W9-BBU-752

SPIEGEL&GRAU

BY REBECCA STOTT

Darwin's Ghosts
The Coral Thief
Ghostwalk

DARWIN'S GHOSTS

For Kate and Anna,
and for Dorinda

Once grant that species [of] one genus may pass into each other . . . & whole fabric totters & falls.

Charles Darwin,

NOTEBOOK C

Masterpieces are not single solitary births; they are the outcome of many years of thinking in common, of thinking by the body of the people, so the experience of the mass is behind the single voice.

Virginia Woolf,

A ROOM OF ONE'S OWN

Preface

I grew up in a Creationist household. As a child, I often thought about Charles Darwin; I wondered who he was and whether he knew, as my grandfather and the other preachers alleged, that he had been sent to earth to do Satan's work. It seemed an odd reason to make a man, I thought, but then, in the scale of things, perhaps no more odd than the story of God and Satan tormenting Job or the angels who appeared in Sodom and Gomorrah, no more strange than the pillar of salt that Lot's wife was turned into, or the Four Horsemen of the Apocalypse. I also wondered if, as Satan's man, Darwin might have hooves and scales. But generally it wasn't a good idea to ask questions about such things.

One hot summer's day, when I was around the age of nine or ten, knowing that I could not ask about Darwin or his ideas without being reprimanded, I went looking for him in the pages of the family *Encyclopaedia Britannica*. The house was empty—my preacher father was away from home and my mother was out gathering my younger brothers and sisters for the evening prayer meeting—but I still felt fear as I eased out the volume marked *D* from the shelves. I knew I could be getting myself into serious trouble.

But the page where Darwin should have been was missing. Along the gap there was a perfectly straight stub: the page, my father told me much later, had been razored out by my grandfather sometime in the 1950s. When the encyclopedia volumes had arrived in their wooden crate, my grandfather had summoned the family to the sitting room in their Brighton house to admire them; during this ceremony he had picked up the *D* volume and taken a razor to the page, while delivering a sermon about the wickedness of Mr. Charles Darwin.

The missing page only made me more determined to find out what Darwin had really said. Because we had only a small collection of carefully selected books on the shelves of the family home—including several morality tales such as *The Story of Mary Jones and Her Bible*—I had already discovered the transgressive pleasures of the school library. There, a few days later, I found another encyclopedia set, and in a stolen moment between lessons, I read as quickly as I could the definitions of evolution, animal-human kinship, and natural selection, convinced that at any moment I might be discovered and denounced. I struggled to understand the complex ideas on the page. I dared not ask questions, however, even of my teachers, for fear that news of my scientific interests might be revealed at a parent-teacher evening. Questions multiplied in my head. I began to daydream about half-animal, half-human forms, molten landscapes and prehistoric worlds.

When my parents later joined a moderate Anglican church and developed more permissive views, later still when my father had lost his faith and my mother had allowed us to work out our own beliefs for ourselves, when, as a teenager, I had the freedom to pursue my own intellectual curiosities unchecked, I continued to feel the simultaneous magnetism and frisson of danger when I wandered, as I often did, back to library shelves containing books on Darwin or evolution or genetics. I still feel it.

Certain curiosities, perhaps especially those that arise out of childhood prohibition and transgression, are not sated by a lifetime's reading and thinking. Evolution opened up a new way of seeing the world for me that was quite different from the one I had grown up with, but not necessarily any easier to understand or any less odd or extraordinary.

Years later, I wrote a book about the young Darwin. I came to admire him for his doggedness, for his rebelliousness, and for the range and brilliance of his imagination, as well as for the way he had stuck to his guns and kept on pursuing answers to his questions about the origin of species even though he knew he would be denounced as a heretic.

At the same time, I became preoccupied with the shadowy figures behind Darwin, his predecessors, the less well-known rebels who, I realized, had asked similar questions about the origin of species before him, in some cases a long time before him, and reached similar conclusions. What kinds of risks had they taken? What price had they paid for their curiosity? Why had understanding nature's laws and the origins of species been so important to them that they had been prepared to challenge intellectual or religious orthodoxies and thus risk their reputations and sometimes even their freedom? I knew they must have been audacious as well as clever.

Priests and bishops denounced Darwin's predecessors; police agents spied on them. They locked their ideas away for fear of bringing disgrace on their families. They deferred publishing. They searched out like-minded men and women and safe places to ask the questions about the origins of time and of species that pressed upon them. They went underground. All of them went on gathering evidence just the same, convinced that species were not fixed and that they had not all been created in seven days.

Many of Darwin's predecessors were called infidels. The word has its origins in the fifteenth century and means, in its broadest sense, those who do not believe. By the eighteenth and nineteenth centuries, when religious leaders and university professors called Darwin's predecessors infidels, they believed them to be as dangerous as the infidel soldiers of the Crusades; anyone who promoted a theory about the mutability of species, they declared, was an enemy of Christianity because such an idea contravened the sacred truth set out in the Bible. By the early nineteenth century it seemed there were so many infidels around—mostly radicals who were promoting atheism as part of a reformist agenda—that evangelicals wrote books with titles like *The Young Man's Guide Against Infidelity* to warn young men how to recognize infidels in public places and how to steer clear of their dangerous traps and snares.

We cannot accurately call these men evolutionists, although I have done so in this book as a kind of shorthand. Even Darwin did not call himself an evolutionist. The word "evolution," which means literally an unfolding or unrolling, did not come into common usage to mean the mutation of species through natural selection until the second half of the nineteenth century. Before then, there had been no common phrase to describe the idea of species transformation or descent with modification. In early-nineteenth-century France, Jean-Baptiste Lamarck and others used the term "transformism." In 1832, the British geologist Charles Lyell, deeply opposed to the theory of species change and scornful of the French, who were all infidels as far as the British were concerned, labeled the idea "transmutation," a word drawn from alchemy; he denounced it as both wrongheaded and heretical. Everyone knew that alchemists held nonsensical notions of magic and unicorns and making gold from lead. Transmutation was preposterous, Lyell declared in forty pages of close refutation in the second volume of *Principles of Geology;* it was a castle in the air.

In July 1837, Charles Darwin, who had been reading and rereading Lyell's *Principles of Geology* on the *Beagle* and had become convinced of the truth of species change through his own investigations, began the first of what would become a series of what he called his "transmutation of species" notebooks. In 1847 he used the phrase "us transmutationists" in a letter to his friend Joseph Hooker to make his allegiances clear and to try to persuade Hooker to do the same. He did not think there was anything ridiculous about such theories, but he knew he would be ridiculed for holding them until he could find a way of proving them to be true beyond all reasonable doubt. He knew that what he was doing was dangerous.

When Darwin finally published *On the Origin of Species by Natural Selection* in 1859 and began to steel himself against the first waves of reprobation, he became increasingly preoccupied with his predecessors. He determined to put together a lineage, a line of intellectual descent, to bring these people out of obscurity, but he failed to uncover much information. He was, he wrote to a friend, a poor scholar of history. He was also afraid of getting that history wrong or of not doing his predecessors justice. He did what he could with the little knowledge he unearthed, adding a preface called "An Historical Sketch" to the third edition of *Origin* in which he listed thirty men who had published evolutionary ideas before him. By the time he revised the list for the fourth edition, the list had swelled to include thirty-eight men. The sketch was just that, a rather hesitant document, the best that Darwin could do in the time he had and perfectly adequate for what was needed. However, as a man who took intense pleasure in reading the biographies of the men and women he admired, he must have remained perpetually curious about the lost lives of his predecessors. This book, *Darwin's Ghosts,* is dedicated to both him and them.

As I began to search for these lost forebears, I found men of science who were not on Darwin's list, men who lived in medieval Basra or in Renaissance Italy and France who had been called "proto-evolutionists" but about whom Darwin would have known nothing. I found others on Darwin's list who should not have been there. To my frustration, I found no women who published evolutionary ideas before *Origin*.

I have not included all the people who might have claims to a place in this book. Of those whom I have chosen, some have their own chapters; some, whose stories are particularly entangled with others, share chapters; others figure briefly in chapters that belong to others. All who appear here are pathfinders, iconoclasts, and innovators, men whom Darwin would have claimed as kin even if he thought their pursuits and ideas misguided or fanciful. Most of them have regrettably disappeared from our view, obscured by the shadow of Darwin. This book aims to bring them back into visibility.

Contents

DARWIN'S GHOSTS

1

Darwin's List

KENT, 1859

Just before Christmas in 1859, only a month after he had finally published *On the Origin of Species by Natural Selection,* Charles Darwin found himself disturbed, even haunted, by the thought of his intellectual predecessors. He entered a state of extreme anxiety that had the strange effect of making him more than usually forgetful.

It had been a cold winter. Though Darwin might have liked to linger on the Sand Walk with his children to admire the intricately patterned hoarfrost on the trees, he knew he had work to do, letters about his book to answer, criticisms to face.

He had weathered the first blasts of the storm of censure in a sanatorium in Ilkley, where he had been taking the water cure, wrapped in wet sheets in hot rooms, the skin on his face dry and cracked with eczema. Since his return to his family home, Down House, now garlanded with Christmas holly, ivy, and mistletoe by his children, he had braced himself every morning against the sound of the postman's footsteps on the gravel outside his study window. The letters, he lamented to his wife, Emma, came like swarms.

Each new mailbag delivered to Down House brought letters voicing opprobrium, some veiled, some outspoken; a few contained praise. But though some reviewers might be expressing outrage, Darwin reassured himself, hundreds of ordinary people were reading his book. On the first day of sale in November, the entire print run of 1,250 books had sold out. Even Mudie's Select Lending Library had taken five hundred copies. Now his publisher, John Murray, was about to publish a second edition; this time Murray intended to print three thousand copies, and he had agreed to let Darwin correct a few minor mistakes. Darwin was relieved. The mistakes embarrassed him.

As readers and reviewers took up their positions for or against his book, Darwin began to keep a note of where everyone stood on the battleground. "We shall soon be a good body of working men," he wrote to his closest friend and confidant, the botanist Joseph Hooker, "& shall have, I am convinced, all young & rising naturalists on our side."

The letter that launched Darwin into a prolonged attack of anxiety came from the Reverend Baden Powell, the Savilian Professor of Geometry at Oxford, a theologian and physicist who had been forthright in his support for the development theory for some time.* The elderly professor was on the brink of being prosecuted for ecclesiastical heresy. Of all the letters in that day's pile, the one from Powell would be innocuous enough, Darwin assumed. He scanned it quickly, relieved to glimpse phrases like "masterly volume" and a few other words of praise. But Baden Powell was not happy. Having finished with his compliments, the professor launched into a direct attack, criticizing Darwin not for being wrong, not for being an infidel, but for *failing to acknowledge his predecessors.* He even implied that Darwin had taken the credit for a theory that had already been argued by others, notably himself.

This was not the first time Darwin had been accused of intellectual theft, but until now, the accusations had been tucked away in reviews and had been only implicit. How original is this book? people were clearly asking. How *new* is this idea of Mr. Darwin's?

He might have protected himself better from charges of plagiarism,

* The Reverend Baden Powell was the father of Robert Baden-Powell, the founder of the Scouting movement.

Darwin reflected fretfully, if he had only written a preface, as most scientists did when they published any controversial set of claims: a survey of all the ideas that had gone before. It gave the ideas a history and a context. It was a way of showing where the edges of other people's ideas finished and your own began. But he had not done so, though he had planned to. And now he was being accused of passing off the ideas of others as his own.

Darwin's Study, Down House, Kent.
Getty Images

As he sat reading and rereading Powell's letter, Darwin's excuses came thick and fast. He should have included a short preface, he wanted to tell Powell, but his book had been rushed. He had not been at all well. His closest friends, the botanist Joseph Hooker and the geologist Charles Lyell, had been badgering him to publish for years. Then, when Alfred Russel Wallace had sent him that alarming essay from the Malay Archipelago showing that Wallace had worked out natural selection, too, Hooker and Lyell had practically forced him to go straight into print. For months, he had hardly slept for writing. He had never written so fast or for so long. And in all that rush, he had neglected to acknowledge those who had gone before. Besides, aware that he was a poor scholar of history, he had not been confident that he knew exactly who *had* gone before or that he had the skills to describe their ideas accurately and fairly. They wrote in every language under the sun. Some of them were obscure, others mad. It would have taken years.

Darwin had known from Wallace's enthusiastic letters that he was getting close to working out natural selection, but until seeing Wallace's essay he had underestimated the speed at which the brilliant young collector was working. The thought that after all this procrastinating, someone like Alfred Russel Wallace could step in and publish his essay and make a claim to the discovery of natural selection before him was more than he could bear. At that point Hooker and Lyell had intervened, explaining to Wallace that Darwin had first formulated the idea some twenty years earlier. Wallace had been generous. He had given up any claim to being the discoverer of natural selection. He had even written to Hooker to say that he did not mind in the least that Darwin was going to take the credit and that it was right that he do so. He considered himself lucky, he confessed, to have been given some credit.

So Wallace had renounced his claim on natural selection. But now, only a year after Darwin had escaped the Wallace tangle, here was another claimant rising like Marley's ghost from the mailbag—the Reverend Baden Powell. Darwin had forgotten about Powell.

My theory. My doctrine. Darwin had been writing those words for years in his notebooks. But *was* it his alone? He had told Hooker and Lyell that he was not ready. It was all very well for them to urge him into print. After all, *they* were not going to be deluged with disgust and outrage. They were not going to have to explain to their troubled wives; they were not going to have to apologize to and mollify bishops and clerics and bigots or answer plagiarism charges. And now John Murray was about to send another three thousand copies of *Origin* out into the world.

There was no stopping any of it. His theory had not leaked quietly into the public domain as he had planned; it had entered the world as a deluge, like the water pipes in the Ilkley water cure establishment, cold and gushing and unstoppable. He, Lyell, and Hooker had simply pulled the rope and released the valve. And here were the consequences.

Hooker would know what to do. Darwin wrote to invite him to Down House. Bring your wife, he wrote, bring the children. On December 21, 1859, Joseph Hooker's wife wrote to Darwin to say that her husband would be happy to visit the Darwins in the second week of January and that he would bring their eldest son, William, with him. Darwin was delighted. Such a visit would do him tremendous good, he wrote to Hooker, for though the water cure had improved his health, now that he was in the midst of the critical storm, he was, he wrote, "utterly knocked up & cannot rally—I am not worth an old button." The eczema had broken out again. He was sick to his stomach.

A portrait of Darwin in his forties made from a photograph taken in 1854.
Harper's Magazine (1884)

The following day, three days before Christmas, while Darwin was still trying to compose a reply to the Reverend Baden Powell, a third claimant emerged, this time from France. Darwin's butler told him that a parcel had arrived in the evening post. Though the children protested, Darwin left the warm parlor where Emma had been reading aloud to them in the shadow of the Christmas tree and slipped away across the hall to the darkened study to retrieve it.

With pleasure and relief, he recognized the handwriting on the label as Hooker's. The parcel contained an essay by Hooker that Darwin had promised to read and a heavy volume of a French scientific journal, the *Revue Horticole.* Hooker explained in an accompanying letter that a scientist named Decaisne* had written to tell him that a botanist named Charles Naudin had discovered natural selection back in 1852. Darwin, Decaisne wrote, had no right to claim natural selection as his idea.

* The botanist Joseph Decaisne.

Hooker enclosed the volume of the journal in which Naudin's paper on species had been published so that Darwin could judge the claim for himself. Darwin had read and admired Naudin's work years before, but he had entirely forgotten the paper.

Darwin ordered the study fire to be relit. He read and reread Naudin's paper late into the night, struggling with some of the French scientific terms, reaching for his French dictionary, making notes as high winds rattled the windowpanes. Naudin's claim was not a serious threat, he finally told Emma a few hours later. The French botanist had not discovered natural selection. He was quite sure of that.

The following morning, returning to his desk cluttered with the debris of the previous evening's struggle with French verbs and the notepaper with all his scribblings across it, he wrote to allay Hooker's fears, explaining with relief: "I cannot find one word like the Struggle for existence & Natural Selection." Naudin had gotten no closer to natural selection than had the French evolutionist Jean-Baptiste Lamarck, he wrote emphatically. Darwin asked Hooker to pass on his refutation of the claim to their mutual friend Lyell, adding with a touch of embarrassment, "though it is foolish work sticking up for independence or priority." He had nothing against Naudin, after all. He was a good botanist.

Darwin could not decide how best to answer these phantom claimants. Should he let others intervene, as Hooker and Lyell had done with Wallace? Should he write directly to each new claimant or simply ignore them? What was the gentlemanly thing to do? Even through Christmas dinner the question troubled him. While the children were playing, he slipped away to write to Hooker in the afternoon, scribbling, as if struggling with his own conscience, "I shall not write to Decaisne: I have always had a strong feeling that no one had better defend his own priority: I cannot say I am as indifferent to subject as I ought to be; but one can avoid doing anything in consequence." The Reverend Baden Powell was different. He needed answering. The man had a point: Darwin was the first to admit that he should have acknowledged all those natural philosophers who had had the courage to publish evolutionary ideas before him—men such as his own grandfather Erasmus Darwin and the misguided but brilliant Lamarck and others. It was bad form not to have done so. In the rush to publish, he had forgotten them.

Everyone had forgotten them.

All through the holidays, all through the singing and the feasts and the toasts, Darwin struggled to formulate a letter to Powell. He imagined the outspoken professor spluttering his outrage to the fellows of Oriel or at Geological Society meetings. Conversations with Powell opened up in Darwin's head again and again, sometimes angry, sometimes defensive or apologetic. Christmas was no time to be defending one's reputation, he told himself, trying to attend to the family celebrations, to be a good father and husband and to be attentive to his eldest son, William, home from Cambridge for Christmas and for William's birthday on December 27.

Despite his resolutions, Darwin still woke in the night, slipping out of bed so as not to disturb Emma and pacing the floor in his study. How many other predecessors had he forgotten? How many did he simply not know about? He had never been a good historian of science. How would he ever write a definitive list?

Hooker's visit was not to be. Down House seemed besieged from both outside and inside; terrible storms lashed the country. Shipwrecks were reported around the coast; a tornado in Wiltshire uprooted trees, destroyed hayricks, and swept the thatch from the roofs of cottages. Heavy lumps of ice fell in a freak hailstorm, killing birds, hares, and rabbits. As the year turned, nine-year-old Lenny Darwin began to run a fever. When the first flush of spots appeared, Emma urged Darwin to write to Hooker to put him off his visit. Darwin wrote sadly to his friend, repeating his wife's words of warning: "Lenny has got the Measles & it is sure to run like wild-fire through the house, as it has been extraordinarily prevalent in village. If your boy Willy has not had measles, I fear it will not be safe for you to bring him here."

In the first week of January 1860, as the measles spread first to twelve-year-old Elizabeth and then to eleven-year-old Francis, and having been unable to talk to Hooker, Darwin resolved to write to Powell and to draft a historical sketch just as he had planned to do years before. The timing was good: the American botanist Asa Gray was organizing an authorized American edition of *Origin* and he wanted a preface from Darwin. Darwin talked aloud to himself, resolving to put it all straight in the American preface by adding a full historical sketch, reminding himself that the idea of species mutability was not his. Not even the idea of the descent with modification was his. It belonged to Lamarck and Maillet, and further back it was probably in Buffon and even in his grandfather's book *Zoonomia.* He had never claimed that descent with modification was his idea, though of course Powell thought that he had. But natural selection—the idea that nature had evolved by selecting the fittest to survive—*was* his. No one, not even Wallace, had discovered natural selection before he had, or at least put all the ideas together in such a way as to make it explain so many large groups of facts. He owed it both to himself and to his predecessors to explain what was his and what was theirs.

It was only when he began to write his letter to Powell on January 8 that Darwin suddenly remembered that he had started writing a list of his predecessors several years earlier. He went to find it. The embryonic historical sketch was in the drawer where he had left it, in the file with the big still-to-be-published full manuscript version of the species book. The list was not finished, of course; it was just a scribbled catalog of predecessors with notes. But it was there. He had started it back in 1856, knowing that his species book would have to have one. And—it made him blush again to see the scale and extent of his own forgetting—there was the Reverend Baden Powell in the catalog, properly acknowledged and praised.

So he wrote to Powell. "My dear Sir," he began,

> my health was so poor, whilst I wrote the Book, that I was unwilling to add in the least to my labour; therefore I attempted no history of the subject; nor do I think that I was bound to do so. I just alluded indeed to the Vestiges & I am now heartily sorry I did so. No educated person, not even the most ignorant, could suppose that I meant to arrogate to myself the origination of the doctrine that species had not been independently created. . . . Had I alluded to those authors who have maintained, with more or less ability, that species have not been separately created, I should have felt myself bound to have given some account of all; namely, passing over the ancients,

—and here Darwin had to glance again at his earlier catalog so as to remember the names, and some of the spellings—

> Buffon (?) Lamarck (by the way his erroneous views were curiously anticipated by my Grandfather), Geoffry St Hilaire [*sic*] & especially his son Isidore; Naudin; Keyserling; an American (name this minute forgotten); the Vestiges of Creation; I believe some Germans. Herbert Spencer; & yourself. . . . I had intended in my larger book to have attempted some such history; but my own catalogue frightens me. I will, however, consult some scientific friends & be guided by their advice.

Darwin read back over the letter to check the tone. His glance snagged on the clause: "my own catalogue frightens me." That was overly candid, perhaps, and a touch histrionic. But candor might well disarm Powell. And after all, it was true: the catalog did frighten him. Those scribbled names on the sheet of paper frightened him.

Predecessors? Who were they? Most of them were dead. Their names slipped from his memory. Why could he not remember the name of the American evolutionist?* Exhausted by the very idea of writing a historical sketch, he folded up the letter to Powell and handed it to Parslow, his butler, for the post.

It seemed as if his work would never be done. He felt the burden of censure heavy on his shoulders now that he was back in the study, stoking the fire, feeling the heat agitating the itching on the dry and flaking skin of his face. He had placed himself at the mercy of all his readers as soon as he had gone into print—the priests, the theologians, the reviewers, the letter writers. Four days before his book had been published, an anonymous reviewer in the *Athenaeum* had denounced *Origin* and declared it too dangerous to read. Darwin wrote to Hooker the following day: "The manner in which [the reviewer] drags in immortality, & sets the Priests at me & leaves me to their mercies, is base. . . . He would on no account burn me; but he will get the wood ready & tell the black beasts how to catch me."

And there in the light from the fire, Darwin remembered the heretics who had been burned in the marketplaces of England. Burned because they kept mass or because they did not keep mass. Burned because, even

*It was Samuel Steman Haldeman (1812–1880), an American taxonomist and polymath.

under torture and starvation, they would not recant. Even his close friends would turn against him now that he had gone into print. Their priests and bishops would expect it of them. This was the final reckoning, the taking of sides. He had warned the naturalist Hugh Falconer on November 11 that when he read *Origin*, "Lord how savage you will be . . . how you will long to crucify me alive." "It is like confessing a murder," he had admitted to Hooker back in 1844 when he had finally summoned the courage to tell his friends about his species theory for the first time.

Over the next three weeks, as winter deepened, a cold spell iced over the lakes and rivers of Britain and high winds returned, whistling around Down House and rattling the windowpanes, Darwin's list grew. There had been only ten names on the list he had sent to Powell, he told Emma, "and some Germans" whose names he had also forgotten. Now, as the predecessors came one by one out of the shadows and into the clear light of his own prose, his fears began to subside. Not only did he come to feel their presence as a kind of protection, a shield from charges of intellectual theft, but he began to think of them as allies, as fellow outlaws and infidels. He read and reread their words, increasingly reassured by his new knowledge. Now, if pressed, he could define exactly where his ideas had been preempted and where they were entirely new.

He admired them. He stopped forgetting their names.

On February 8, Darwin sent the first version of his "Historical Sketch" to America for the authorized American edition, a corrected and revised version of the first (pirated) version. Darwin's list had almost doubled in length since he had assembled the first tentative ten names for Powell in mid-January. There were eighteen names on this new list published in the summer of 1860. Darwin's catalog of predecessors was now, he was sure, as definitive as he could make it. He sent the same version of the "Historical Sketch" to Heinrich Georg Bronn in Heidelberg, who was translating *Origin* for the first German edition of 1860.

Eighteen predecessors. A good number. But still a relatively small one.

Meanwhile the hostile reviews of *Origin* were becoming more overtly aggressive. The gloves were off. "The stones are beginning to fly," Darwin wrote to Hooker, and he reassured Wallace that "all these attacks will make me only more determinately fight." To Asa Gray he wrote: "I will buckle on my armour & fight my best. . . . But it will be a long fight. By myself I shd. be powerless. I feel my weak health acutely, as I cannot work hard."

There were still other important evolutionists yet to step out of the shadows to claim some of Darwin's glory.

On April 7, 1860, his favorite journal, the *Gardeners' Chronicle,* carried an article by a man he had never heard of named Patrick Matthew, a Scottish landowner and fruit farmer. Matthew claimed that he had discovered natural selection back in 1831, twenty-eight years before Darwin. There was no beating about the bush. This was a direct accusation: Darwin had no right to claim natural selection as his own, Matthew wrote. By way of proof, he republished numerous short extracts from his original book, unpromisingly entitled *Naval Timber and Arboriculture.*

Darwin was horrified that such an attack should be rehearsed in the pages of his beloved *Gardeners' Chronicle.* Moreover, Matthew's claim to be the discoverer of natural selection was a strong one. Seriously alarmed, Darwin sent for the book and was reassured to find that the passages in question were tucked away in the appendix of what was a very obscure and specialist book. Nonetheless he determined to be a gentleman.

A week or so later Darwin sent a letter to the *Gardeners' Chronicle.* "I freely acknowledge that Mr. Matthew has anticipated by many years the explanation which I have offered of the origin of species, under the name of natural selection," he wrote. "I think that no one will feel surprised that neither I, nor apparently any other naturalist, has heard of Mr. Matthew's views, considering how briefly they are given, and that they appeared in the Appendix to a work on Naval Timber and Arboriculture. I can do no more than offer my apologies to Mr. Matthew for my entire ignorance of his publication."

Darwin's response took the wind out of Matthew's sails. Flattered and mollified, the fruit farmer published his final word on the matter in the *Gardeners' Chronicle* on May 12: "To me the conception of this law of Nature came intuitively as a self-evident fact, almost without an effort of concentrated thought. Mr. Darwin here seems to have more merit in the discovery than I have had; to me it did not appear a discovery."

Matthew had conceded the throne, but he retained his claim to an important place in Darwin's list.

Eighteen names became nineteen.

That same May, Charles Lyell sent Darwin a paper on natural selection by a Dr. Hermann Schaaffhausen published in 1853; nineteen names became twenty.

In October 1860, an Irish doctor named Henry Freke sent Darwin a pamphlet he had published in 1851 describing animals and plants evolving from a single filament. The pamphlet was, Darwin told Hooker with some relief, "ill-written unintelligible rubbish." But if Darwin was to play by the rules of the game, even eccentric Henry Freke had a claim to a place in the list.

Twenty names became twenty-one.

By the time Darwin revised the "Historical Sketch" again for the third English edition of *Origin of Species* in late 1860, his list of predecessors had grown to include thirty men, including his own grandfather. New claimants included Patrick Matthew, Henry Freke, Constantine Rafinesque, Robert Grant, Dr. Schaaffhausen, and Richard Owen.

Putting the poison-tongued Oxford naturalist Richard Owen on the list gave Darwin particular pleasure. Owen had written a spiteful and envious review of *Origin* in April 1860. "Odious," Darwin had called it. Owen had not even had the courage to sign his name to it, he complained; instead he had taken cover in anonymity, although Darwin's friends had later rooted out his identity. Owen had also sneered at Darwin's failure to include a list of his predecessors. So putting Owen on the list was for Darwin a way of getting even, a way of ridiculing Owen's philosophical inconsistencies and contradictions. In the new version of the "Historical Sketch," he quoted Owen's extraordinary claim of 1852 that he had discovered natural selection, allowing himself a touch of scorn: "This belief in Professor Owen that he then gave to the world the theory of natural selection will surprise all those who are acquainted with the several passages in his works, reviews, and lectures, published since the 'Origin,' in which he strenuously opposes the theory; and it will please all those who are interested on this side of the question, as it may be presumed that his opposition will now cease."

Robert Grant, who was also new to the list, was Darwin's old mentor at Edinburgh. Now impoverished and mocked for his views, he was teaching at the University of London. Reading Darwin's *Origin* had prompted Grant to finally publish his evolution lectures and to remind Darwin that he had published articles on evolution in Scottish journals all through the 1820s. Darwin disliked Grant's radical political views and wanted to distance himself from them, but he knew he would have to include him in the list if he was to stick to the rules of gentlemanly behavior.

There were demotions, too. In 1860, Darwin took one name off the list: Benoît de Maillet, the eccentric Frenchman who had worked up a theory of animal-human kinship in Cairo in the early eighteenth century. In his savage review of *Origin,* Richard Owen had implied that Darwin was as foolish as the deluded Maillet, who had believed in mermaids. That was more ridicule than Darwin could bear. He took a pen and put a line through Maillet's name.

By the fourth edition of *Origin,* completed in ten weeks in 1866, Darwin's list had swelled to no fewer than thirty-seven names. Since the publication of the third edition, he had found another eight European evolutionists in an article published back in 1858 by his German translator, Heinrich Georg Bronn, which he had not been able to read until Camilla Ludvig, the Darwin family's German governess, translated it for him. Darwin no longer had the time or the patience to test each of the claims individually, so he placed all eight new names inside a single footnote.

And then in 1865, just as Darwin was completing the final amendments to the fourth edition of *Origin,* the ancient Greek philosopher Aristotle stepped out of the shadows as a claimant. James Clair Grece, a town clerk and Greek scholar from Redhill, wrote to Darwin claiming that he had found natural selection in Aristotle's work, ideas recorded in lecture notes scribbled in Athens two thousand years earlier. He had translated the passage into English for Darwin as proof. Darwin had read Aristotle at school. He admired him above all other naturalists, he told Hooker—even more than Linnaeus or Cuvier. But he knew so little of his work, and he was not going to learn Greek at this stage in his life. So in every version of the "Historical Sketch" he had written so far, he had simply "passed over" the "ancients," apologizing for the limitations of his knowledge.

The passage Grece sent was from a book that Darwin did not know, and, given that Grece's translation was pretty incomprehensible and that he was reading the words out of context, it was difficult for him to tell whether it really was an ancient Greek version of natural selection, as Grece claimed. But Darwin was prepared to give the clerk the benefit of the doubt because he admired Aristotle; he was the first man to have looked closely at animals and the structures and connections of their bodies—all animals, right down to the sea urchins and the oysters and the sponges. And he had done all of that close observation and dissection without microscopes or dissecting tools or preserving spirits.

With no time to ask abroad or test the claim, Darwin placed both Aristotle and Grece together into the same footnote destined to appear in the fourth edition of *Origin*.

Aristotle was now the first man on Darwin's list and the last man to enter it. Darwin was delighted to add Aristotle to his list but wished he could have said more, explained more about how the Greek philosopher might have come to understand species and time more than two thousand years earlier. Instead he had to make do with a footnote.

The next time Grece wrote to Darwin it was not about Aristotle but about a pig.

It was November 12, 1866. Darwin's morning mailbag had doubled if not tripled in size since the publication of *Origin*. People continued to write to him from all over the world. They offered him facts like gifts, as if he were now the sole chronicler of all nature's strange and peculiar ways, as if he were the owner of a great factory of facts, grinding them out in the millstones of his brain to make something that might be labeled "nature's laws." People sent him facts about the tendrils of climbing plants, the valve structures of barnacles, the mating habits of hummingbirds. He collected them all and filed them away.

This morning was no different. Darwin reached for the first letter from the top of the pile that his butler had arranged on his desk. The envelope was postmarked Redhill in Surrey. He tried to recall who he knew there, who might have sent the letter. Inside the envelope, he found a letter from Grece and a cutting from the *Morning Star* dated November 10, 1866. Grece explained that he was sending an oddity of nature for Darwin's files in case it might be of use in the future. The newspaper headline read "FREAK OF NATURE," and the article described a pig that had apparently sloughed off its entire black and bristly skin from snout to tail in one mass in a single night, revealing underneath an entirely new mottled pink body. The pig was, the journalist recorded, apparently unperturbed by its night adventure and was eating as hungrily as before, oblivious to the scores of visitors who had flocked to see it. The owner had pinned the discarded skin to the door of the pig's sty with a notice that read "Do not touch." No natural philosopher, the letter writer complained, had yet been to see the pig. He encouraged Darwin to do so. He might be able to make sense of the unusual occurrence.

"You may recollect me as having some year or two since pointed out to you a passage from Aristotle," Grece wrote, "shewing that 'Natural Selection' was known to the ancients." Grece was claiming his due, Darwin realized, as if having been placed in a footnote with Aristotle in the fourth edition of *Origin* were not reward enough. By 1866, Darwin was weighed down with a sense of the debts he owed to the hundreds of naturalists who sent him things. "Should you like to see the animal," wrote Clair James Grece, town clerk of Redhill council, railway enthusiast, chronicler of the local sloughing practices of pigs, "it is on the premises of one Mr. Jennings, a baker, in Horley Row about one mile north of the Horley Station of the London and Brighton railway. A fly might not be procurable at that station, so that you might prefer to alight at the Redhill Station, where vehicles are readily obtainable, and whence it is about four miles to the southward."

—

By the time Darwin's "Historical Sketch" appeared in the fourth edition of *Origin*, it had been ten years in the making. Of the distribution of nationalities of these evolutionists, fourteen were British, nine French, six German, two American, one Italian, one Russian, one Austrian, one Estonian, one Belgian, and, if he were to count Aristotle, one an ancient Greek. A reviewer might easily have thought that Darwin was making a point about British superiority in the biological sciences. Yet only Darwin knew how little design there had been in the composition of the "Historical Sketch." Only he knew the way in which certain names had been shoehorned in at the last minute and how doubtful he was about the status of some of those claimants, particularly the most recent additions.

Yet Darwin found the final distribution of nationalities pleasing. There were only nine Frenchmen as against fourteen British. Now he had finally proved once and for all that evolution was not an exclusively French idea, that it was not the spawn of French revolutionaries, part of a conspiracy to bring down the church and government and all social hierarchies. It was just as much the discovery of British clergymen, doctors, fruit farmers, and gentleman naturalists working away with microscopes in houses in the British provinces.

Darwin looked at the gaps in the list, too. That enormous gap between the first person on his list and the second—the Greek philosopher Aristotle and the eighteenth-century French naturalist Buffon—puzzled him. What had happened in that chasm of more than two thousand years? If Grece was right and Aristotle had begun to formulate vaguely evolutionary questions about the history of animals in 347 BC, even if they were only flickers of a vision he could not yet see clearly from his vantage point, what had happened to those embryonic ideas? Where had they disappeared to? Religious repression was too easy an answer; there were always freethinkers in a population of people, however repressed, however much they lived under the eye of censoring priests. There must have been transmutationists in that gap of two thousand years, he reflected. Perhaps they had disappeared beyond all historical record.

Something else about the Aristotle footnote troubled Darwin long after the fourth edition of *Origin* had found its way into the bookshops.

He could not see how anyone in ancient Greece, even the great philosopher, could have foreseen natural selection. There were no microscopes and so no way of studying single-celled organisms. There were no taxonomic theories to work with or against, so there was no way of understanding the various families of animals or the relationship between the plant and animal kingdoms. There were no systematic anatomical or dissection methods and no way of preserving body parts during examination. There were no studies of the effects of plague or population statistics. No libraries. Surely there were only superstition and sacrifice and vengeful gods and the relentless Greek sun turning everything black and fly-infested. How was it possible?

2
Aristotle's Eyes

LESBOS, 344 BC

Dawn. In the port of Mytilene, on Lesbos, a scattering of fishing boats sail back into harbor across the turquoise sea. Once the boats are securely moored, ropes thrown and knots tied, the fishermen lug their morning's catch in woven baskets to a table built into the side of the harbor wall, spilling out a torrent of glistening-scaled silver, gold, and red fish. Here, every morning, under a cloth canopy erected to protect the catch from the heat of the sun, the fishermen sort their fish into piles, separating out octopus from mullet, sardines from bream, making a separate pile for the small fish for which they have no name. Cats crouch under stools and behind old crates; men arrive in carts pulled by donkeys; the bartering begins.

A group of young men wearing finely woven tunics wander unhurriedly through the harbor market among the fishermen and traders. The slaves who accompany them carry rolls of papyrus,* jars, nets, and

* The ancient Greeks did not have books. They wrote on papyrus, on clay tablets, on wood, and on animal skins. Papyrus was the most widely used writing surface. From around 3000 BC the Egyptians manufactured papyrus sheets.

cutting implements. Philosophers from the school across the sea in Assos, these men live in the harbor town but spend their days in the island forests and meadows talking, poking about under stones or in caves, or studying the trees in the petrified forest, always arguing, peering, recording, asking questions of farmers, fishermen, and animal breeders. On tables and rock surfaces they cut open animals and insects and fish and peer inside: cuttlefish, crickets, chameleons, and butterflies. Nothing is too small for their attention.

The fishermen know one of the philosophers as Theophrastus, now in his early twenties, who was born in Eresos, a village on the southeast of the island where his father worked as a laundryman. It was Theophrastus who had brought the philosophers here. He knows Mytilene well, though he has been away in Assos for several years studying in the school there. The older man in the group, clean-shaven and wearing fine clothes and expensive rings, is the great teacher Aristotle. People say he is wise beyond all men, but he is also a Macedonian and even in remote Lesbos the locals keep the Macedonians at a distance. Macedonia is a country to be reckoned with. Philip, the king of Macedonia, has imperial ambitions. He has his eye on the Greek city-states, and beyond them, Persia. And the rumors about Aristotle always connect him with Philip. Some people say the philosopher is Philip's creature; others that he advises the great king; still others that he is a spy. In Lesbos, though, the fishermen say, he seems to be interested only in fish.

The fishermen of Mytilene harbor collect fish for Aristotle. They throw the fish that he has asked them to find for him into large clay bowls filled with seawater that stand in the shade under the fish market table. He wants them alive, not dead, and he pays well for unusual or particularly well preserved specimens or for fish fat with eggs. Theophrastus explains that Aristotle is collecting the names of all the animals on earth. He wants to describe every single living thing, every fish and bird, so as to discover the secrets of nature's patterns. He wants to know how the fish of Lesbos swim, feed, defend themselves, what they eat, whether they sleep, smell, or hear.

The fishermen tell the philosopher everything they know. Aristotle, amused at some of their descriptions, notes how their minutely detailed knowledge of marine life is shot through with myths, magic, and superstition. They tell him that they have seen fish with twenty eyes and octopuses dancing in submarine caves with their arms entwined, and that they live in fear of a whale the size of an island with the head of a man. But they know how to tell one fish from another; they know about spawning and the movement of shoals and mating patterns. Aristotle is interested only in their facts.

Aristotle did not inherit a tradition of natural philosophy; he had no mentors or teachers. Although huntsmen, farmers, pigeon breeders, beekeepers, and apothecaries had built up a body of knowledge about agriculture, stock breeding, hunting, foods, drugs, poisons, childbirth, and dying over centuries, Aristotle was the first to collect animal specimens, the first to describe and record species, the first to think those things worth doing. He was the first to believe that if he looked long and hard enough inside the bodies of birds, bees, butterflies, and fish, nature would reveal its secrets. He was the first to believe that nature had secrets, and that those secrets would answer complex metaphysical questions in addition to physical ones.

From the sky, the island of Lesbos looks like a creature with two pairs of arms reaching out into the sea—one pair reaches southeast toward the coast of modern-day Turkey and the other southwest toward the southern coast of Greece. Each of those pairs of arms enfolds a vast lagoon, like two watery eyes—shallow, clear as glass on a windless day, and teeming with fish and seafood. The island landscape is austere and volcanic, studded with olive groves and fragrant shrubs. To the west, up near the small town of Eresos, where Theophrastus was born, the stumps of a petrified forest of ancient sequoias spike up through white gravel. The climate is wet and hot. Natural springs bubble up through cracks in the limestone rock on the eastern side of the island, and the island's rivers are seasonal: in the summer they disappear; in the winter they run like torrents. Sixty different species of flower are native to the island.

Lesbos, the crossroads between three continents, is also an island of migrants. Thousands of birds stop here on their migration paths to and from Africa; fish, reptiles, and insects, accidental travelers carried in the holds of trading boats from ports on all sides of the Aegean, have made their homes here. The islanders themselves are the descendants of migrants—Persians, Anatolians, and Greeks.

Aristotle was a political refugee when he crossed the sea to Lesbos from the hill town of Assos, in Turkey, where he had recently established a school. While the king of Macedonia was expanding his empire south and taking city after city in his path, it was not safe for Aristotle, as a Macedonian, to return to Athens. He arrived in Lesbos with his pupils and disciples and his young wife, who might also have been pregnant with his first child, at the invitation of his student Theophrastus, the young botanist who had been born here and who described to him the beauty and abundance of the island.

When Aristotle saw the lagoon and the fish, he was enchanted. The philosopher and his students stayed on Lesbos for two years inscribing hundreds of papyrus rolls with descriptions of the wildlife, descriptions that years later he would write up into lectures for his pupils when he returned to Athens. Those lecture notes—expanded and added to—eventually became some of the most influential books about the natural world of all time: Aristotle's *Parts of Animals, The History of Animals,* and *On the Generation of Animals.* They contained the very first systematic and empiricist studies of nature, the very first attempts to decipher nature's codes. All of Aristotle's great philosophical works—his ideas on governance, metaphysics, ethics, logic, and rhetoric—were inflected by the great zoological project that he began on Lesbos.

Aristotle's life can be mapped only through the fragments of letters, myths, stories, and eyewitness accounts that have survived. There are times when he disappears altogether from the historical record. But nonetheless, despite the disappearances, we know enough about the contours of his life to be confident that he was a man of voracious curiosity, that he traveled widely, and that he worked constantly to seek answers to an extraordinarily wide set of questions about subjects that today we would subdivide and label as ethics, art, poetry, cosmology, physics, metaphysics, politics, rhetoric, theater, linguistics, biology, and zoology but which he saw as indivisible from one another.

———

It all began with a sea journey.

From the age of seventeen, Aristotle had lived a charmed life in Athens wandering the colonnades and libraries of Plato's Academy and the streets, marketplaces, papyrus stalls, and gymnasia of the great city. For twenty years, first as a student and then as a teacher, he had been Plato's most challenging and most promising protégé. People said he might even succeed Plato at the Academy.

But as a Macedonian living in Athens, Aristotle was always a *metic*, an immigrant. While the politics of the Aegean were turbulent—and they always were in the fourth century BC—to the locals, the fact of Aristotle's not being an Athenian citizen made him a foreigner from the barbarian north, an outsider. Later writers have claimed that Plato described his clever student as a wild horse, a stallion who needed bridling, a young man who was forever kicking back, kicking up. If Plato did say these things about Aristotle, perhaps he was thinking of that Macedonian childhood, that northernness.

We know that Aristotle felt his foreignness keenly in Athens, for he complained about it to his friend the Macedonian general Antipater years later. "In Athens things that are proper for a citizen are not proper for an alien," he wrote; "it is dangerous [for an alien] to live in Athens." He could not own property or vote; he paid a monthly poll tax and might be conscripted at any time. Although there were many *metics* in the city, Athenians were especially wary of Macedonians because of the king of Macedonia's imperialist ambitions. Macedonian *metics* might be spies working for the king or interlopers or assassins, Athenian citizens whispered to one another; they were not to be trusted. And Aristotle was no ordinary Macedonian. His father had been the court physician and friend of King Amyntas III of Macedonia, father of the barbarian King Philip.

In the fourth century BC, the lands and the sea to the east of Athens were a battleground between empires. To the east of the Aegean there were Persia and Anatolia, modern-day Turkey. Macedonia lay to the north and the powerful Greek city-states to the west. Armies fought for fertile land or to control the ports that were the gateways into new trade routes; the Macedonians, the Persians, the Spartans were all at various times in the ascendant in the Aegean.

In 348 BC, when Aristotle was about thirty years old, Philip of Macedonia laid siege to the important coastal city of Olynthus. Built on the flat-topped hills overlooking the Gulf of Torone and under the protection of Athens, the town was of strategic importance to Philip, but he also wanted to capture two of his troublesome relatives who had taken refuge in the town and were making claims to his throne. Although the Athenians sent troops and ships to protect the Olynthian citizens, they failed to reach the city in time. Philip's troops looted the city before razing all its buildings and selling its entire population—including the Athenian garrison—into slavery.

In the wake of the attack, anti-Macedonian feeling erupted in Athens. Anti-Macedonian slogans were daubed on house walls; Athenians threw rotting food at Macedonians and attacked their property. It was probably not safe for Aristotle to walk alone at night. He put up with the persecution at first, keeping to himself. But he knew he was vulnerable; his status as a respected philosopher would not be enough to protect him. Only fifty years earlier, in 399 BC, the great Socrates had been charged with heresy by a jury of five hundred Athenian citizens, accused, Plato tells us in his *Apologia,* of being "an evil-doer and curious person, searching into things under the earth and above the heavens . . . and teaching all this to others." Socrates had been sentenced to death by a margin of six votes. Refusing the opportunity to escape the city and thus break the Athenian law that had condemned him, he was executed by poison.

As the xenophobia in the streets of Athens thickened in 348 BC, Aristotle finally gave the order to his slaves to begin packing up his belongings. He left the city quickly, boarding a merchant ship down in the port of Piraeus, taking with him his beautiful clothes, his extensive library,* and some of his slaves. He almost certainly did not expect to return. Few places of refuge were open to him: he could not head north to his hometown, Stageira, because Philip had destroyed it on his march into Olynthus; nor could he return to his mother's homeland in Chalcis, on the island of Euboea, because Philip had stirred up the islanders against Athenians and to them Aristotle was an Athenian. The political landscape of the Aegean had turned against him.

* Aristotle was one of the very first book collectors. His library would have been made up of papyrus rolls.

This is where the picture of Aristotle's early life disappears from view for a while. We know that he left Athens in 348 BC and that he arrived in the port of Atarneus, on the coast of the Troad, two hundred miles east, a few weeks or months later. We lose him in between, somewhere in the Aegean Sea. It is probable that he first headed directly north, hugging the coast of Euboea and then Thessaly, making his way to the Macedonian court in Pella, the capital city of Macedonia, the city where he had spent at least some of his childhood and which might have seemed a natural refuge. If he did find a way of getting to the Macedonian court, he would have found Philip, now in his midthirties, seriously disfigured since he had last seen him; he had lost an eye in battle, and the scar stretched across half of his face. Philip was also now a father of two

The seascape of Aristotle's journey.

© John Gilkes

boys—the seven-year-old Alexander, who would become Alexander the Great, and his elder brother, the ten-year-old Arrhidaeus—and he was married to his fourth wife, Olympias, a member of the snake-worshipping cult of Dionysus. The courtiers gossiped that she slept with snakes and that Alexander, already wild and uncontrollable, was the offspring of an alliance between his mother and Zeus, not the natural son of Philip.*

Aristotle's feelings toward Philip must have been complex at this point. Even if he had fond memories of the prince whom he had known as a boy in the Macedonian court, it would have been difficult for him to forgive the king for the destruction of Stageira. Under Philip's orders, the Macedonian soldiers had massacred or enslaved all the occupants of his hometown, and these may have included members of Aristotle's family, his late father's slaves, and his tutors and friends. They had set fire to and razed all the buildings he remembered with affection—the marketplace, the temples, the gymnasia. But he was now a political refugee and he would have to bite his tongue.

Whichever route he took on his great sea journey across the Aegean from Athens to Atarneus, Aristotle would have stopped at ports on many different Greek islands so that the captains could restock supplies, repair their sails, and rest their sailors. He spent a good deal of time in harbors, getting in and out of boats. Day by day, week by week, island by island, he would have noticed the bird and plant life changing around him. Watching the sailors fishing with nets and lines from the side of the boat, Aristotle would have noticed the differences between the fish that came up in the nets, the birds that followed the boats, and the leaves from the different plants that flourished in each new port. He no doubt besieged the sailors and the fishermen with questions. Each set of answers gave him a new set of questions to ask about the natural world: about adaptation, reproduction, species, varieties, diversity. For Aristotle, now at sea, like Ishmael in Melville's *Moby-Dick,* "the great flood-gates of the wonder-world swung open."

Aristotle's boat docked at the port of Atarneus a few weeks later. At some point since he had fled Athens, either one or both of his protectors, Philip of Macedonia or Proxenus, his uncle, had secured him a new posi-

* Even at this young age, Alexander attracted mythical stories.

tion as a political adviser in this foreign court. Aristotle's new patron and protector, Hermias, was one of the most interesting figures in this fourth-century-BC Aegean theater of power. Hermias had been born a slave and had worked as a private secretary for a successful banker who had spent his money on extending his territories along the coast of Asia Minor opposite Lesbos; he had even been sent to study at Plato's Academy in Athens. When his master died, Hermias inherited a wealthy coastal princedom, protected by a small army of mercenaries; now he wanted to establish a court that would replicate something of the life he had experienced in Athens. Plato, assuring him that all just and successful rulers needed philosophers in their courts to advise and temper the power of the king, had already sent him two philosopher brothers from the academy, Erastos and Koriskos, to give him advice on governance and ethics.

Aristotle flourished in Hermias' court. He married Hermias' young niece Pythias* soon after he arrived, a marriage that indicates the high esteem in which his patron held him. Now Aristotle was more than a court philosopher to Hermias; he was kin. Soon he was also father to a daughter, named after her young mother. He had a new kind of power here, an opportunity to apply his emerging ideas about politics and governance to the concrete day-to-day problems that pressed upon his patron and to help shape the ruler's relationships with his subjects.

In return for effective political counsel, Hermias gave Aristotle and his companions a school in Assos, some one hundred miles away, a small coastal town built on raised ground that looked across the sea to Lesbos. As news of the school spread, young men from settlements inland came to Assos to attend classes. Aristotle was still a Platonist, but in his new Academy he was striking out for himself, shaping and defining new questions about the natural world.

It was in Assos that Aristotle began the project that would become his great book *The History of the Animals,* an attempt to describe the habits of as many different species as possible and to use those descriptions to try to discover key patterns and designs amid all the apparent diversity he had seen on his great sea journey and continued to see on the

* Some say niece, some say courtesan or even adopted daughter. The uncertainty is perhaps revealing.

coast of the Troad. He took his students out of the lecture theater and down the road to the rocky seashore. He showed them how to collect detailed facts about plants, lizards, and birds, how to find birds' nests, how to observe animals in their natural habitats, how to dissect a snake or a seabird to observe its internal as well as external structures.

Nobody had asked questions and gathered information quite like this before; certainly no one in Athens had been interested in studying the shapes and functions of birds' beaks or the stamens of plants or the life cycles of cicadas. As Aristotle and his students collected along the rocky shoreline of Assos, wading in and out of rock pools, watching seabirds nesting, mating, and feeding, they talked incessantly, sketching out arguments and counterarguments about nature's patterns and rhythms. Start with the facts, Aristotle would insist, then work up and out to general principles. Accept nothing as truth unless you see it with your own eyes. The younger men, such as Theophrastus or Coriscus' son Neleus, accepted these new principles and methods easily, enthusiastically collecting beetles, describing plumage patterns, and learning to dissect. It was always more difficult for Aristotle to persuade his older Athenian companions, still preoccupied with Plato's abstractions, of the worth of this fact-collecting work.

There had been a revolution in Aristotle's way of seeing and thinking in the months since he left the Academy. Plato's world had been one of abstracts and ideal forms, concentrated not on the here-and-now, the materiality of nature, but on another world entirely, *up there.* Aristotle's new world, in contrast, consisted of observable facts, a world of *down here.*

Nature had usurped his childhood gods; it astounded him. *Philosophy starts in wonder and wonderment,* he would tell his students as he gathered them to watch a chick hatch from its egg or a spider weave its web. Astonished by the immense variety of species and body parts he had seen on his sea journey, Aristotle was searching for evidence of the unity and regularity of design and structure at the heart of all this apparent diversity. There were continuities as well as discontinuities with Plato's philosophy in this zoological work—Aristotle was, like Plato, searching for the irreducible form in things. But Aristotle had become what we would now call an empiricist. And this empiricism began in exile from the Academy, on the seashores and lagoons of Asia Minor.

—

After two years teaching in the school in Assos, Aristotle, his wife, and his companions sailed across the nine-mile-wide channel to the island of Lesbos, where they stayed for a further two years. Theophrastus brought him to the island as Aristotle's zoological project gathered pace, describing the fish in the lagoons, the petrified forest up in the hills, the natural springs that gushed up in cracks in the rocks or in lush moss-lined caves. The Greek poets Alcaeus and Sappho, who had once lived there, described the island as wet and heavy with fruit, flowers, grapes, and olives. Alcaeus wrote of the "bloom of soft autumn," and claimed to have "heard the flowery spring coming," the "chirring of summer cicadas, the long-winged widgeon flying overhead, harlequin necks outstretched." Sappho described an orchard where "cool water babbles through apple-branches, and the whole place is shadowed by roses, and from the shimmering leaves the sleep of enchantment comes down; therein too a meadow, where horses graze, blossoms with spring flowers and the winds blow gently."

During the two years in which Aristotle lived on Lesbos, he favored the largest of the two lagoons for fish observation, perhaps because it bordered Pyrra, one of the five main cities of Lesbos, built high up on a hill, a citadel above a tiny natural harbor. Pyrra was destroyed by an earthquake only a hundred years after Aristotle walked its streets and marketplaces; the whole town disappeared into the lagoon. Today you can still see stumps of columns sticking out of the lagoon, half a mile from the salt flats where fresh sea salt is harvested and where Aristotle watched fish spawn and cuttlefish change color.

For two years the beautiful sea lagoon at Pyrra became Aristotle's natural laboratory. He would have come to know the time of day by the coming and going of the fishermen's boats, by the length of the shadows, by the sounds of the men at work on the salt flats, by the calls of the birds, or by the depth of blue in the water. His zoological questions multiplied here, they became more focused and precise; he had begun to question the relationship between stasis, change, and diversity, between form and function. He assigned specific natural philosophical questions to particular students. Why had the starfish in the lagoon multiplied to such an extent that they had become a pest to the fishermen and yet the scallops had all but disappeared? Why did sea urchins spawn in midwinter here but almost nowhere else in the Aegean? Why were there so many varieties of cuttlefish in the lagoon and yet no octopuses at all? Why were there no parrot wrasses, or any kind of spiny fish, or sea crawfish, or the spotted or the spiny dogfish; and, why did all the lagoon's fishes, except only a little gudgeon, migrate seaward to breed? Why was the chameleon common to the Asiatic coast and found all around the lagoon and yet never to be found on the Greek islands farther west or on the Greek mainland?

The philosophers set to work to investigate nature's ways. All around the bay below the citadel of Pyrra, the fishermen had built walkways and piers where they fished with rod and line and tended simple fish farms. On land, on the harbors and near the markets, they built holding bays or cisterns to keep their finest and most valuable fish alive and protected from the scorching sun. Lying on the platforms built out over the lagoon, Aristotle could observe the fish breeding and feeding in their natural environment. Here, either in the cool of the covered cisterns or out on the wooden walkways in the cool of the early mornings or as the sun began to dip, he watched them for hours at a time, making careful notes or dictating to a slave and formulating questions to discuss with Theophrastus and the other students.

In the introduction to *Parts of Animals,* Aristotle passionately defends this orderly observation of particulars; one senses that it was difficult to convince students who had been trained on abstract verbal dueling that investigating earthworms or insects was worth doing. "If any person thinks the examination of the rest of the animal kingdom an unworthy task," he wrote in that introduction,

> he must hold in like disesteem the study of man. For no one can look at the primordia of the human frame—blood, flesh, bones, vessels, and the like—without much repugnance. Moreover, when any one of the parts or structures, be it which it may, is under discussion, it must not be supposed that it is its material composition to which attention is being directed or which is the object of the discussion, but the relation of such part to the total form. Similarly, the true object of architecture is not bricks, mortar, or timber, but the house; and so the principal object of natural philosophy

> is not the material elements, but their composition, and the totality of the form, independently of which they have no existence.

In defending his methods, Aristotle reminded his students of what Heraclitus had said to the strangers who hesitated to enter when they found him warming himself by the kitchen fire. He told them not to be afraid, for gods were present even in the kitchen. Nothing was too low for the philosopher. Every aspect of nature's realm was worthy of study. Every animal opened up to investigation revealed something natural and beautiful. Nothing was without design, nothing without purpose. Instead of speculating on the things in the heavens you cannot know, he told his students, use your eyes and hands and senses to investigate what *is*, not what *might be.*

The more Aristotle and his students collected facts, the more those facts challenged the worth of the old classification methods. He came to understand that you could not describe one group of animals as feathered and another as unfeathered, one group as wild and another tame, or call some animals walkers and others fliers. Take an ant, for instance, he would explain, or a glowworm. If you stick to the old categories, you will have to put them in both the winged and unwinged groups. Now Aristotle was preoccupied not with dichotomies but with gradations, species that looked almost identical and yet had one or more quite different body parts.

From his descriptions we know that Aristotle would lie for several hours on a small platform at the lagoon watching the giant goby (*kobios*) and the intertidal blenny (*phucis*) as they swam about together in the shallows. The two fish are almost impossible to tell apart; they have the same salt-and-pepper markings, body shapes, and upturned eyes. Lifting them out of the water, Aristotle examined their fins, noting that the goby's fins are like suckers and the blenny's fins are like rays. But though the design of each is different, they work in exactly the same way, he argued. Both are perfect designs to enable the fish to hold their positions on the muddy floor of the lagoon when high winds whip the waves into peaks. And by dropping scraps of meat into the water he would show his students how the two fish eat different things, too. *So why are two fish so similar and yet so different?* one of his students might have asked. Aristotle's answer was that these two fish had been given limbs and parts that were the most useful designs for their own particular environments. Nature had provided those already perfected forms. Nature itself was perfect.

Once the pile of collated facts about animals had grown substantial, Aristotle became more confident. He was certain the cosmos had not been created by a god or gods; it had not come into being nor would it pass away; it was eternal. And if there was no beginning to the world, it could not have been designed; there had been no maker. It was ruled and driven from inside itself, not from the outside. Some aspects of the body parts of animals served a purpose; others were just the outcomes of a process. All around him he saw the evidence of modes of adaptation that ensured the preservation of species: the different shapes of the beaks of the Lesbos birds he studied; the varying shapes and lengths of birds' legs. But he had also become certain that species, though subject to minor changes and adaptations, were as eternal as time itself. Certain small animals might be spontaneously generated in mud or decaying matter, but this was no proof of species change. While he knew of the existence of hybrids, some fertile, others not, this, too, did not challenge his conviction that species were eternal and that they had not come into being through the chance collisions of atoms, as the Greek philosopher Democritus had once argued.

But the work was frustrating, too. Aristotle would find a pattern or a rule and then he would find an exception to that rule. "Oviparous fish as a rule spawn only once a year," he noted down with satisfaction, but then a fisherman would bring him a new fish from the lagoon to dissect and he would have to add: "but the little *phucis* or black goby is an exception, as it spawns twice."

As he worked, Aristotle became increasingly preoccupied by zoological anomalies. On Lesbos, as the scale of the project grew, the students began to specialize. Theophrastus was studying the plants of the lagoon and the surrounding hills. Aristotle himself was concentrating on the fish. The two men would have exchanged facts and questions across the boundary between their two specialties, just as Darwin and his friend the botanist Joseph Hooker were to do two thousand years later in Darwin's study. How can an animal be defined? Aristotle would ask his students. What makes an animal *absolutely different* from a plant? And what makes humans *absolutely different* from animals? What are the essential differences? And the students repeated back the formula that Aristotle had taught them: humans differ from animals because they have reason; animals differ from plants because they perceive and feel; plants have nutritive and reproductive faculties only. It was a beautifully simple formula.

But in the lagoons of Lesbos and in the sea around its shores, Aristotle found animals and plants that did not fit his definitions. He called them "dualisers" or "borderliners." The fishermen of Mytilene brought him plant-shaped creatures that appeared to be sensitive to touch and animal-shaped creatures that were rooted to rock—sponges, sea squirts, jellyfish, sea anemones, pinna, and razorshells, each of which had both animal and plant characteristics. The sponges looked like plants—they had branches and roots. They did not move. Yet the fishermen insisted they were sensitive to the touch. They appeared to straddle the two kingdoms. They seemed to suggest that nature moved in a continuous gradation from the nonliving to the living, and from plants to animals.

Instead of turning away from these exceptions to his general theory, Aristotle kept coming back to them. The sea sponges were the most baffling of all the organisms Aristotle studied. Inert, rooted to rock on the seabed, their body shapes resembled a branching tree or sometimes the delicately laced underside of a mushroom. They looked like plants. But under closer examination in a terra-cotta pot full of seawater, Aristotle was certain that he could see excretion going on through the sponge's pores. And plants do not excrete. When his categories broke down like that, Aristotle knew he was up against something both troubling and important. He was, he wrote in his notes, *at a loss*. He could *see* the sea sponge, he could turn it over in his hands, he could dissect it, he could draw and map every tiny orifice of its body, inside and out, he could watch it ingesting plankton under water, but he could not *make it out*. It did not fit. And his eyes were not strong enough to see even farther inside the sponge's pores, where there might be more clues to be found.

Sometimes Aristotle was at a loss because he had reached the limit of what he knew or what it was possible to know—he was limited by what his eyes could see; he had no microscope. And sometimes he was at a loss because the thing in front of him—dissected, minutely examined, flayed—just did not fit the patterns he believed to be fundamental to all biological structures. Sometimes both things happened at once. Some sea creatures were simply uncategorizable. When Aristotle was at a loss, the language he used in his notes wavered a little. He did not try to force round pegs into square holes; he was always honest about the difficulties in front of him. He repeated certain words in those moments of unknowing: *phainetai*, which means "what appears" or "what is evident," or *dokei*, which means "what seems" or "what is thought to be the case." There were sometimes major difficulties in his deciding: "The boundary and the middle, between the non-living and the animals," he wrote, "escapes our notice [*lanthanein*]"; "as regards certain things in the sea, one would be at a loss [*diaporēseien an tis*] to know whether they are animals or plants"; "how to classify them is unclear."

In his fact collecting on Assos and later on Lesbos, Aristotle relied heavily on the local knowledge of huntsmen, beekeepers, fishermen, sponge divers, and herdsmen as well as on his students. Hermias, no doubt fascinated by Aristotle's new work, sent him to elderly and experienced men from the mountains or from the coastal villages who could bring him quickly to the bodies of hunted deer before they began to decay or could hunt down certain wild animals for him or bring him a beehive. Aristotle forged relationships with those men by asking endless questions and by respecting their knowledge.

Aristotle could sail out to sea with the sponge divers, but he could not follow them to the seafloor. Sponge diving was skilled and treacherous work. Many of the older sponge divers were deaf because their eardrums had burst. Some were paralyzed, their bodies twisted and contorted from surfacing too fast. They spoke of extraordinary fish and monsters for which they had no name. So, out at sea on the sponge divers' boats, Aristotle watched the men work.

A single sponge diver took his place at the prow of the boat, with a long rope tied around his waist, a heavy stone in one hand, and a sharp cutting implement in the other, scanning the waves for the familiar shapes of the sponges hundreds of feet below, his mouth full of white oil that he used to light up the darkness of the seabed, his comrades cheering him on. Aristotle would watch the lean body first squat and then arc through the air, disappearing deep into the water, weighed down by the rock, the rope slithering after him into the depths. He would see the diver resurface gasping for air minutes later, clutching a basket full of sponges scythed from the rock on the seafloor. On deck, the resting divers trampled the sponges with their bare feet, rinsing them with seawater and threading them onto a length of rope to be washed again at the back of the boat before the sponges were dried out in the sun and sacked up ready for market.

No one asked the sponge divers questions as Aristotle did. He asked them about touch, about the effect of shadows falling across sponge beds. He needed them, he said, to be absolutely precise in their answers, because he could not go where they went, or see what they saw. Do the sponges sense you coming? Can they hear you? Do they live longer if you cut or pull them off the rock? Do they grow again if you put their severed roots in seawater? And of course, the most pressing question of all: Are they plants or animals? One wonders whether Aristotle shared his philosophical puzzles with the sponge divers and fishermen. And if he had found a way of explaining his project and its philosophical importance, what might those ideas have meant to the sponge divers of Lesbos, who depended in all their dangerous daily work upon rituals and sacrifices, prayers and myths, who leaned on their gods both as protection and as explanation?

The local people had myths to explain everything. They said that the trees in the petrified forest up at Eressos had been turned to stone by Zeus' thunderbolt; that the enormous bones the farmers sometimes plowed up were the remains of a giant race of humans and animals that had lived long before man and that had been wiped out by jealous gods. Aristotle refused to accept mythical or supernatural explanations for the anomalies he found in the rock pools and nets and caves and under the ground. These were superstitions and folktales, he said. They were not stories for philosophers. He refused to accept any explanation that he could not verify with his own eyes.

But if there were no gods doing the making, one of the sponge divers might have asked Aristotle, how did life come to be? And Aristotle would have told him that there was no beginning, no origin. The world in all its variety has always been and would always be. "Coming to be and passing away," he wrote, "will . . . always be continuous and will never fail." He had come to see the living world as a series of cycles in a serenely perfected and inextinguishable world. "For becoming starts from non-being and advances till it reaches being and decay goes back again from being to non-being," he wrote. That is the great beauty of it. No egg is without a parent. Everything comes out of something that has lived before. There can be no ultimate origin or beginning.

Sometimes, by way of providing a useful counterargument to his own ideas, he might have told his students about the bold propositions of his predecessors, the pre-Socratic philosophers. There was Anaximander of Miletus, who, speculating on the origin of life three hundred years before, had proposed that all life had originated in water; that in the early years before life began, the sun's heat had interacted with the primal mud of the earth to generate sea creatures. Some of these creatures eventually progressed to a chrysalis stage, Anaximander suggested, which hatched open to produce the first primitive humans who had clambered out of their fish-germinated chrysalis shells and onto dry land. It was a theory of great ingenuity, Aristotle might have added, but of little worth.

Then there was Empedocles of Acragas, the poet, who claimed a hundred years earlier that before humans there had been myriads of strangely shaped creatures suspended in some vast primal soup—disembodied heads, limbs, eyes—and that these body parts came together in hundreds of different strange combinations, thrown together randomly, chaotically: men with the heads of oxen, oxen with the heads of humans. The unviable combinations of body parts perished over time, Empedocles believed, while the combinations that worked survived. Aristotle thought Empedocles' words worth quoting—but "Empedocles was wrong," he declared in *Parts of Animals*. Species could not have been formed by random collisions and accidents. Democritus was wrong, too, he insisted, to declare a hundred years earlier that forms had come into being as a result of random interactions of atoms.

Such stories were wild and unverifiable, Aristotle told his students; they were no better than myths, no more verifiable than Deucalion's stories of the flood. Nature worked according to laws that were already in operation. It was not chaotic, nor was it random, as Empedocles and Democritus had argued. Nature produced animals perfectly adapted to their environments with all the body parts they needed to survive and reproduce. Nature was perfectly designed. You had only to look at the design of human teeth to see that. How could such a structure have come about randomly, by chance or collision? Tales of species' mutability, stories of gods and monsters and divine interventions, magical transformations and animal-human hybrids, he repeated, were tales told by soothsayers and priests. They were not stories for philosophers.

Aristotle was an intellectual risk taker and maverick. He admired the ideas of the natural philosophers who had gone before him and collected their theories. He cited them respectfully in discussion. But he accepted little as truth without having witnessed it. He might have entertained the idea of inherited characteristics or of modification with descent if someone had presented him with the proof. In mulling over the borderline creatures, he was considering zoological puzzles that would lead to evolutionary answers thousands of years later. But for Aristotle, the reason for the blurring here on the border between the animal and plant kingdoms was not that all species had evolved from common primitive aquatic life-forms, but that nature was arranged along a scale and that the blurring he saw here signified the point of transition in that scale: "For nature moves continuously from lifeless things to the animals through things that are alive but not animals," he wrote. What he observed did not topple his absolute conviction of the fixity, regularity, immortality, and order of animal forms.

Aristotle saw corals spawn and dogfish mate, he saw the stages of the development of a chick in its egg, he saw bees dance and chameleons change color, but he did not see what we would call evolution or natural selection. Darwin was mistaken in his belief that Aristotle was the first man to publish evolutionary speculations; he was misled by James Clair Grece, the enthusiastic classicist and town clerk of Redhill, who had made an error. Grece had come across a passage of the *Physics* in which Aristotle seemed to be promoting what looked to him like a protoevolutionary theory. In fact, in this passage, Aristotle had been summarizing the ideas of Empedocles in order passionately to rebut those ideas. Grece mistook Empedocles' ideas for Aristotle's. He translated the passage and sent it to Darwin insisting, because he believed it to be true, that Aristotle had adhered to a theory of species change. Darwin, poor classical scholar that he was, believed James Clair Grece's claim and added Aristotle to his "Historical Sketch."

Aristotle was extraordinary and radical in his understanding of nature's laws. He understood the principle of change to be at the heart of the natural world; he refused supernatural or mythical explanations of natural phenomena; he understood that sea and land had changed places through deep time; he saw the conceptual importance of borderline organisms; he even grasped the principle of analogous parts—that the human hand was like the fin of a fish or the wing of a bird; he understood animal-human kinship; he saw that species are continuous and gradated. But he did not believe that species had transmuted from earlier forms. In Aristotle's world, all species were fixed within a world of unlimited duration. Individuals might change, they might evolve and metamorphose from nonbeing to being and back to nonbeing again—in that repeated arc of birth, growth, death, decay—but for Aristotle, species were beautifully, unchangeably fixed for all time. Flesh might bloom and decay, but the shape of the human body, the cicada body, the fish body—all those shapes and functions remained unchanging. Aristotle's nature was steady-state and beautifully, eternally continuous.

Aristotle was the first to see gradation in nature. He saw nature shading almost imperceptibly in those dark seabed forms he studied in the lagoon of Lesbos with the sun beating on his back, the sponges and sea squirts and jellyfish shading from animal to vegetable and from vegetable to animal. But while later philosophers might be able to convert Aristotle's "unbroken sequence," his *scala naturae,* into the Great Chain of Being, which dominated eighteenth-century science, and which would make it easier to think or see in evolutionary ways, Aristotle did not see evolution. He held too many beliefs that were irreconcilable with evolutionary ideas. Although he had broken with Plato's conception of ideal forms, nonetheless he believed that knowledge could be based only on what was fixed and not in flux. Only the fixity of individual forms through the eternity of species makes Aristotelian nature and thus knowledge possible.

Aristotle's world, a seascape of enormous variety of color bounded by the shores of the Aegean, was a world of flux contained within fixed shores. Perhaps given the flux and unpredictability of his own migrant life, living as he did in the shadow of Philip of Macedonia and his soon to be imperial son, Alexander, at the mercy of a theater of war, a world in which the landscape of his childhood had been leveled in a single siege, he was bound to believe in a form of deep unchangeability.

Aristotle's philosophical landscape differed profoundly from ours. He had different explanatory stories within which to work, different orthodoxies, different predecessors, different myths and belief systems to question or confirm. There was no orthodox account of creation in ancient Greece defended and policed by an established Church. There was no official Church-pronounced age of the earth. There was no heresy to guard against, no inquisition to fear. But there was no philosophical consensus, either. When the ancients argued about creation, their narratives were not inflected by stories of a man and a woman in a garden or a god moving across the face of the waters; their arguments turned on different questions, about whether the universe had been shaped by gods or atoms colliding chaotically or was the outcome of a design deep within all animal and plant forms.

Aristotle never returned to the intensive zoological work he had undertaken in Lesbos except to write up his lecture notes and to add more facts to the files he had already gathered. He never experienced such focused scientific work again. Philip summoned him to teach his son Alexander, who would become Alexander the Great; teaching the clever, unbridled, and energetic young man was an experiment in itself, important and challenging work. While Alexander pushed eastward, taking territory after territory right up to the borders of India, in one of the most aggressive imperialist campaigns of history, Aristotle was safe to continue to teach in his lyceum in Athens. When Alexander died, Aristotle was forced into exile from Athens a second time, dying of natural causes at the age of sixty-two in Chalcis, a town built on a peninsula of an island and thus almost entirely surrounded by the sea.

After Aristotle died and his manuscripts and library had been stored in a cellar in Scepsis for safekeeping, where they remained underground for several hundred years, the landscapes of those quarrels and disagreements shifted again, away from the arguments that Aristotle had made for a regulated, ordered, and eternal nature. A young man named Epicurus opened a new school in Athens in 306 BC called the Garden promoting a philosophy of ethics that returned to the ideas of the atomists, arguing, as Democritus had done, that different worlds come to be and pass away as the product only of the chance collisions of atoms, not of a controlling god. Instead there was a controlling principle at work, among plants and animals, a struggle for survival that resulted in numerous adaptations and modifications as well as extinctions.

Epicurus' writings have survived only through the retelling of Lucretius in *De rerum natura* and through the fragments of a thirty-seven-volume manuscript called *On Nature* that surfaced in the eighteenth-century excavations of a vast philosophical library in a seaside villa in Herculaneum, a library that had been engulfed by volcanic ash and lava in AD 79. Epicureanism took a short but significant hold on Roman life and ideals, but as Christianity swept across the ancient world, these notions were outlawed and denounced as pagan and materialist by Christian emperors as they closed down the academies of Greece.

Darwin did not read Epicurus or Aristotle. The Greeks frightened him a little. They seemed to be from a different world entirely. If he had been able to read "the ancients," as he called them, he might have recognized in parts of Aristotle's writings and in corners of Epicurus, Democritus, and Empedocles the glimmerings of thoughts and questions that were remarkably similar to his own. Instead he was misled by a town clerk in Redhill who managed to persuade him that Aristotle was an evolutionist, and so Darwin put Aristotle at the top of his list of predecessors. But although Darwin was, like Aristotle, looking closely at the intricate patternings of plumage on pigeons and studying the habits of worms and asking similar questions about what he had documented, he was battling with different orthodoxies, different gods, negotiating different philosophical landscapes and belief systems. Perhaps Darwin was right to be a little frightened by the gulf of time that stretched between him and the man who wandered the shores of the lagoons of Lesbos in the fourth century BC. Perhaps Aristotle was always beyond his comprehension not only because Darwin did not read ancient Greek, and thus always had to rely on others to translate and understand the words and concepts that Aristotle used, but because their ways of seeing and understanding the workings of time and of nature were so immeasurably different.

3

The Worshipful Curiosity of Jahiz

BASRA AND BAGHDAD, 850

In the ninth century AD, the port city of Basra sprawled across a great web of palm-fringed canals; it was a city of water and waterways, flanked by the desert on one side and the great river Tigris on the other. The sparkling minarets of three mosques rose above streets lined with shops, markets, suqs, and warehouses. At the western side of the city, the mud brick of houses gave way to rich pastureland and palm groves and finally to the Desert Gate, where the city met the desert at the famous Mirbad, a great open space where the Bedouin caravans halted.

Here date croppers had once dried their dates on vast racks laid out in the sun, but by the ninth century it was the busiest part of the city, the place where Bedouin traders, migrating to Basra to sell their animals or their jewelery and to restock their water supplies, kept their camels and sheep. Open to the sky outside the city walls, the Mirbad must have been breathtaking: a sea of camels, the brightly colored woolen clothes of the Bedouins, the flapping of cloth in the wind, the sounds of men singing, poets reciting, camels and mules braying, snake charmers and magicians, smoke drifting from cooking fires, the smells of cooking meats and spices, of animal dung.

It was a meeting place of dialects and cultures. Scholars walked here from the city; philologists came to study language in use and to search out interesting grammatical constructions and unknown words. Collectors and anthologists came to record the poetry that the Bedouin poets recited here, performing on platforms to crowds of dedicated followers. The bookstalls, bookbinders, and stationers' suqs spread out like a labyrinth from the marketplace, with books bound in gleaming leather piled on low tables or arranged on shelves in booths. Here in the dust of the Mirbad, on the edge of desert and city, as well as in the mosques, libraries, and observatories of the city itself, scholars transcribed, translated, and recorded. An oral culture was passing into a written one.

Among the crowd of philologists and lexicographers listening to the recitations of the Bedouin poets in the Mirbad in the early ninth century was a young man of striking ugliness, a man with eyes so protruding that his friends called him al-Jahiz, or Goggle-Eyes.* Jahiz, who may have been of part African descent, had grown up in the streets of Basra, working as a boy selling fish on the banks of the canals to bring in money for his family. Though his family was not socially well connected, he attended the elementary Qur'an school and the mosques where eminent scholars gathered to teach; later he educated himself in the bookstalls of the suqs as well as by listening and talking to the Bedouins in the Mirbad. Jahiz would become one of the most prolific and versatile writers of the Abbasid Empire, producing more than two hundred books during his lifetime, works on literature, biology, zoology, history, rhetoric, psychology, theology, and polemics.

In the extraordinary *Book of Living Beings* (*Kitab al-Hayawan*), probably written between 847 and 867, Jahiz produced the first extensive study of animals published in the Islamic world and came close to a theory of evolution and natural selection that would not be matched for another thousand years. Some Islamic nationalists, journalists, and Internet bloggers now claim Jahiz as Darwin before Darwin, the real inventor of

*His full name was Abu 'Uthman 'Amr ibn Bahr al-Kinani al-Fuqaimi al-Basri.

evolution; some even allege that Darwin stole Jahiz's work and passed it off as his own.

But Darwin had never heard of Jahiz and could not have plagiarized his ideas. He did not read Arabic, and there were no English translations of *The Book of Living Beings* for him to read; there is no complete English translation to this day. But had he been able to read Arabic, he would no doubt have been enthralled by Jahiz's book and might have recognized Aristotle's influence in its pages. Jahiz admired Aristotle's *The History of Animals,* which he had read in Arabic translation, but he was also certain that the Arabic people, and particularly the Bedouins of the desert, knew a great deal more about animals than the great Greek philosopher had. *Living Beings* was his attempt to bring ancient Greek and Arabic knowledge together in one place, to turn oral into written knowledge, and to demonstrate at the same time that everything has a purpose and a place in the great scheme of nature and that everything proves the existence and wisdom of God. *The Book of Living Beings* is a strange and beguiling jumble of insights, speculations, and facts about animals; more than a miscellany, it is virtually an encyclopedia. Through seven volumes, Jahiz weaves together pre-Islamic poetry, passages from the Qur'an, comic stories, and passages of philosophy, metaphysics, sociology, and anthropology. Its elaborate, meandering, and complex prose is very different from the sparse, precise sentences of Aristotle's *The History of Animals.*

Ideas flower in certain places and at certain moments in time. Jahiz came into adulthood in the midst of the extraordinary Abbasid translation movement. While Europe slumbered fitfully through the Dark Ages, in the eighth century the Abbasid caliph Mansur, the third caliph of the dynasty, built the great round city of Baghdad on the rich agricultural land of what is now central Iraq.* By his grandson Ma'mun's time, Baghdad had become the center of the largest and richest empire in the world through an agricultural revolution supported by new irrigation, engineering, and drainage technologies and a network of trade routes that

*The Abbasid dynasty had only recently—AD 750—come to power after overthrowing the Arab-Syrian Umayyad caliphs.

stretched from North Africa through Spain to the borders of France, across India and Afghanistan and the Persian Empire, to the edge of western China.

While Vikings' longboats were invading Britain, while European feudal dukes ruled over small courts and controlled bands of mercenaries, while Christian priests quarreled over the Book of Revelations and the anticipated date of the end of the world, in Baghdad the Abbasid caliphs and the courtly elite paid scholars to expand medical knowledge and to develop chemical and alchemical methods for the manufacture of goods, mathematics for the calculation of taxes and the administration of the empire, astronomy for mapmaking, astrology for casting horoscopes, and engineering and physics for irrigation, agriculture, and navigation.

The caliph Ma'mun, the Abbasid court propagandists repeated as they spun richly colored myths extolling the glories of their ruler's intellectual quests and patronage, had a direct line not only to God but also

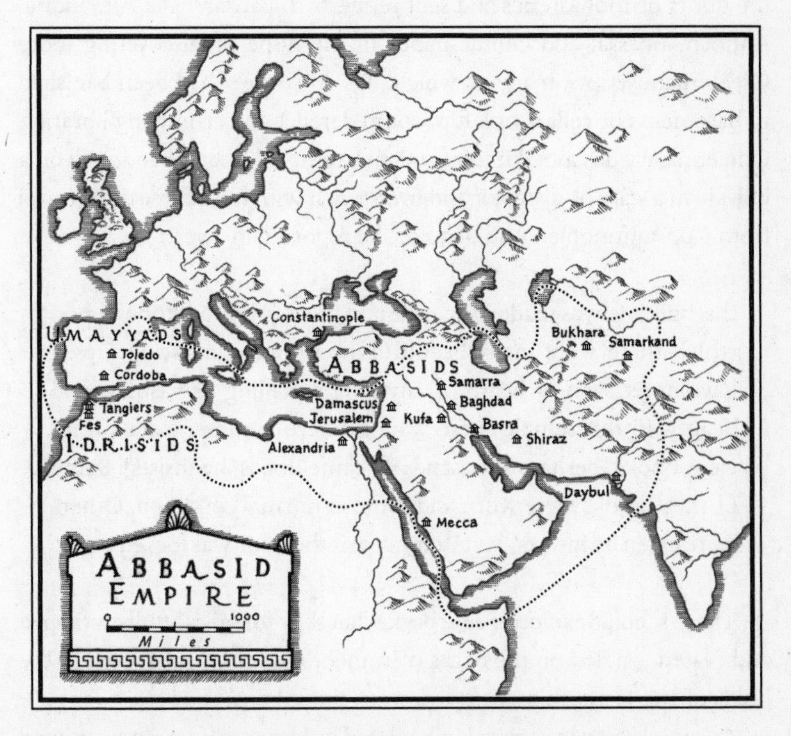

The extent of the Abbasid Empire in the ninth century AD.

© John Gilkes

to the great Aristotle. A man with pale skin, a flushed face, and bloodshot eyes had once appeared to the sleeping caliph, they said; when asked to identify himself, the apparition at the end of the bed replied simply: "I am Aristotle." Before dawn he had persuaded Ma'mun that the rationality of the Greeks and the revelation of Islam could be braided together. But when the caliph awoke and ordered his eunuchs to bring him all of Aristotle's works, the eunuchs told him apologetically that the libraries of Baghdad held only a very small number of Aristotle's manuscripts, and these were in Greek and not Arabic; the remainder of Aristotle's works were lost, locked away in moldering libraries and basements in Byzantium, Syria, and Alexandria.

For decades Ma'mun and the many wealthy courtiers and administrators of Baghdad, compelled by the desire to rediscover and translate lost knowledge, sent out emissaries to hunt for ancient Syriac and Greek manuscripts in Syria, Palestine, and Iraq. Scholar-explorers knocked on the doors of monasteries and sent requests to patriarchs in Alexandria, Antioch, Edessa, and Gundeshapur in the hope of discovering more Greek manuscripts, many of which, like Aristotle's, had been banished to basements or cellars or left to rot in derelict and crumbling libraries. One emissary described finding a Greek temple library abandoned on a hillside in a state of disrepair and overgrown with trees three days' travel from Constantinople. "And behold," he reported in awe,

> this building was made of marble and great coloured stones, upon which there were many beautiful inscriptions and sculptures. I have never seen or heard of anything equalling its vastness and beauty. In this temple there were numerous camel loads of ancient books [perhaps thousands of camel loads, he insists]. Some of these books were worn and some in normal condition. Others were eaten by insects. . . . After my exit the door was locked again.

These scholar-explorers sent back what they found carefully wrapped and crated, carried on the backs of camels and mules or stacked in the holds of merchant ships, across deserts, mountains, and seas.

As camel caravans arrived in Baghdad bringing manuscripts quarried from the locked mountain monasteries of Syria and Alexandria or carrying dusty scholars from all over the Abbasid Empire, the caliph Ma'mun established what became known as the House of Wisdom, in Baghdad, an institute for the translation, preservation, and pursuit of knowledge centered on an extraordinary library. Wealthy patrons built elaborate palaces, libraries, and gardens in Baghdad and lavishly endowed hospitals, but they displayed their wealth most ostentatiously in competing to commission translations as a demonstration of their sophistication and their pious dedication to the expansion of knowledge.

Each translator living and working in Basra or Baghdad spoke several languages, and though many were lowborn, they were richly paid for their work.* Many were Greek-speaking Syriac Christians. Sometimes friends or family members worked together in small translation dynasties. Within a century they had translated almost all the ancient Greek secular scientific and philosophical texts in existence, including all of Aristotle's works, translating them from Pahlavi and Greek usually first into Syriac and then into Arabic. Once rendered into Arabic, Greek wisdom became gradually assimilated into a body of distinctly Arabic-Islamic knowledge. These translations were commissioned to enrich, cultivate, and empower the Abbasid elite, but they also provided the means by which an extraordinary body of knowledge came to be rediscovered and channeled into Europe, which would bring about what we call the Renaissance.

The invention of paper further revolutionized the circulation of new and revived knowledge. The Abbasid Empire depended on a vast administrative and intelligence machine, which in turn depended upon the production of immense quantities of leather and parchment and, later, paper for correspondence, intelligence, and record keeping. By the middle of the ninth century, Baghdad's mills produced enough paper for the use of all the secretaries and administrators in the War Office, the Office of Expenditure, the State Treasury, the Office of Correspondence, the Office of Letter Opening, the Caliphal Bank, and the Post and Intelligence Office. Cheap paper meant that books could be produced more easily. More than a hundred shops selling paper and books lined the Stationers' Market in the southwest of the city. By the middle of the tenth century, ship

*Some reports suggest that a number of translators received as much as the modern equivalent of $40,000 a month.

mills—floating paper mills powered by the river's current—were moored on the Tigris. Booksellers rented books to their customers and procured books for the expanding libraries of affluent and educated citizens.

Jahiz, the enigmatic author of *The Book of Living Beings,* surfaced from this wealthy, intellectually and scientifically curious culture; he emerged from the streets of Basra and Baghdad as Rabelais did from the streets of Paris or Joyce from Dublin. But it is hard to excavate him from that history. What we know of him comes from his own beautiful and tantalizing descriptions of conversations, his descriptions of bird breeders and Bedouins, the glimpses he gives us of places, streets, and rooms. On every page of *The Book of Living Beings,* his voice is hauntingly alive and engaging, reaching us across more than a thousand years. Like Aristotle, he disappears now and again from the record.

Even as a very young man, Jahiz was an extraordinary observer, attentive, insatiably inquisitive, his ear attuned to new languages and to new turns of phrase. As a young man growing up in Basra, he watched the kaleidoscopic world around him with an anthropologist's eye, looking out for differences in behavior and manners, memorizing and recording those differences. A good Muslim, he spent much of his time at the mosque, not just for worship but also for debates and lectures. He mixed with the *masjidiyyun,* whom he satirized, savants who gathered in the congregational mosque of Basra to discuss all manner of questions, mostly secular, honing their conversational and dialectical skills and discussing theological and philosophical issues. There he heard lectures on grammar and lexicography and participated in political, philosophical, and ethical debates.

The dialectical competitions that were staged in the mosque and the conversations he heard in the Mirbad probably shaped Jahiz's writing style and method. In the pages of *Living Beings* he moves between different positions or theories, equivocates, raises voices and lets them speak, orchestrating them as though he were a referee rather than a participant. His prose swaggers brilliantly from one subject to another, often making no obvious connection between them, and then, when he is done with wandering, suddenly returns to his original subject. His long sentences tangle and twist and turn back on themselves, meandering down and around and back again as intricately as the canal systems of Basra and as multivoiced as the Mirbad itself.

Jahiz was an *adib,* a writer of the *adab,* elegant collations of knowledge assembled for the amusement and edification of the cultivated man; both encyclopedic and conversational, didactic and dialogic, they were threaded through with anecdotes and aphorisms drawn from politics, ethics, religious thought, etiquette, literature, and science. In Jahiz's prose we hear the voices of the ninth-century Basra mosque and marketplace talk about the political, religious, and moral questions that concerned them. In turning Basra conversations into words on pages, Jahiz sometimes made a kind of poetry.

When he was not listening to the poets in the Mirbad or debating with the savants in the mosque, Jahiz would be reading or walking through the warren of streets of paper shops, pen sellers, bookbinders, book copiers, ink sellers, and booksellers in the Mirbad, shops that were multiplying by the day in Basra as the young philologists, poetry collectors, and translators produced more and more books for sale. Jahiz read every book he came across wherever he happened to be. While most readers would be content to rent a single book and return it within a few days, as was the custom, Jahiz would rent whole shops for the night and stay in them alone until morning so that he could read everything at hand without interruption.

When Jahiz left Basra for Baghdad around 815, he almost certainly traveled by water up the main artery of the canal and then up the great Tigris. At first he was retained in Baghdad in an official capacity as an adviser and apologist for the government, writing letters and reports; later he worked for several different powerful politicians and judges. Although he lived in Baghdad for the next forty years and produced most of his books there, he returned to Basra regularly, and he also traveled to Samarra when the seat of the caliphate moved there in 836, riding on mules or on boats navigating the great network of waterways that threaded across the country like veins, always on the lookout for new knowledge on the journey in conversation with his fellow travelers.

Although the court at Baghdad provided him with a living and with intellectual freedom, conditions were not always safe there. Patrons were plentiful but some allegiances dangerous. Petitioning a patron, entering his entourage, and maintaining a place there was always difficult because competition for the finest patrons was fierce. Scholars and writers had to be both rhetorically skillful and canny. The life of a writer was lucrative but precarious.

The streets and palaces of Baghdad gave Jahiz inspiration for new books. Although he had seen many domestic species of animals in the marketplaces and harbors of Basra—chickens, camels, donkeys, dogs, sheep—in Samarra and Baghdad he could wander in the zoological gardens of the Abbasid caliphate. Here he saw lions, tigers, and giraffes—animals imported from Africa and Asia, animals that walked the banks of the Tigris with their native keepers. Some say the great zoological garden in Samarra, with its ornamental rose gardens and fountains, contained a thousand animals and plants of hundreds of different species shipped in from every corner of the empire; visitors described waterwheels being turned not by oxen but by ostriches. Jahiz, like so many of his peers, was dazzled and mystified by the diversity of all these exotic species, by plumage patterns and colors, by the different structures of necks and beaks and claws. He talked to the keepers, asking questions about diet, behavior, and reproduction. Why, he asked, did some animals seem to adapt easily to their new home, while others sickened and died?

It was in a library in Baghdad that Jahiz almost certainly read Aristotle's nineteen volumes of zoological work for the first time, now partially translated into Arabic. Sometime around 846–47 he hatched the idea of writing *The Book of Living Beings,* an Arabic reworking and rectification of Aristotle. It was an attempt to gather together all the facts known about animals from every possible source, to understand and to capture in writing the totality of God's creation and man's place in it. He began to collect facts, anecdotes, and knowledge. He may have believed he could come to an end, that he could catalog everything, get to the bottom of everything, but the list of animals he wanted to write about kept on growing and his thoughts about animals kept on expanding. Inevitably the book, seven volumes long, was still unfinished when he died twenty years later.

There was no other book in the ninth-century Abbasid Empire like Jahiz's *Living Beings*, no compendium of animal knowledge, except Aristotle's *The History of Animals.* That is not to say that there was no knowledge of animals in pre-ninth-century Islam. Quite the reverse. A successful civilization such as that of the Abbasids depended on a sophisticated and extensive knowledge of animal breeding, animal husbandry, hunting methods, medicine, migration patterns, and animal diet. The nomadic Bedouin tribesmen, proud and self-sufficient, were the custodians of this knowledge in the Abbasid Empire. They passed it down from parent to child, preserving it in songs and stories and in their poetry, which was prized as the highest of the art forms. When they came to the marketplaces of the great Abbasid cities to sell their camels and sheep, scholars like Jahiz came to consult with them.

In assembling *Living Beings,* Jahiz was not concerned with argument or theorizing. He was concerned with witnessing; he promoted the pleasures and fascinations of close looking and told his readers that there was nothing more important than this. We can approach a better understanding of God's creation only if we examine closely, he insisted. Here and there amid the close looking there are visions, glimpses of brilliant insight and perception about natural laws, but the overt purpose of *Living Beings* was to persuade the reader to fulfill his moral obligation to God, an obligation enjoined by the Qur'an: to look closely and to search for understanding.

If certain historians have claimed that Jahiz wrote about evolution a thousand years before Darwin and that he discovered natural selection, they have misunderstood. Jahiz was not trying to work out how the world began or how species had come to be. He believed that God had done the making and that he had done it brilliantly. He took divine creation and intelligent design for granted. So did all the people he talked to about animals—the Bedouins and the hunters and the animal trainers in the zoological gardens in Baghdad. There was, for him, no other possible explanation.

What is striking, however, about Jahiz's portrait of nature in *Living Beings* is his vision of interconnectedness, his repeated images of nets and webs. He certainly saw ecosystems, as we would call them now, in the natural world. He also understood what we might call the survival of the fittest. He saw adaptation. Like Aristotle, he believed in spontaneous generation—he had seen flies emerge from the flesh of dead animals. None of those things was remarkable or controversial in his time. His aim in writing *Living Beings,* he told his readers, was to prove to them that the world of animals around them was interconnected, mutually dependent, that everything had its place in the great web and that it was possible to account even for the presence of harm and danger in the world as a sign of God's generosity and blessing.

Jahiz's web of interdependencies foreshadows Darwin's vision of the "entangled bank," the exquisite and poetic extended metaphor that Darwin used to distill the key ideas of natural selection and to conclude *On the Origin of Species:* "It is interesting," Darwin wrote,

> to contemplate an entangled bank, clothed with many plants of many kinds, with birds singing on the bushes, with various insects flitting about, and with worms crawling through the damp earth, and to reflect that these elaborately constructed forms, so different from each other, and dependent on each other in so complex a manner, have all been produced by laws acting around us. . . . Thus, from the war of nature, from famine and death, the most exalted object which we are capable of conceiving, namely, the production of the higher animals, directly follows. There is grandeur in this view of life, with its several powers, having been originally breathed into a few forms or into one; and that, whilst this planet has gone cycling on according to the fixed law of gravity, from so simple a beginning endless forms most beautiful and most wonderful have been, and are being, evolved.

Jahiz had his own version of the entangled-bank metaphor. In Book 2, he described the creatures gathered around a desert fire. "All you need do is light a fire in the middle of a clump of trees or in the desert and watch the various insects that converge on it," he wrote; "then you will see creatures and shapes that you never imagined God had created. Moreover the creatures that come towards the fire vary according to whether the fire is in a clump of trees, the sea or the mountains." To demonstrate this principle, he lit fires on the canal banks of Basra, in the courtyards of palaces, and in mountains and forests in the company of philosophers, caliphs, patrons, or the Bedouins he traveled with.

It is not difficult to imagine Jahiz a few miles outside Basra or Baghdad in the desert plains with a group of Bedouins sitting around a fire in the dark, close to a tangle of tents that have been erected from sticks, ropes, and rugs. Beyond the circle of the fire, the darkness presses; the Bedouin camels and mules are present only as brayings in the night when they are startled by a desert fox. Above them, stars of different degrees of brightness make patterns on the great arc of the night sky, patterns that the Bedouins use to map the desert. In boxes and in nets, Jahiz and his companions catch and count the flying insects and listen for the sound of desert foxes, or watch for the tracks of mice, snakes, and lizards that the fire draws out of the night. Sometimes there are as many as seventy different species in the circle of a fire, Jahiz shows his companions, but the combination is always different. You can sit by a night fire by the canals of Basra or by the great ocean or in the desert or in the forest, and the assembly will always be different. God has made it that way—every animal and every insect has its place in the great web.

When one of the Bedouins, pointing to the stinging insects, scorpions, and snakes in the trapping nets, asks why God has created so many bad animals among all of the useful ones, the men agree that everything is connected, good and bad, that nature requires opposites for balance, that man is only one animal, like the mosquito, in the great web. They discuss how everything depends upon everything else for its food: toward the top of the chain is the wolf, then the fox, then the hedgehog, the adder and other snakes, the sparrow, the grasshopper, the hornet's brood, the bee, the fly, and finally the mosquito. "Each species," Jahiz wrote in *Living Beings,* "constitutes a food for another species. Each animal cannot do without food, and so must hunt. Thus each animal feeds upon weaker species than itself. Similarly each animal is the food for a species that is stronger than itself. Divine Wisdom has decreed that some species are the source of food for others, and some species are the cause of death for others." Hunting and defense are mutually coexistent in these webs. Jahiz wanted his reader to see that something that might appear noxious might not actually be noxious if it was comprehended properly. Everything had its place and function.

A lion eating the entrails of a cow from a later illustrated manuscript of Jahiz's *Book of Living Beings*.
Biblioteca Ambrosiana, Milan

It is in passages like these that some historians of Islamic science have claimed that "we can see the germs of Darwin's . . . natural selection." We can; but if this is natural selection, Jahiz's natural world is governed not by Darwin's savage nature but by a divinely perfected and balanced universe. While there are certainly glimpses of an understanding and perception that might have led Jahiz to evolutionary speculations, the vision of nature in his work is profoundly different from that of Darwin.

Jahiz reminded his readers that the keys to all the answers to the great metaphysical questions—life and death, good and evil—are to be found in the natural world. Everything around them was a system of signs. He told them that all they had to do was to look closely and to use their reason, that everything they saw in the natural world—miraculous, interconnected, interdependent, diverse—was proof of the existence of a creator; he wrote his book so that

> every man endowed with reason may know that God did not create His creation to no purpose, and did not abandon His creatures to their fate; that he overlooked nothing, left nothing without his distinctive mark, nothing in disorder or unprotected; that He makes no mistakes in His wondrous farsightedness and no detail of His dispositions fails him, nor yet the beauty of wisdom and the glory of the powerful proof. All of that activity extends to everything from the louse and the butterfly to the seven celestial spheres and the seven climates of the globe.

Like Aristotle before him and Darwin after him, Jahiz depended on animal trainers, beekeepers, pigeon trainers, and animal breeders; he depended on the Bedouins above all. Everything that he had read in Aristotle's zoology, he claimed, was already known by Bedouin traders and already contained in the great Arabic poetry that passed by word of mouth, from father to son, from philosophers to poets and from poets to philologists, in deserts and marketplaces and menageries across the empire. "We rarely hear of a statement by a philosopher on natural history," he wrote, "or come across a reference to the subject in books by doctors or dialecticians, without finding an identical passage in Arab or Bedouin poetry, or in the everyday wisdom of those who speak our language and belong to our religious community."

In the Wild Beast Park of Baghdad, the caliphal menagerie reached via the Perfume Market, Jahiz talked to the elephant and giraffe keepers and asked them why their animals did not mate in captivity. He watched elephants, lions, giraffes, and leopards, and he asked the animal trainers why the crocodiles that had been imported from India had died in the Tigris. He traveled to a gazelle farm outside Basra to talk to a farmer about breeding. In the bird markets of Basra and Baghdad he talked to pigeon breeders from Antioch who described the time and money they spent in breeding and training their birds. These pigeon breeders and trainers were valuable men in the Abbasid Empire; trained pigeons carried important reports, letters, and instructions between geographically distant cities in one of the first systems of pigeon post. Jahiz was impressed by how much they knew and how skilled they were: "They carry [the pigeons] on their backs after transporting them by boat, shut them up in houses, separate and reunite them at the appropriate times, take the females far away from the males, put the males with other females, and are at pains to avoid the ill effects of close inbreeding. . . . They are less solicitous for the wombs of their own wives than for those of hen pigeons that yield thoroughbred chicks."

Yet while Jahiz clearly admired Aristotle's observations and depended on them, he did not always trust Aristotle's eyes or very much of what the Greek philosopher had to say about fish, believing he had relied too heavily on the claims of sailors and fishermen. He explains:

> We have not devoted a separate chapter to fish and other denizens of salt and fresh water, canals, rivers, swamps and streams because for most of these species we have been unable to find key passages of sufficient exactitude to inspire confidence. . . . The only material available is information provided by sailors, and sailors are not renowned as respecters of the unvarnished truth. The stranger the story, the more they like it; and moreover they use vulgar expressions and have an atrocious style. . . . Aristotle devoted much space to the subject, but I could find no evidence in the book beyond his own assertions. I once said to a sailor: "Aristotle claims that whenever fish eat anything they swallow some water at the same time, owing to their voraciousness and the size of their mouth aperture." His own answer was to say: "Only someone who has been a fish or been told by a fish can know for certain." . . . That sailor must have been a dialectician, and have prided himself that he understood the causes of things to have answered me so.

However, while Jahiz, unlike Aristotle, may have been humbled by the unknowability of fish, like Aristotle he became increasingly preoccupied with anomalous organisms in the natural world. Everywhere in *Living Beings,* Jahiz defends and celebrates hybrids and crossbreeds, creatures that are interstitial or transcategorical in some way. He lost no opportunity to point out when creatures did not fit comfortably or easily into one category but extended across several: birds, for instance, that were both herbivorous and carnivorous. After a long preface about the usefulness of written as opposed to oral knowledge, the first book of *Living Beings* begins with a sixty-page disquisition on hybrids, crossbred birds, giraffes and mules, marriages between people of different races, and the eunuch, an important figure in the Abbasid court and in the military, whom he describes as having a greater life expectancy and superior intelligence to other humans. Books, he pointed out, were also hybrid creations, collages, mixtures, compilations, yet they were profoundly important as a way of bringing different races and peoples together under one commonwealth.

Like Aristotle, Jahiz attempted to generalize about species and about their positions and function in the natural world and found it increasingly difficult to do so the more he investigated. Hybrids, anomalies, and organisms that seemed to be both animals and plants preoccupied him precisely because they forced orthodox taxonomies to collapse and opened up new questions about how to produce a coherent taxonomy of creation. Had he pursued his investigation of anomalies as Darwin did later with his barnacles, for instance, he might have reached different conclusions about how species had come to be, explanations that might have centered on the idea of descent with modification. Instead, Jahiz used the collapse of taxonomies to question categories of high and low in creation, to show, again, how the All-Powerful Creator had fashioned a world of endless complexity and sophistication in which every organism had its place and depended on everything else.

It was the scientific curiosity of the world Jahiz lived in, a curiosity enjoined by the Qur'an, that taught him to value observation and empiricism and that led him to the writings of the great philosopher of zoological empiricism, Aristotle. But Jahiz could never take his patrons or his safety for granted. Eighteen years after he had arrived in Baghdad, and while he was still compiling *Living Beings,* the extraordinarily centralized Abbasid caliphal power was beginning to move into a stage of slow decline. Every day, postal envoys brought news of insurrections and disturbances in the provinces, and of occasional riots in Baghdad. The caliphs, fearing that the increasingly powerful Turkish slave-soldiers would stage a coup or an uprising, had moved their entire court up to Samarra some years before, seventy-five miles north of Baghdad, taking their administrators, military leaders, and savants with them. Baghdad had become too small for a garrison complex.

Jahiz's principal patron at Baghdad had until now been the vizier Zayyat, a wealthy Arabized Persian whose family had traded oil and who had managed to retain power through three caliphates. It was Zayyat to whom Jahiz had addressed and dedicated *Living Beings,* because Zayyat was funding his great book. But in 847 the caliph al-Wathiq died and was succeeded by his brother al-Mutawakkil, whom Zayyat had slighted. Zayyat's main rival at court, the chief *cadi* (judge) Ahmad ren Abi Du'ad, seized the opportunity to discredit Zayyat with the new caliph. As part of a campaign of retribution against the powerful men who had once threatened his authority, the caliph ordered his Turkish soldiers to arrest and imprison Zayyat. For six weeks the vizier was tortured by being placed inside an iron maiden, a wooden cylinder lined with spikes. All his houses in Samarra and Baghdad, his furnishings, his libraries, his slaves and singing girls and his warehouses of grain, raisins, figs, and garlic—everything was confiscated. The great libraries were broken up, the books dispersed. Zayyat died in prison from the wounds inflicted during his torture.

Jahiz, fearing for his life, fled to Basra and went into hiding, but he was discovered and brought back to the court by soldiers under orders of the *cadi*. For two years he worked as a propagandist for the chief judge, Ibn Abi Du'ad, and his son, Muhammad, but when the balance of power shifted again among the ruling elite, Ibn Abi Du'ad and his son were removed from high office. They were arrested days later, their estates confiscated, their libraries sold. Ibn Abi Du'ad was so badly tortured that he was partially paralyzed; he died three years later in 854, a few days after his son.

The new caliph, Mutawakkil, put an end to the court-sponsored inquiry into the nature of the Qur'an that had been conducted by the last three caliphs and pronounced that the Holy Book was uncreated rather than created, as his predecessors had maintained, promoting a form of monotheism that incorporated the Bible, the Gospels, and the Torah. At the same time he ordered all Christians to wear yellow sleeves on their outer garments and decreed that all newly built churches and synagogues be demolished. The rewards for writers and poets in the new court were good, particularly for poets who defended and justified the actions of the caliphate and who were prepared to participate in the work of the secret police.

The aging Jahiz, now in his seventies, forever adaptable, enjoyed an Indian summer of prosperity under Mutawakkil. He began by proposing a book on the primacy of Islam against the Christians. He wrote to tell the new vizier, al-Fath ben Khaqan, about this proposed book and petitioned him to secure a position in the court. Fath was a Turkish aristocrat, part of the growing and influential Turkish praetorian guard. He had an enormous library in his palace and a thriving courtly salon and was intimate with the caliph. Fath replied to Jahiz's petitions with enthusiasm and flattery. "The Commander of the Faithful has taken a tremendous liking to you," he wrote. "He rejoices to hear your name spoken.

Were it not that he thinks so highly of you because of your learning and erudition, he would require your constant attendance in his audience chamber to give him your views and tell him your opinion on the questions that occupy your time and thought." He urged Jahiz to finish his book. "You will be receiving your monthly allowance: I have arranged for you to be credited with the arrears and am also having you paid a whole year in advance. There is a windfall for you."

To ensure the favor of his new patron, Jahiz put aside his *Living Beings* again to finish a long treatise titled "On the Merits of the Turks," praising the pride, courage, and military prowess of Fath's people and incorporating an epistle he had composed much earlier on the subject. He chose his patron well. For seven good years he had access to one of the finest libraries in the world and to one of the finest salons. He began to publicize *Living Beings*. But in 861, Fath and the caliph were both violently assassinated. Jahiz, now suffering from a paralysis that was making it impossible to move without help, was allowed to return to his hometown of Basra, where he died in 869 at the age of ninety-four, sixty years more than the average life expectancy for a man or woman living in the Abbasid Empire at that time. According to popular lore, he was crushed to death when a wall of books fell on him.

Jahiz died in the first year of the great Zanj uprising, which took place in the marshes south of Basra. The Zanj people, African slaves in origin, lived in shantytowns out on the marshes, either removing topsoil from the alluvial plains to render the plains fertile for agriculture or working in the salt mines. Their rebel leader was an eloquent and charismatic poet who had taken refuge in Basra only a year before. On a September Friday in 871, the Zanj rebels, joined by Arab horsemen from the desert, marched into the city through the great marketplace of the Mirbad and set fire to the market stalls, the mosques, and the bridges; they rounded up the citizens and massacred hundreds of them by sword outside the palace. All the animals of the Mirbad died, engulfed in the flames, and every structure in the Mirbad was razed.

It took the caliphate army two decades to quell the Zanj uprising, which spread north. In 893, the historian al-Tabari reported that in an attempt to maintain public order after the Zanj had been defeated, "the authorities decreed in Baghdad that no popular preachers, astrologers, or fortune-tellers should sit and practise their trade in the Friday mosque. Moreover, the booksellers were sworn not to trade in books of theology, dialectical disputation or philosophy." When the traveler Ibn Jubayr came to the city in 1184, he found Baghdad "a shadow of itself, a washed out ruin or the statue of a ghost."

Any Arabic translations of Aristotle's works that survived the later sacking of the great cities and libraries of Islam by the Mongol hordes in 1258 were scattered throughout Europe. But the Catholic Church, long obsessed with rooting out and destroying the works of infidels abroad during the Crusades and tracking down heretics within Europe through the work of inquisitors, continued to be deeply suspicious of pagan ideas. In 1210, as the great walls of Notre Dame's western façade rose to transform the Paris skyline, the Bishop of Paris ordered that all the members of a small sect in Champeaux be burned at the stake for preaching pantheism and materialism inspired by reading Aristotle. "Neither the works of Aristotle on natural philosophy," the bishop decreed, "nor their commentaries are to be read at Paris in public or private. This we forbid under penalty of excommunication." But it was too late to stop the spread of Aristotle's ideas. Within twenty years, the pope, under pressure from the professors of France, insisted that the bishop reverse his decree. The ancient Greek's rhetorical arts and his knowledge, the professors claimed, were an essential part of the war against heresy.

When Constantinople fell in 1453 to the Crusaders, a stream of refugee scholars fanned down through Italy pushing handcarts of rescued manuscripts containing the combined knowledge of Arabic and Greek science and philosophy—Aristotle, Averroes and Avicenna, Euclid, Ptolemy, and Plato. Many settled in Florence, where they established new networks and took in students, accelerating a flow of knowledge that had been crisscrossing Europe for centuries.

In 1423 a Florentine bookseller and book agent named Giovanni Aurispa returned to Venice from Constantinople with 238 complete Greek manuscripts. Within decades these had been translated into Latin, printed up on one of the many new printing presses in the Italian cities, and were on sale in the booksellers' stalls of Florence and Milan, where a young artist named Leonardo da Vinci bought three translations of works by Aristotle for his rapidly increasing library.

4

Leonardo and the Potter

MILAN, 1493; PARIS, 1570

Sometime in 1493, a family of Italian peasants pushed a handcart into the courtyard of a grand but faded palace in the center of Milan and asked to speak to Il Maestro, Leonardo da Vinci. Inside the Corte Vecchio the sculptor chiseled away at a colossal clay mold of a horse, twenty-four feet high, that lay on its side in the dusty half-light of the old palace ballroom, a statue commissioned by Leonardo's patron, the great Duke of Milan, Prince Ludovico il Moro, to commemorate his father. Assistants and pupils crossed quietly between rooms carrying paint, clay, and wax; nearby, pupils stretched canvases, sketched, or mixed paints.

The visitors from the mountains waited, fascinated by the hum of activity in the atelier and hoping to get a glimpse of the legendary horse or the flying machine. Once owned by a great Milanese dynasty, the palace—fortified, towered, and moated—had fallen into disrepair by the time Leonardo, in search of larger premises, moved in there. The colonnaded pillars in the courtyards were peeling, the corridors drafty, the floors bare, the carved library shelves dusty and empty of books, the frescoes faded. But for Leonardo, who needed high ceilings, good light, rooms for his assistants, stables for his horses, space for his flying machine, stage sets, and mechanical designs and models, a study for his books, a laboratory for his experiments, and shady courtyards in which to draw, no building could have been better. It even had a section of flat roof from which he could launch his smaller flying machines when they were complete.

When the tall, handsome man with the mane of dusty brown hair emerged from the dark interior, the peasants told him they had been traveling for days, pulling their heavy cart over mountain passes and across pitted roads. In the mountains, people talked of Il Maestro and his mighty horse, they told him; people whispered that he collected the rocks called petrifications. So they had brought him red rocks famous in their mountains, rocks with strange markings and shapes, flecked with oyster shells and corals, some as big as a hand. How did the shells get to the tops of the mountains? they asked the sculptor as he examined their gifts in the sunlight of the courtyard. The priests maintained that they were carried there by the great flood described in the Bible, they told him, but the astrologers insisted that they had been made by magic deep in the rocks on nights when the stars lined up in particular patterns. Leonardo shrugged. He had no idea, he told them.

For months after the peasants had returned to their mountain village, well paid for their trouble and their gifts, Leonardo puzzled over how the seashells had washed up in the mountains. He turned over the red rocks in the light of a candle, peering into the mystifyingly entombed and clustered shells and consulting his books, struggling with his Latin to find an answer to the question the peasants had framed. As the illegitimate son of a nobleman and a peasant woman, Leonardo lacked a formal education, but like many artists and engineers of his day, he had fashioned himself into an intellectual. And like others in the community of intellectuals in which he was accorded equal status, he was searching for answers to the great mysteries of creation. Like the humanists around him, Leonardo was returning to the ancient texts that had flooded into Italy after the sack of Constantinople searching for answers. He had made himself a library.

Impatient to understand the way everything worked, and scornful of the dominance of the Church and of alchemy and necromancy, Leonardo had taught himself to read and write Latin; he wandered the booksellers' shops of Milan looking for translations of ancient Greek manuscripts, now being translated into Latin and spun out by the thumping printing presses invented only fifty years earlier by Gutenberg and now pressing ink onto paper in all the major Italian cities. Leonardo arranged his growing collection of books in the Corte Vecchio near a desk by an open window, books and notebooks that helped to channel the noise in his head, the perpetual rush of ideas and questions. By 1493 his library contained thirty-seven books, expensive printed volumes that included the Bible, the Psalms, books by Ficino, Ovid, Livy, Aesop, Petrarch, and Pliny, and three newly printed Latin translations of Aristotle. A decade later it would contain three times as many volumes.

Leonardo added the red rocks to the others in the collection of natural objects he had gathered from trips into the mountains. Beyond the rooftops of Milan, the mountain range of Le Grigne tantalized him; he would stand at the top of the towers of the Corte Vecchio or up on the roof of Milan Cathedral gazing north, running his eyes along the perpetually changing colors of the snowcapped mountain peaks as the sun moved and shadows brushed across them, wondering about those shapes and how they had come to be. He remembered a moment in his childhood when, walking in the mountains, he had looked down into a cave mouth and had been "astounded," petrified, he wrote, both by "fear and desire—fear of that dark, threatening cave; desire to see if there was some marvellous thing within." He painted striations of rocky landscapes glimpsed distantly through the Corte Vecchio windows, from the veering crags of *The Virgin of the Rocks* (ca.1483) to the framed, sunlit, always out of reach mountains of *The Last Supper* (ca.1496).

On several occasions, sketchbook balanced on his knee, Leonardo traveled by mule from Milan to Lecco up into the mountains along the Carraia del Ferro, the old Iron Road used for centuries to transport ore from the mines of Valsassina. He told himself and the duke's envoys that he was studying watercourses for canalization projects, but he was also, as always, engaged in numerous investigations about the way water—deluges, watercourses, rivers, streams, and storms—had carved and chiseled and polished the landscape above and below ground through unimaginable stretches of time. Leonardo scrambled down into caves and mine workings so that he could draw the shapes and layers of rock, in places as molten as petrified water. He sketched the whirled and flurried patterns in the exposed rock outcrops—the signs of ancient flows and upheavals; he noted the lakes and rivers he saw there and the "fantastical things" he saw in the copper and silver mines; he described fish in the high lakes, hot springs at Bormio, and the ebbs and flows of the Fonte Pliniana. He talked to the mountain people, recording their knowledge of their landscapes. "In the Chiavenna valley," he wrote in his notebook, "are very high barren mountains with huge rocks. Among the mountains you see the water-birds called *marangoni*. Here grow firs, larches and pines. There are deer, wild goats, chamois and terrible bears. It is impossible to climb without going on hands and feet. The peasants go there at the time of the snows with huge devices to make the bears tumble down the slopes." In all his descriptions, he repeated the phrase "I myself have seen it" again and again, invoking the Aristotelian imperative he lived by: never trust a fact unless you have seen it with your own eyes.

But when the peasants came to the Corte Vecchio with their questions about the oyster shells in the mountain rock, Leonardo could not answer them. It drove him back to his books and notebooks. He knew he was not the first to be vexed by the problem of inland shells: the Greek philosopher and poet Xenophanes and the geographer Herodotus had concluded in the fourth and fifth centuries BC that mountain shells had been left behind by marine or alluvial inundations in some ancient unimaginable past. Leonardo scanned the pages of his copy of Aristotle's *Metereology* for further speculations, but Aristotle had nothing to say about petrifications.

But in Milan there was never enough time to attend to anything, never any quiet to pursue trains of thought. There was always the horse; there was always the duke's envoy asking about the horse; there were the duke's parties to design entertainments for, pupils coming and going, the flying machine to engineer, and new commissions for paintings and frescoes.

Two years later, as he worked both on *The Last Supper* in the refectory of the monastery of Santa Maria delle Grazie in Milan and on the horse in the Corte Vecchio, Leonardo was still distracted by a host of connected scientific and artistic questions, still taking notes, still puzzling about water, shells, and rock. War with France interrupted his work and his studies. The duke, despairing that his brilliant engineer and sculptor would ever finish anything, reclaimed the bronze he had assigned for the horse to build cannons. When the French army invaded Milan in 1499, the soldiers used the unfinished clay horse for target practice, fracturing it into pieces. Leonardo fled the city, eventually returning to his hometown, Florence. Over the following years, the fresco of *The Last Supper* in the Milan monastery crumbled, cracked, and faded.[*] In 1517, the diarist and travel writer Antonio de Beatis lamented that the mural was "beginning to spoil," and forty years later Vasari saw "nothing visible but a muddle of blots."

Leonardo da Vinci's drawing of Neptune and his horses, ca. 1504–5.
The Royal Collection © 2011 Her Majesty Queen Elizabeth II/The Bridgeman Art Library

Leonardo saw mutability everywhere. He drew swirls in rock, and later, when those swirls reminded him of the curls of hair on the head of an old woman or the eddies in a mountain torrent, he drew the head, the

* It was vandalized by Napoleonic soldiers in the early nineteenth century and narrowly escaped the bombs of the Allies in the summer of 1943.

rock, and the river next to one another in his notebooks. He looked at the profile of an old man and saw the jutting features of an apparently still molten mountain ridge. Sometimes everything seemed to him to be in the process of transformation. When he saw horses fighting he saw male warriors, their mouths open, foaming, teeth clenched. When he traced the arteries of a dissected body he saw trees and root systems; when he drew trees he saw veins. He believed, as the Greeks did, that the patterns and structures of small things mirrored the same patterns and structures in the universe as a whole. But, for Leonardo, if the world was made up of repeated patterns, none of them was a static archetype or the blueprint of a divine designer. Leonardo's patterns and shapes were all on the move, passing continually through ripeness to decay. Nothing, he knew, remained untouched by time. Even the apparently solid masses of mountain ranges were passing constantly through processes of putrefaction and regeneration, driven by water that was ever flooding, gushing, eroding, silting, slicing, leveling, blocking—destroying and remaking landscapes in an endless cycle. He pushed his vocabulary to its utmost limit, forever collecting words to describe the actions of water on rock.

If rocks were no more than forms caught temporarily in time, then the same was true of living beings. For Leonardo, all forms dripped, flowed, and metamorphosed like Ovid's creatures. As he sketched a face or a body in motion, other forms seemed to be constantly pushing through. He drew grotesque and misshapen bodies in states of transformation, feeling for points where the edges broke down, where something else seemed to be surfacing. The body of the duke on his monumental clay horse: Where, Leonardo asked himself, did the man end and the horse begin? Life was continuity and flux.

In October 1503, after a summer spent completing the portrait of Lisa del Giocondo and making excursions into the Pisan hills, Leonardo moved into the large disused refectory of the monastery of Santa Maria Novella so that he could begin working on a new commission, the first major public work he had attempted since *The Last Supper.* His task was to decorate a wall of the Council Hall on the first floor of the Palazzo

Vecchio with a huge battle scene to commemorate the Battle of Anghiari, in which Milan had triumphed over the Italian League led by the Republic of Florence in 1440. The battle had become a symbol of the city's power and ambition.

It was winter, and the roof and windows leaked. Workmen mended them and, at Leonardo's instruction, papered them up, perhaps to keep out the light or to ensure privacy. Niccolò Machiavelli, the Florentine statesman and scholar, sent over a long description of the battle copied out by his assistant, but the scene was already taking shape in Leonardo's mind. Ten years earlier he had written instructions on "How to represent a battle," noting down images as vividly as if he were describing a recurring dream. The battleground was for him a place of intense mutability: "First you must show the smoke of the artillery, mingling in the air with the dust thrown up by the movement of horses and soldiers," he wrote. "The air must be filled with arrows in every direction, and the cannon-balls must have a train of smoke following their flight. . . . If you show one who has fallen you must show the place where his body has slithered in the blood-stained dust and mud. . . . Others must be represented in the agonies of death, grinding their teeth, rolling their eyes, with their fists clenched against their bodies and their legs contorted. . . . There might be seen a number of men fallen in a heap over a dead horse." Always horses merge with men. His preparatory sketches juxtapose human, lion, and horse heads, eyes and nostrils dilated, teeth grinding, fury making them one. He called battle a "Pazzia Bestialissima," a most bestial madness.

The battle scene that took shape on the paper suspended by ropes in a frame on the refectory wall in early summer 1505 might have been a scene from Ovid's *Metamorphoses;* a great tangle of human and animal body parts, it was impossible to see where one body ended and another began. In Rubens's extraordinary watercolor copy of Leonardo's fresco, painted in 1603, horses and men merge as the soldiers fight for possession of the standard.* Everywhere in the sketches for the battle scene, Leonardo dissolved the edges between man and animal and pressed fur-

*There are several copies of Leonardo's lost *Battle of Anghiari:* an anonymous copy in oils painted on a wooden panel is called the *Tavola Doria*; there is a painting by Lorenzo Zacchia dated 1558, and the Rubens watercolor, which is in the Louvre.

ther into the mutability of species than he had yet dared, further into the scaled, furred, feathered, and watery mutability of his dreams. On the left he drew the enemy leader, monstrous in his wrath, turning into a centaur: his skin, clothed in ram's fleece, is turning into fur and scales; the ram's horn on his breastplate, a symbol of Mars, has become a second head; his helmet is the coils of a snake, his shoulder a giant xenophora shell. Leonardo drew spirals everywhere—in hair, the ram horns, shells, helmet snake coils, even in the curls of the ram's fleece. Beneath his paintbrush, the repeated spirals of terrestrial and aquatic animal forms became the eddies of a deluge wiping out all distinctions and leveling everything.

Between finishing the huge wood-framed patchwork of a drawing for the scene in the refectory in midsummer and moving into the Council Hall to begin work on the fresco of it with his assistants, Leonardo left Florence and headed for the promontory of Piombino, where he stood on the waterfront to watch the wind whipping the sea into torrents. Years later he remembered the violence of the sea and wind: "of the winds at Piombino," he wrote, "gusts of wind and rain, with branches

Rubens's watercolor copy of Leonardo's fresco of *The Battle of Anghiari*, ca. 1604.
Louvre, Paris/The Bridgeman Art Library

and trees blown into the air. The emptying of rainwater from the boats." He had conjured a storm into the battle; painting it even seemed to summon storms. "On 6 June 1505, on Friday," he recorded with drama and precision, "at the stroke of the 13th hour [about 9:30 A.M.] I began to paint at the Palazzo. At the moment of putting down the paintbrush the weather changed for the worse, and the bell in the law-courts began to toll. The cartoon came loose. The water spilled as the jug which contained it broke. And suddenly the weather worsened, and the rain poured down until nightfall. And it was as dark as night." It was a kind of omen. As he applied the paint to the thick primer he had pasted to the wall, pocked with holes from the imprints of the drawing, the paint began to drip. He and his men suspended large burning charcoal braziers from ropes close to the painting to save what they could, but only the lower part survived intact. The upper part ran and bled, the colors intermingling, the edges fluid, the wall awash.

Leonardo lost patience with the project. He broke his contract with the city fathers and abandoned his already crumbling battle scene, turning instead back to birds and to flight—to his flying machine projects and to his unfinished sketches of *Leda and the Swan*.

Three years later, in the spring of 1508, in the quiet and privacy of the Palazzo Martelli on the Via Larga in Florence, where Leonardo was one of several guests of the intellectual and art patron Piero di Braccio Martelli, Leonardo, now in his sixties, returned to the shells in the red rocks. Distracted and vexed by lengthy legal cases that his half-brothers had begun against him in an inheritance dispute, he had determined to sort through his boxes and crates of manuscripts, overwhelmed by the great weight of unfinished investigations—he called them *casi,* or cases—he had collected. He came upon the notes he had made on the red shell-flecked rocks of Verona and remembered the arrival of the peasants at the Corte Vecchio and the urgency of their question.

Now, all these years later, he had begun to wonder whether the presence of shells in the mountain rock might be an answer to the set of questions about geophysics and hydraulics that plagued him and that now seemed increasingly entangled: questions about tides and currents and the atmospheric, geological, and erosive effect of water on the surface of the earth. It all needed careful, systematic investigation. It needed quiet thought. Aristotle, he remembered, perpetually moving from investigation to investigation, had gotten only so far with his pursuit of the same set of questions. So with his red, shell-tombed rocks laid out on his desk in front of the open window, his copy of Aristotle probably open beside him, the pages of his earlier notes scattered about, he began the Leicester Codex, a notebook entirely dedicated to geology and the physics of water. It was a book he had been thinking about all his life.

He began his investigation of the shells-in-the-mountains mystery by rebutting some of the existing explanations, his language inflected by weeks of writing legal defenses and letters to lawyers. Were the shells really made in the rocks by the action of stars or the moon, as some alchemists and astrologers claimed? That was superstitious nonsense. "And if you should say that these shells have been and still constantly are being created in such places as these by the nature of the locality or by potency of the heavens in these spots," he scratched out in mirror writing (the back-to-front writing he used in his notebooks), "such an opinion cannot exist in brains possessed with any extensive powers of reasoning."

Could the shells instead have been washed there by a universal deluge, as the priests claimed? Or even have made their own way to the higher ground? The cockles and oysters simply could not have reached the tops of mountains carried by the rain that fell in forty days, he reasoned. And even if the flood had been universal, they certainly could not have propelled themselves from the sea along the new seafloor so far in that time. He bought live cockles from the Florence fish market and tipped them out into a long container filled with seawater and sand in his rooms, noting down that no cockle moved more than eight feet a day: "With such a rate of motion," he declared, "it would not have travelled from the Adriatic Sea as far as Monferrato in Lombardy, a distance of 250 miles, in forty days."

Had they been drowned by the flood and then washed up here? Leonardo carefully chipped away pieces of rock with his sculpting tools. Young and old oyster shells clustered together there, attached to the rock and to one another, just as he had seen them grow on the shores of the Mediterranean. These oysters had not been washed there; they "had been born there," he wrote; they had grown on the very mountain rock in front of him and later they had been entombed while still alive, calcified in sedimentary rock.

The answer was therefore much simpler than the one the priests told about the flood. It had an austere poetry to it: "The peaks of the Apennines once stood up in a sea in the form of islands surrounded by salt water," he concluded, "and above the plains of Italy where flocks of birds are flying today, fishes were once swimming in large shoals." The mountain rock had once been a seafloor. Where birds now fly, fish once swam.

Piece by piece, always assuming that nature acted as it had always acted because there was no good reason to believe otherwise, Leonardo puzzled out the problem of the oyster shells, consulting every book in his now much larger library and in the libraries of his patrons. He reread Aristotle and Theophrastus on geology; he consulted his copy of Pliny's *Natural History;* he studied Avicenna and Averroes from medieval Latin sources. He knew that Aristotle understood about the upheavals that turned seabeds into mountains; the Greek philosopher had described that process nearly two thousand years earlier with a breathtakingly confident statement about the endlessness of time:

> It is therefore clear that as time is infinite and the universe eternal neither Tanais nor Nile always flowed but the place whence they flow was once dry: for their action has an end whereas time has none. And the same may be said with truth about other rivers. But if rivers come into being and perish and if the same parts of the earth are not always moist, the sea also must necessarily change correspondingly. And if in places the sea recedes whilst in others it encroaches, then inevitably the same parts of the earth as a whole are not always sea, nor always mainland, but in process of time all change.

Like Aristotle, Leonardo scorned supernatural explanations of natural phenomena, dismissing as "ignoramuses" all those who claimed that celestial influences had placed the shells in the rocks. He trusted his own eyes over and above the pronouncements of priests or alchemists or astrologers. And in this he played a dangerous game. At the beginning of the sixteenth century, those who rejected the opinions of priests might find themselves under surveillance—or, worse, on the lists of the inquisitors, even imprisoned or tortured.* He learned to keep his own counsel and taught himself to mirror-write to protect his notebooks from prying eyes. "He had a very heretical state of mind," the painter Giorgio Vasari wrote of him fifty years later. "He could not be content with any kind of religion at all, considering himself in all things much more a philosopher than a Christian." He was also dismissive of the claims of alchemists to be able to transmute materials, although he did depend on the chemical expertise recorded in alchemical formulas. He made enemies. He had to be careful not only to veil his intimacies with the men he loved, but also to guard his ideas and practices, for assistants and rivals might report him to the authorities as a heretic at any moment.

Throughout his life, Leonardo was plagued by questions about water and its movements through and across the earth and the trails and hollowings that it left behind. But like most polymaths, he was easily distracted; he abandoned project after project, notebook after notebook, in order to follow a new thought or to return to an old one. His range of reference and the scale of his curiosity were both his strengths and his weaknesses. His notebooks are full of brilliant ideas, connections, and inspired questions and plans, but few lead to any complete theory. His leaping frustrated him, too. "My concern now," he wrote in one notebook, "is to find subjects and inventions, gathering them as they occur to me; later I will put them in order, putting together those of the same kind. So, reader, you need not wonder, nor laugh at me, if here we jump from one subject to another."

But though later scientists, like Charles Lyell, working more patiently and systematically three hundred years later, using a mass of fossil and mineralogical evidence and microscopes, might take the problem of shells in the mountains toward an answer that suggested an earth his-

*The Church's powers of surveillance and inquisition became even stronger after the sequence of decrees and doctrinal definitions issued by the Council of Trent between 1545 and 1563.

tory that was millions of years older than had been previously believed and thus pave the way for an understanding of evolving species as well as evolving earth surfaces, Leonardo was doing something rather different in Florence in 1510. Certainly, he wanted to refute the authority of the priests and to disprove the theory of a universal deluge. He took the great age of the earth for granted. But he wanted more than anything else to prove that the earth worked like the human body, to prove what many Renaissance philosophers and scholars believed to be true: that the earth contained a soul, an *anima mundi,* that permeated the cosmos and connected all living beings within it.

"Nothing originates in a spot where there is no sentient, vegetable and rational life," he wrote.

> Feathers grow upon birds and are changed every year; hairs grow upon animals and are changed every year, excepting some parts, like the hairs of the beard in lions, cats and their like. The grass grows in the fields, and the leaves on the trees, and every year they are, in great part, renewed. So that we might say that the earth has a spirit of growth; that its flesh is the soil, its bones the arrangement and connection of the rocks of which the mountains are composed, its cartilage the tufa, and its blood the springs of water. The pool of blood which lies round the heart is the ocean, and its breathing, and the increase and decrease of the blood in the pulses, is represented in the earth by the flow and ebb of the sea; and the heat of the spirit of the world is the fire which pervades the earth, and the seat of the vegetative soul is in the fires, which in many parts of the earth find vent in baths and mines of sulphur, and in volcanoes, as at Mount Aetna in Sicily, and in many other places.

Underwater courses were always veins to Leonardo. When he traveled down into caves, he was traveling inside an enormous body that lived and breathed, seeped, and flowed. The earth was alive for him; it had veins and arteries; it was as sentient as the human body, and as vulnerable. He had brilliantly expressed this vision of the earth in the legendary

Mona Lisa, painted in Florence in 1503 only months before he began the first sketches for the *Battle of Anghiari:* La Gioconda is depicted on a balcony overlooking a complex geological landscape of flowing waters, demonstrating an entire hydrological cycle. The veins and waterways of the landscape mirror the veins and waterways of the sitter's body.

Leonardo's mind sought connection between all the workings of nature, between the macrocosm and the microcosm. Landscape and bodily surfaces concealed, depths revealed. To come to know more than his predecessors, he saw that he had to make himself a tunneling man. He had to go underground. It is no coincidence that while he was writing his pages on geophysics in Florence, he was also undertaking a series of secret human dissections in the hospital of the Santa Maria Nuova, a few streets away, tunneling beneath skin, feverishly searching out the secrets of the vascular system, aware that his eyesight was beginning to fail, drawing page after page of diagrams and notes.

Leonardo passionately rebutted Christian explanations for the presence of shells in mountains. He understood the idea of strata; he knew that the gaps between layers of strata represented thousands of years of time and that this meant that the earth was incomprehensibly older than the Church proposed. But he did not contemplate species *evolution* as we now understand it. He did not ask why, for instance, there were no human remains entombed with the oysters in the deepest levels of time. He was interested in different ways of explaining nature's laws. What the shells in the rocks of Verona told him was that the mountains had once been seabeds, and that provided proof of his deep conviction that the land and sea were in a continual process of rise and fall, like the lungs of a body. "The body of the earth, like the bodies of animals," he wrote in the Leicester Codex, "is interwoven with a network of veins, which are all joined together, and formed for the nutrition and vivification of the earth and of its creatures." Microcosm and macroscosm; connectedness across the cosmos; common patterns and structures not fixed in geometrically structured shapes but forever moving. It was a way of seeing shaped by the idiosyncratic collection of books in his library, by ideas quarried from Greek philosophy, from Renaissance humanists like Ficino, Aristotle, Averroes, and Avicenna; but however widely he roamed among the Greek philosophers or Renaissance humanists, there in his library or up in the mountains or inside his own head, moving between metaphysics, geology, physics, and hydrology, Leonardo was never able to answer to his own satisfaction the question buried in the shell-laced red rocks.

Leonardo was not the only Renaissance scholar-artisan to have spent his life returning to the great mystery of fossils and pondering the creation and origin of species. In France, only fifty years or so after Leonardo was buried in the Church of St. Florentin, in France, a Huguenot potter named Bernard Palissy was working on a similar set of questions about fossils, also using art as a means of detection. His patron was an extraordinary woman of Italian birth, the great Catherine de Médicis.

On the north bank of the Seine, in the district of Saint-Germain-l'Auxerrois, the classical columns and pilasters of the Queen Mother's magnificent new palace-retreat rose against the skyline in the midst of spectacular gardens along the river. The palace, named the Tuileries after the old tile kilns, or *tuileries*, that had once stood on the site, was the most elaborate royal building scheme in all of western Europe, but, like so many of Catherine's projects, it was, in 1570, still unfinished. When Catherine's architect died that year and an astrologer told her that she would die in Saint-Germain-l'Auxerrois, she found herself no longer in such a hurry to move her household there. Instead, she began designing another palace for herself, the Hôtel de la Reine, built within Paris's medieval walls in the northwest of the city. Though the Tuileries remained unfinished, the Queen Mother still held banquets and festivities there and brought her visitors and ministers to walk in the gardens.

Hidden away in the back of the Tuileries Gardens, in a series of artisans' ateliers, masons and carpenters chiseled away at the pilasters of columns or architraves; in one of the workshops Bernard Palissy was slowly piecing together a spectacular terra-cotta and ceramic grotto for the regent's beloved garden. Inspired by classical models, grottoes had become the rage across Europe, but Palissy had promised the Queen Mother that her grotto would be the finest in Europe.

The potter, who proudly bore the title of Worker of the Earth and Inventor of Rustic Ware to the King and the Queen Mother, never stopped working or talking. His clothes sweat-stained and his hands rough and covered in scars, he was a ball of energy, always quick to tell his visitors about the sacrifices he had made for his art—the terrible weeks when he was forced to tear up his own floorboards and take an ax to his last chairs to keep his kiln burning; his periods of semistarvation; the denunciations he suffered from the mouths of his neighbors; his wife's persecutions; the perpetual burdens of domestic and familial responsibilities; his endless clashes with the police, with debt collectors, and with the Inquisition; his time in prison accused of heresy and Protestant agitation. He owed his life to his first patron, he would recount, the constable of France, Montmorency, who had secured his release from prison and guaranteed his future safety. And now here he was in Paris, potter to the Queen Mother.

Catherine de Médicis, Italian by birth, an orphaned Medici heiress raised in a series of wealthy Medici courts in Florence under the protection of the pope, had been married at the age of fourteen to the French king, Francis I; after her husband's death, her young son took the French throne as Francis II; when he died a year later, she became regent to her second son, Charles IX, who was only ten years old when he came to the throne. Catherine's power, wealth, and interest in the arts made her the most important patron in Europe, with several hundred artists, sculptors, architects, masons, and landscape designers listed in her payment books in 1570. She had brought the principles of the Florentine and Medici classical revival to France, and for decades wealthy French aristocrats and intellectuals competed to follow her lead. On her travels around the country, she searched out provincial artists and brought them to her court or poached them from the courts of wealthy aristocrats. She found her Huguenot potter in Saintes working in his atelier, already designing grottoes and ceramics for the Duc de Montmorency, constable of France, her late father-in-law's closest friend. The visionary grotto, she insisted, must be built for her in Paris.

Month by month from 1570 when the work on her garden began, Catherine de Médicis, marveling at her potter's miraculous art, watched her grotto take shape in a partially underground chamber in her garden that measured roughly five by forty feet. Made entirely of terra-cotta, its longest walls had niches and columns that were cast from molds of rocks and shells. On the second story, terra-cotta human figures metamorphosing into rock and shell stood between windows. Opposite the entrance, like an altarpiece, a fountain shaped like an enormous crag gushed water from its spouts, which fell down over the rock to the pool at its base lined with water-spouting ceramic fish, seals, reptiles, amphibians, and crustaceans. All of these sea creatures were cast from life, the Queen Mother was proud to tell the foreign ambassadors and envoys who visited here. Each and every one of them had been cast from real frogs and animals caught in the rivers and ponds of France.

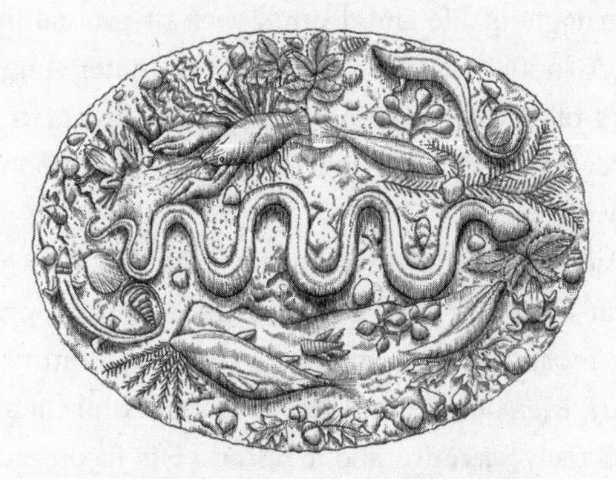

Drawing of a Palissy plate: a theatrical display of interdependence, mutation, and predation.
Victoria Sawdon

Palissy had made versions of these metamorphosing human-rock-shell figures before. In his earliest printed work, *Architecture and Order* (*Architecture et ordonnance*), written while the potter was imprisoned in Saintes in 1563, he described a sequence of six herms he had built (classically inspired head and torso figures that become an inverted triangle beneath the waist), each gradually turning into plants, stones, cockle shells, and then rock as they become sequentially turned back into the rock that had spawned them. Palissy's grotto figures were part human, part animal, part rock; they issued from a mind that had become a cauldron of classical, natural-philosophical, pagan, Protestant, and alchemical ideas.

By 1570, the Queen Mother's grotto maker had become something of an institution in Paris. Palissy's plates, miniature versions of the great grotto, had become collectors' pieces. The visitors who came to his atelier came not to see the grotto—which, unfinished, was still for the Queen Mother's eyes only—but to commission one of his extraordinary plates displayed in cabinet after cabinet across Europe.

Beyond the purse of all but the richest buyers, and bought as gifts to be given in elaborate gift exchanges between aristocrats, Palissy's plates were some of the most unusual and bizarre productions of Renaissance France. Oval, brightly colored, and highly glazed, they depicted pondbeds writhing with lizards, crustaceans, snails, snakes, and fish; each creature, interacting and reacting with others, is shown netted together in a staged struggle of life and death, a web of mutual predation and dependency. A spotted slate-gray viper takes center stage in most of these theaters of nature, creating danger and threat; frogs, fish, and crayfish scatter outward in all directions, both in the pond water and onto its rocky edges.

Bernard Palissy encouraged patrons to visit the public rooms of his workshop, but he allowed no one into the interior rooms where he worked. Like most other artisans in the sixteenth century who had labored for years, even decades, to discover a successful chemical process, Palissy was fiercely secretive about his art. His income depended on keeping a monopoly on production. Forever in fear of imitators, he revealed his process only to his sons, who were also sworn to secrecy. Late in his life he made the mistake of promising to reveal it to his son-in-law, Charlemagne Moreau, as part of his daughter Marie's wedding dowry. When Palissy procrastinated, perhaps regretting his promise after Marie had died, his son-in-law took him to court and, after ten years of court cases, finally smashed up Palissy's kiln, atelier, and house in Sedan in revenge.

We know some of the secrets of Palissy's process not because he revealed them, but because his followers did. The potter used live specimens—he kept toads, frogs, lizards, and snakes in jars in his workshop, collected for him from the ponds and ditches in the fields around Paris. To make a cast of an animal, he would first take it to the brink of death by immersing it in urine or vinegar; then he would grease its skin before embedding it in flattened plaster and posing it to make it seem alert. Once the plaster had set, the potter made a clay impression from the plaster. The casts were man-made fossils. The animal actors were then placed onto a rough-hewn plate or basin cast from a rock mold. Using needles, palette knives, and other tools Palissy worked the scene, integrating the ceramic interstices of earth and animals, water and vegetable, until the drama was both at its peak and permanently sealed, suspended in time.

These strange plates were not the random creations of the potter's fevered imagination, nor were they the freak productions of a culture obsessed with newfangled things. In these ceramic theaters, Palissy was raising the curtain on what he believed to be nature's deepest secrets: a miraculous process of decay and regeneration set in a borderline ooze where new life was constantly being generated. Palissy's plates were philosophical objects, gifts, conversation pieces, things to be contemplated and discussed, at a time when scientific debate and speculation took place in the courts and salons of princes and dukes as well as in the halls and corridors of universities.

Palissy thought of himself as something of a prophet. He was convinced that, through a lifetime of suffering comparable to the labors of Hercules or the trials of Job, nature's secret laws had been revealed uniquely to him and that it was his duty to pass this knowledge on. He began a lecture series; he took care to label all the objects in his museum, and over decades he rarely changed the staging arrangements of the pond theater plates. His sense of mission also explains the risk he took in returning to Paris after the political tide in the city turned violently against the Huguenots. He was a man possessed.

Palissy believed that ponds, salt marshes, caves, and wells were nature's kilns or cauldrons. Here, in rotting warm water, feces, and mud, nature turned dead things into new life through the congelation of various salts, waters, and minerals. "There are a great many kinds of ponds [*mares*] both natural and artificial," he told his lecture audience;

> several call them *claunes.** In some places they are a shallow trench, dug on a slope, so that the rain water flows into the trench or pool, and oxen, cows and other cattle may easily enter or leave to

* Shallow pools covered with a clay or claylike residue.

> drink in it, and these are dug only on the hanging side. . . . They are warmed by the air and the sun and by this means generate and produce many kinds of animals. And a great quantity of frogs, serpents, asps and vipers always gather around these *claunes* in order to feed on the frogs. There are often also leeches in them, so that if oxen and cows remain some time within these pools they cannot avoid being bitten by the leeches. I have often seen serpents lying curled up at the bottom of these ponds.

Like Leonardo da Vinci, Palissy believed that the earth, like the human body, pulsed with life and with rising and falling fluids. But unlike Leonardo, he was convinced that these fluids were life-generating, like semen. They ran through the veins of the earth, surfacing in cracks and holes, where they were warmed by the sun into new life. The words Palissy used to explain these processes—"putrefaction," "congelation," "generation," "exhalative," "evaporative," "germinative," "transmutation"—he took from the language of alchemy in general and from the works of the German-Swiss alchemist Paracelsus in particular. "Transmutation" was the most important word of all for alchemists, the idea around which all other ideas turned and the goal of all practice. The alchemist was constantly searching for ways to "transmute" matter into a new state by applying fierce heat or cold or by mixing up new combinations of fluids and gases. Like Leonardo, Palissy was scathing about the claims of many alchemists of his time, but he believed in transmutation; he was convinced that alchemists had simply misunderstood the process and agency by which it happened. Nature's secret generations happened in salt and water and rot, he insisted in all his writing, not in fire.

Palissy's attraction to ponds was not unusual in the Renaissance. Numerous Renaissance natural philosophers, drawing on alchemical ideas and conceiving of nature as a sorcerer practicing secret arts, were fascinated with the aqueous and mineral worlds, places where all the elements of alchemy mixed in a cauldron or crossing place. In Renaissance art and philosophy, frogs, salamanders, and snakes are shape-shifters; as transitional creatures they were endowed with magical qualities beyond our rational imaginings. Throughout the sixteenth, seventeenth, and early eighteenth centuries, alchemical concepts pervaded many aspects of what we would now call science, from ideas derived from Aristotle about the human body being made up of a combination of four elements—earth, water, fire, and air—to new experimental chemical and industrial knowledge.

Speculation about spontaneous generation had preoccupied natural philosophers for centuries, from Aristotle and Theophrastus to Lucretius and Ovid to van Helmont and Harvey. Palissy may have claimed to be working in isolation, uninfluenced by the theories of others; he may have claimed he was uneducated, a man, like Leonardo, without letters; but despite these declarations, like Leonardo, Palissy had a library and he was a reader. His published lectures and treatises make frequent references to the writing of Greek and Roman natural philosophers and to those of his own day, particularly the two great Renaissance Frenchmen Pierre Belon and Guillaume Rondelet, both of whom he read in French and in Latin.

Like Leonardo, whose work he may have encountered through the Italian Renaissance astronomer, mathematician, and physician Jérôme Cardan, Palissy rebutted the priests' explanation that fossils were washed by Noah's flood to the tops of mountains by making the same case that Leonardo had made seventy years earlier: "I maintain that shellfish . . . are born on the very spot," he wrote, "while the rocks were but water and mud, which since have been petrified together with these fishes. . . . I have explained to you above that these fishes were generated in the very place where they have changed their nature, keeping the form that they had when they were alive." He believed, as Leonardo did, that mountains were once seabeds.

Ideas and theories such as these about natural philosophy, medicine, and alchemy gushed from Palissy like the waters of his fountains. An opinionated, highly intelligent polymath with no formal classical education, he boasted of his lack of book learning: "I am neither Greek, nor Hebrew, nor Poet, nor Rhetorician, but a simple artisan hardly at all learned in the arts of language." Everything he knew, he insisted, was based on his own observations and experiments, on years of hard looking and practice, rather than on books and theories. Sometimes so many people came to listen to his disquisitions that his atelier resembled a salon more than a workshop.

But in his salon, Palissy was walking on broken glass. He was a Huguenot living through the bloody days of the French wars of religion, and despite the patronage of the Queen Mother and his own intellectual bravado and range of knowledge, he always had to be careful about what he said and who might be listening. In his provincial hometown of Saintes, he had been constantly under surveillance and had been imprisoned and questioned for periods of time. Many of the Huguenot men and women he worked with had disappeared or been tried or executed for heresy. For the moment, he and his family were protected by the patronage of the Queen Mother; while the grotto was under construction and his famous plates were becoming collectors' pieces, fetching high prices from the aristocracy of Europe, his religious beliefs were tolerated.

Palissy was lucky. Throughout her regency, Catherine de Médicis had insisted on religious toleration from her princely sons and ministers, particularly toward the large number of Huguenots who lived in France. Many of the most valued craftsmen who worked on her projects were Huguenots, she insisted. But she could not control her Catholic dukes and their brigand hordes, so again and again, violent sectarian battles and massacres were played out in the woods and fields and villages of France in cycles of attack and counterattack, massacre and reprisal.

Unfortunately for Palissy, Catherine's patience with the Huguenots ran out in 1567 when a Huguenot army tried to capture the royal family and the seventeen-year-old king, Charles IX, her son, at the Surprise of Meaux. Catherine, furious and at her wits' end, returned to Paris and kept her own counsel, biding her time. While her beloved potter was building her visionary grotto, she was planning her revenge on his people—she is even said to have planned the details of her coup in the grotto itself. Feigning conciliation, she arranged a marriage between her daughter and the Huguenot Henry III of Navarre. When aristocratic Huguenots came to Paris in their thousands to celebrate the wedding in 1572 and an assassin wounded one of the prominent Huguenot admirals, Catherine urged her son to strike against the Huguenots instead of risking an uprising in Paris. "Then kill them all. Kill them all," Charles is said to have answered. Within days, at least thirty thousand Huguenots died on the streets of Paris and throughout France, some murdered by soldiers, most by mobs. Henry of Navarre converted to Catholicism in order to avoid being killed. Palissy and his family survived, escaping Paris for the Protestant town of Sedan, no doubt tipped off by the queen's soldiers.

Perhaps Palissy had grown used to these oscillating periods of danger and safety, or perhaps he had acclimated to the refugee life, or perhaps his disputes with his brother-in-law about the fate of his grandchildren forced him to make more money; but given the level of danger in Paris after the St. Bartholomew's Day Massacre, and given his public profile, it is surprising to find that he was back in Paris in 1575, not keeping a low profile but pasting posters onto walls announcing the opening of an academy and advertising a series of lectures on alchemy, medicine, and the natural sciences. He had also established a small museum in a room next to his atelier that shelved his enormous collection of petrifications; these objects would later be used as evidence for his increasingly eccentric theories and demonstrations.

In these Paris lectures, Palissy unveiled his grand theories. The quantity of all the substances on the earth—minerals, metal, rock, and earth—was fixed and absolute, he claimed; it could never be increased or decreased. But within these fixed quantities, all these materials were in a constant state of flux—seas to mountains, mountains to seas, cloud to rock, rock to water. Here he was absolutely in line with Leonardo, but he went further: metals and rocks and minerals, he argued, "grow" over vast amounts of time into new and increased quantities of themselves through the action of generative water in the moist, humid, enclosed womb of the earth. He described a brilliant system that explained many natural phenomena from petrification, fossilized shells on mountaintops, and hydrologic cycles to the nature of fertilizers. Fossils were, according to Palissy, fish, shells, and mud that came from lakes that had since congealed because the lakes' waters had at one point blended with generative salt.

Since the terrors of the St. Bartholomew's Day Massacre, when so many of his friends and neighbors had died on the streets, Palissy had become a man possessed. He could have disappeared into rural anonymity or recanted his ideas, but instead he returned to Paris to lecture. For nine years, the Huguenot potter held the attention of the Paris intelligentsia. He was a kind of elderly eccentric Scheherazade—each day that he entertained and fascinated his audiences with his philosophical ideas, his life would be prolonged a little longer. Hundreds attended his lectures in the Tuileries Gardens: physicians, including the king's man Ambroise Paré, canons, jurists, humanists, dukes. The museum was always open, each of its labeled objects conveying the sense of urgency that characterized everything Palissy did. "I have set up a cabinet," he wrote in the preface to his published lectures, *Admirable Discourses* (*Discours admirables*), "in which I have placed many admirable and monstrous things which I have drawn from the bowels of the earth, and which give reliable evidence of what I say, and no one will be found who will not admit them to be true, after he has seen the things which I have prepared in my cabinet, in order to convince all those who do not believe my writings." In his writing, as in the labels in his museum, he impels individuals to see, touch, listen, and smell. In *Admirable Discourses,* he implores: "Look . . . look . . . look at the slates! . . . Don't you see?" And then: "I have put this rock before your eyes." The objects speak for themselves, he insists: "Consider this great number of shellfish I have put before your eyes, which are now all reduced to stone . . . and look at all these kinds of fishes I have put before your eyes. . . . They were once alive."

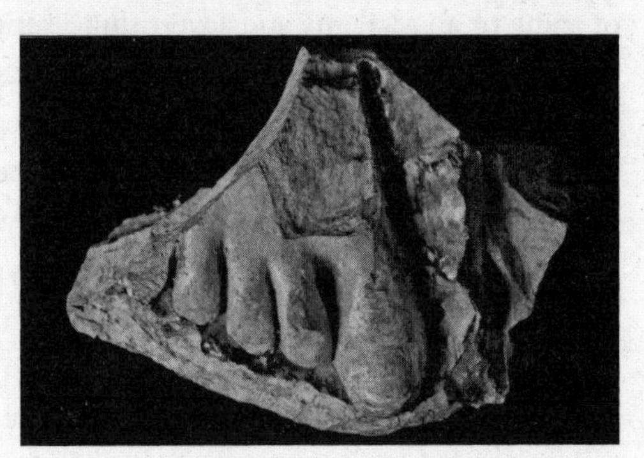

Fragment of a human foot cast from life, from Palissy's Paris grotto.
Louvre, Paris

In July 1585, the Treaty of Nemours granted Protestants six months to convert to Catholicism or leave the country. Palissy fled Paris again and went into hiding in a Protestant refuge near Saint-Germain-des-Prés on the rue des Maretz, also known as the Little Geneva and now called rue Visconti. Here he was arrested with several other men, and another round of exhausting sentences and reprieves began. In January 1587 a judge decreed that the three men be flogged, that their books be burned in the marketplace, and that they themselves be banished from France. Palissy again went into hiding, this time in the Faubourg Saint-Germain, very close to his atelier. He was arrested again a few months later and imprisoned with a group of elderly Huguenot women who were all sentenced in 1588 to be "hung and strangled and their bodies reduced to ashes, imposed on them for heresy." The sentence was quashed again, despite Palissy's objections. He was ready to die, he told his warders. At the age of eighty, Palissy was moved to the Bastille, where the prison warder, determined to test the old man's mettle, told him he would be burned alive immediately if he did not convert. Palissy still refused. He died a year later from "misery, need, and poor treatment."

Palissy's books disappeared from the shelves of Paris libraries, only to be rediscovered again 150 years later in another garden in the city, the mighty Jardin des Plantes, by the great naturalist the Comte de Buffon, who was looking for writings on fossils. Until then Palissy's ideas about the origins of life had left no trace. When auditors arrived in Paris in 1589 to assess the famous Tuileries Gardens as part of the estate of the recently deceased Queen Mother, they found the grotto in serious disrepair and not worth inventorying, its clay animals broken and scattered about the grotto and the surrounding gardens. In the nineteenth century, excavators working in the foundations of the Louvre found fragments of clay animals, ostrich feathers, lion heads, fish, crabs, and salamanders buried deep in the soil of the palace gardens among ceramic human faces covered in shells and leaves turning slowly back into rock.

5

Trembley's Polyp

THE HAGUE, 1740

The Count of Bentinck's summer residence* once stood in spectacular gardens on the edge of sand dunes that stretched down from the outskirts of The Hague to the North Sea at the port of Scheveningen. Within a series of elaborately patterned parterres, gardeners clipped trees and cone-shaped and tunneled hedges into carefully orchestrated symmetries. Exotic flowers bloomed in carved stone pots among classical statues. An artificial hill called Mount Parnassus rose against the skyline, netted with a symmetrical system of paths. Complicated latticework tunnels led from one part of the park to another through a system of ornamental ponds teeming with exotic fish. Everywhere in this garden, nature had been sculpted, ordered, restrained, symmetrized, and framed.

In the summer of 1740, in a large study looking out over the orangery, the count's two young sons—six-year-old Antoine and three-year-old

*The surviving house at Sorgvliet is now the country residence of the Dutch prime minister, and its once extensive estates have been engulfed by the spread of The Hague.

Abraham Trembley and the Bentinck boys
at work in Trembley's study at Sorgvliet.
Abraham Trembley, Memoires *(1744)*

Jean—and their thirty-year-old Genevan tutor, Abraham Trembley, peered into glass jars arranged along the window ledge. The count and countess were in the midst of an acrimonious divorce; the boys' mother had returned to Germany, where she was already living openly with her lover, and, as their father spent much of his time with his lawyers in The Hague, this summer, the fascinating Monsieur Trembley had become the Bentinck boys' entire world. The boys and their tutor were together night and day. Trembley taught the boys to read and write and to become fluent in French, Latin, and English; the young man worked hard to make everything they learned seem like play. He was passionate about teaching. He made sure the boys spent as much time outside as inside, catching aphids or caterpillars by day, moths in the evening, using the microscope, taking thermometric measures, speaking in French, declining Latin verbs. Trembley carefully prepared object lessons for them, using the study of a moth or a community of insects living on a single tree or the germination of seeds or the arrangements of a beehive as an opportunity to teach French, logic, morals, religion, history, science, and mathematics. "So far as is possible," he wrote, "it is advisable that ideas should be preceded in the mind by curiosity—by the kind of curiosity that excites and sustains attention and thereby helps to cause objects to be seen in a precise manner and to be observed with pleasure." To this end he had turned his study into a laboratory.

The boys had spent the morning wading through the ditches and ponds of the estate collecting water creatures and had returned to inspect their treasure. Under the magnifying glass and in the light of the sun, the water teemed with bright, moving green forms, some like flecks of grass, others resembling the tufts of dandelion seed heads. The boys, their fingers pressed up against the sides of the glass, called the flecks plants, but Trembley, raising his magnifying glass to the jars, was not so sure.* He shook the glass slightly; in the agitated water the creatures contracted to the size of a grain and then slowly extended their arms again. "I was surprised," he recalled later, "and this surprise only served to excite my curiosity and redouble my attention." The creatures in the jar moved by themselves and were sensitive to movement. These were characteristics of animals, not of plants, he reminded the boys.

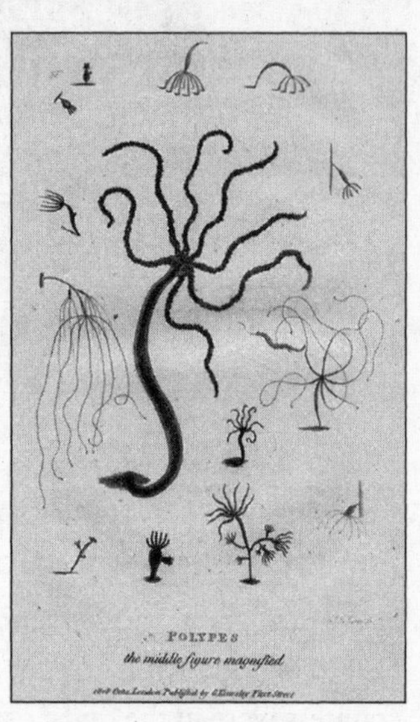

An early-nineteenth-century print of the polyp (hydra) showing its varied movements. *Copper engraved antique print by J. LeKeux, published in* Zoological Lectures, *delivered at the Royal Institution, 1809*

A few days later the two excited boys called their tutor to show him the creatures walking like caterpillars across the inside of the jar. These strange creatures were about to make philosophical history; they were about to upset what eighteenth-century naturalists believed to be natural laws. And they were about to make

* The freshwater polyp is what we now call a Hydra.

Trembley's name famous across Europe and, despite his resistance, embroil him in the spread of atheism.

That summer, Trembley's natural history interests lay with moths and aphids. Inspired by the French natural historian René Réaumur's dazzling new volumes on insects, which he had read the previous summer, Trembley had begun to collect the caterpillars of carpet moths, sealing them in glass powder jars with pieces of woolen cloth to watch them pupate, hatch, and fly. In the book that had become Trembley's bible, Réaumur had called on all natural philosophers to collect facts about insects. Facts—not speculations—were needed, he insisted, if nature's laws were to be fully understood.

Every week Trembley corresponded with his brilliant nephew, Charles Bonnet, who lived in Geneva. The young men were close. Both were pious Huguenot second-generation refugees who had grown up in the small Protestant republic of Geneva and attended the Academy there. Their families had fled France for Geneva either shortly before or just after the St. Bartholomew's Day Massacre in 1572. Bonnet, though ten years younger than Trembley, was intellectually more precocious and philosophically inclined. His natural curiosity had already brought him to the notice of his professors, although his father wanted him to study law.

Trembley and Bonnet had been writing to each other for some time, sharing their natural history discoveries and honing their observation and description skills. Bonnet had devoured every page of Réaumur's insect volumes as soon as he had access to the library in the Geneva Academy, and at the age of fifteen he had written directly to the Parisian insect master, sending him detailed and imaginative descriptions of the behavior of caterpillars, antlions, and spiders. He and Réaumur had been in correspondence ever since. A third friend of Trembley's, Pierre Lyonet, thirty-four years old in 1740, also of Genevan-French descent, lived a mile away in The Hague, where he worked as a lawyer. When he was not preparing legal cases, he spent his time dissecting insects. All three men were obsessed with insects, and with insect sex. All three were in correspondence with Réaumur and with one another.

In Geneva, Bonnet, now nineteen, bored with his law studies at the Academy, had determined to try to solve the mystery of aphid reproduction, a mystery that Réaumur himself and numerous predecessors had failed to crack. Despite many efforts and experiments, he had not yet found a male aphid, he complained in volume 3 of his insect book, nor had he seen any act of aphid coupling. He was completely mystified. He challenged all young naturalists to solve the mystery of aphid reproduction. Lyonet also set up aphid experiments in his rooms in The Hague at the same time, but with his legal work, he could not watch his aphids day and night as Bonnet could do in his student rooms in Geneva.

It was young Bonnet who made the first breakthrough. On May 20, 1740, he placed a single newborn aphid on a branch of an evergreen shrub called the spindle tree inside a glass flask sunk in a container full of earth. His task was to guard and testify to her virginity. Always quick to make dramatic classical analogies, he compared his aphid to the virgin Danae in Greek myth, who was imprisoned by her father in a tower because he had been told her child would be his murderer. Between May 20 and June 24, Bonnet watched the newborn aphid moving around her glass tower through every hour of the day to ensure that no one tampered with his experiment and thus ensure that she remained a virgin. "I have been an Argus more vigilant than the one of the fable," he wrote, referring to the giant watchman of Greek mythology who was reputed to have a hundred eyes.

Astonishingly, Bonnet's imprisoned female aphid not only survived her first molts but also actually gave birth on June 1. Over the following twenty-three days she produced ninety-four more aphids. In July, Bonnet sent his meticulous records and tables of findings to Réaumur by letter. He was not the first human to have witnessed a so-called virgin birth, a form of asexual reproduction that requires no fertilization by the male (now known as parthenogenesis),* but he was the first to record proof of it through repeated experiments, crowning seventy years of investigations and experiments by others working in different entomological traditions.

*The term "parthenogenesis" was coined in 1849.

In July of that year, Réaumur read Bonnet's letter and findings to the assembled members of the French Academy of Sciences, established in 1666 to encourage and protect the spirit of French scientific inquiry. The audience, though respectful of the young man's meticulous observation and recording skills, agreed that they could not confirm the discovery of virgin births in aphids until the experiments had been repeated and witnessed by others. Réaumur wrote to his friend the doctor Gilles Augustin Bazin in Strasbourg and to Abraham Trembley in The Hague to ask the two men to repeat Bonnet's experiments with several different species of aphids. Trembley accepted the task, proudly explaining to the Bentinck boys, as they angled the microscope arm onto the glass tower that housed their own imprisoned aphid, that they were now engaged in a serious investigation that might put their names into the annals of science.*

Within weeks, he and the Bentinck boys, Bazin, and Lyonet had all verified Bonnet's discovery; they had all witnessed aphid virgin births for themselves. In August 1740, Réaumur sent Bonnet a letter of congratulation. Not only was Bonnet the first human to have both witnessed and recorded a virgin birth; he had toppled one of the central premises of eighteenth-century science—the belief in the universality of sexual reproduction. "These are assuredly observations of great importance in natural history," Réaumur wrote to Bonnet, "since they teach us that the law of coupling is not a general law." He appointed Bonnet to a prestigious place as a corresponding member of the Academy.

For the rest of the summer, the carpet moths, Bonnet's extraordinary aphid discovery, his own aphid verification experiments, and the boys' lessons occupied all of Trembley's attention. In late September he and the boys watched the first two carpet moths hatch; proudly he sent Réaumur a detailed description of the moths' life cycle and enclosed some specimens. But for Réaumur the carpet moth was old news. He had al-

* The angled-arm microscope that Trembley used for the polyp experiments was one that he was probably already using for insect observation.

ready studied the genus, he replied apologetically, and had already published a description of its generation in detail. Trembley and the boys dejectedly released the moths and caterpillars, but Bonnet's aphid discovery had given the young tutor a desire to test other supposedly natural laws: "A fact such as the one which aphids presented could only inspire in me a great deal of distrust of general laws," he wrote. "I felt strongly that Nature was too vast, and too little known, for anyone to decide without foolhardiness that one or another property was not found in such and such a class of organised bodies."

Meanwhile, all through the summer, the water creatures had continued to thrive in their jar on the windowsill, largely unobserved—"Almost the whole of the month of September 1740 passed without my giving them the least attention," Trembley wrote—until one morning he noticed that the tiny green creatures had all congregated on the sunlit side of the jar. When he turned the jar away from the sun, the water creatures migrated slowly back toward it. Could these minute beings actually be sentient? Curious and still baffled, he told the boys that for the next few weeks this creature—whatever it was—would become their new object lesson. Using a range of microscopes, they would investigate everything about it: what it ate, how it reproduced, how it moved. They would describe in detail everything they observed, and when they had recorded everything that was knowable, they would send a report to Réaumur, and then perhaps he would grant them the same honor he had granted Bonnet.

The investigation began. First, he told the boys, they must determine one way or the other whether the creatures were plants or animals. Might those "tentacles" be branches or roots rather than "arms"? The only way to answer this question was to cut the creature in half, for only plants could survive being cut in two. "It was on the 25th November 1740 that I cut the first polyp," Trembley recalled. "I put the two parts in a flat glass, which only contained water to the height of four or five lignes. It was thus easy for me to observe these portions of the polyp with a fairly powerful lens. . . . The instant that I cut the polyp, the two parts contracted so that at first they only appeared like two little grains of green matter at the bottom of the glass." The creature did not die. To Trembley's astonishment, "the two parts expanded on the same day on which I separated them. . . . I saw it move its arms; and the next day, the first time I came to observe it, I found that it had changed its position; and shortly afterwards I saw it take a step."

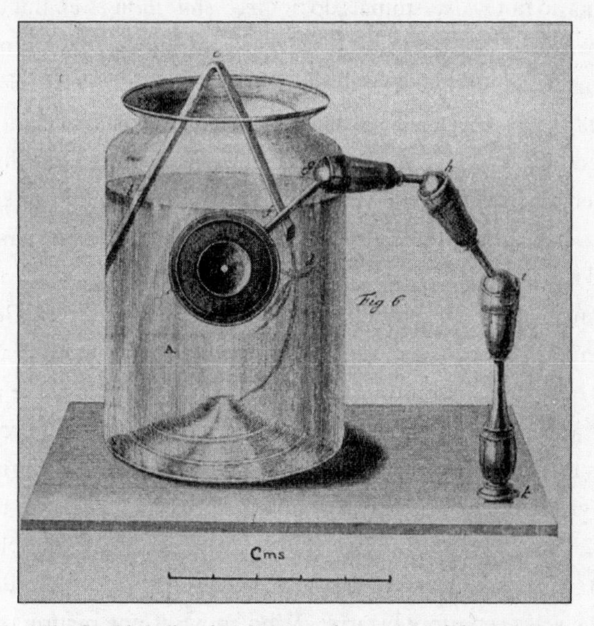

The flexible ball-and-socket-armed structure of the aquatic microscope that Trembley invented in 1745 allowed him to observe the polyps in movement and from every angle.
Abraham Trembley, Memoires *(1744)*

It was important not to jump to conclusions, Trembley warned the boys. They must watch and record everything just as Bonnet had done, just as Réaumur insisted. They might only be witnessing a feeble remnant of life, or perhaps they had cut off a nonvital section like a lizard's tail. It was too early to say. But over the next few days, as the boys sketched diagrams of what they had seen, the two halves of the polyp continued to move vigorously, one with the original set of the hornlike crown of arms and the other without.

Now Trembley was losing sleep. But there were stranger transformations yet to come. After nine days, the hornless, armless half of the cut creature began to sprout what looked like small arms. "Throughout the day I continually observed the points," he wrote; "this excited me extremely, and I awaited with impatience the moment when I should know with certainty what they were." Within a further week neither he nor the boys could tell one half from the other.

Plants do not *walk*. Animals do not *regenerate themselves*. But the pond creatures did both. Alarmed, Trembley wrote directly to Réaumur, summarizing what he and the boys had seen and enclosing some early sketches. "I saw these parts walk, take steps, mount, descend, shorten, lengthen, and make so many movements, which up to the time I had only seen made by animals. I did not know what to think," he exclaimed. Now he needed the philosophical big guns. He needed others to help unravel this enigma.

Meanwhile, on December 18, 1740, Bonnet wrote from Geneva to tell Trembley that he had discovered a male aphid. It was rampantly sexual, he wrote, "perhaps one of the most ardent that there is in Nature. It appears to me that it does nothing except have intercourse as soon as the day arrives." Had they been wrong about the virgin births? Surely virgin births and sexual coupling could not coexist in the same species? There were only two possible explanations that Trembley could suggest. Might aphids mate invisibly *inside the womb*? he asked. The alternative was even more bizarre: "Who knows if one mating might not serve for several generations?" he asked his friend. They should rule nothing out. Nature was proving more extraordinary than anyone had anticipated. Still he told Bonnet nothing about the polyp experiments.

On receipt of Trembley's letter and drawings in January 1741, the incredulous French professor urged Trembley to send fifty of the polyps in a sealed jar to him in Paris as soon as he was able. Trembley packed up a jar and sent it with a servant on horseback to Paris, a journey of seven days. Only now that Réaumur had taken his claims seriously enough to send for the polyps did Trembley write to tell Bonnet about his "little aquatic being."

The creatures arrived in Paris on February 27 dead, suffocated in their glass jar by a too-tight seal of Spanish wax. Réaumur, frustrated, wrote to ask for more, suggesting that this time Trembley use cork as a sealant rather than wax and asking permission to read the letter to the savants in the French Academy of Sciences. Trembley, determined to make his experiment transportable, sent the count's servants out on horseback on a trial run for miles across the dunes carrying the creatures in various bottles with different sealants. While he awaited their arrival, Réaumur read Trembley's letter to the Academy at three consecutive meetings, on March 1, 8, and 22. Anticipation ran high. On March 16, Trembley, satisfied that he had now found the right traveling arrangements, sent off another batch of twenty water creatures from The Hague to Paris; this time they survived.

A few days after he had dispatched the second batch of animals to Paris, Trembley received an excited congratulatory letter from his nephew: his "little aquatic Being ought to be regarded as one of the greatest marvels that the Study of Natural History can offer," Bonnet wrote. "One can say that you have discovered the point of passage from the Vegetable to the Animal. What you have reported is close to an enigma, which I do not know how to decipher. But I easily console myself when I see that the clever men here, even learned men, such as our Professors to whom I have been very pleased to show your letter, are confounded." Despite Bonnet's advice to Trembley not to get caught up in metaphysical speculation and always to follow Francis Bacon's rule of thumb to avoid hypothesis and stick to collecting facts, Bonnet's mind had begun to spin. His was not the only one.

As soon as the new jars of aquatic creatures arrived in Paris, Réaumur repeated the experiments following Trembley's detailed instructions—cutting and waiting and watching through the microscope. The creature—Réaumur named it a polyp, meaning "many armed"—regenerated itself again and again. Exhilarated by the philosophical questions the experiment raised, questions already being discussed heatedly in salons and coffeehouses across Paris, Réaumur repeated the demonstration over the three following days both to the "entire academy" and "to the court and the city" in Paris. The microscopic polyp had become a circus animal.

Back in The Hague, Trembley and his young protégés spent the spring of 1741 developing more experiments for their pond polyps. They learned to produce a polyp with seven heads, provoked a polyp to "swallow" other polyps, grafted the halves of two different polyps together, and turned one inside out. Trembley had scores of experiments planned. He wanted to learn everything knowable and measurable about his "little machine."

In April 1741, Bonnet, already struggling with strained eyes, began a new series of aphid experiments in order to answer Trembley's question about how many generations of aphids might be produced without coupling. This time he watched through night and day for three months as nine generations of aphids emerged. When he had asked Trembley to solve the aphid enigma, his friend had made a valiant effort and then, defeated, had asked "Who knows?" in his letter of reply. "If this excellent friend had been able to foresee all the evil that his 'who knows' did to my eyes," Bonnet wrote, "I am very sure that his tender friendship for me would not have permitted him to express it. It was, however, on this simple 'who knows' that I undertook a new study which was much more laborious than the preceding one. I was young and full of ardour: it seemed that these two words reduced to nothing all my previous work."

By the end of the summer, the polyp had usurped the place of the aphid as the philosophical object of the season in Paris. In August, Réaumur wrote to Trembley to report that in Paris, "never did an insect cause so much uproar as do the polyps." The reports of the Academy of Sciences for 1741 described the discovery in sensational theatrical language: "The story of the Phoenix who is born from the ashes as fabulous as it might be, offers nothing more marvellous. . . . From each portion of the same animal cut in 2, 3, 4, 10, 20, 30, 40 parts, and, so to speak, chopped up, just as many complete animals are reborn, similar to the first. Each of these is ready to undergo the same division . . . without it being known yet at what point this astonishing multiplication will cease." The popular French entomologist Gilles Bazin described the upset in his novel *Letters from Eugène to Clarice on the Subject of the Animals Called Polypes* (*Lettres d'Eugène à Clarice*): "A miserable insect has just shown itself to the world and has changed what up to now we have believed to be the immutable order of nature. The philosophers have been frightened; a poet told us that death itself has grown pale."

But Réaumur also wrote to tell Trembley that he was encountering skepticism everywhere among the natural philosophers of Europe. Everyone who had not seen the polyp regenerating with his or her own eyes remained an *incrédule,* he reported—an unbeliever.

More of a philosopher than his uncle, Charles Bonnet was quickly alarmed by the potential metaphysical meanings of the polyp experiments. What did it all mean? If this simple creature could be cut up into infinite parts and regenerate itself from these parts, then where was its soul? Was Descartes then right after all to argue that the universe and all living organisms were like clocks governed by natural laws? Like many Christian natural philosophers in the seventeenth and eighteenth centuries, Bonnet was always looking for opportunities to refute this apparently reductionist and potentially godless way of seeing the world. Every time he admired a complex organism or looked at the patterns on the wing feathers of a bird under the microscope, it was for him further evidence of the miracle and benevolence of God's design. But the design was looking increasingly incomprehensible. Under the microscope, nature was looking more and more various, bizarre, and wayward. Its codes were becoming indecipherable.

Increasingly aware that Trembley's polyp was being taken up by materialists (there is only matter—there is no supernatural controlling force) and atheists (there is no God), Charles Bonnet, now three months into his aphid vigil and a little high-strung, wrote to his professor Gabriel Cramer in Geneva in June: "All that I ardently wish is that my poor insects will not be degraded too much, and I have reason to fear it terribly. I beg you, Sir, strive not to let [the insects] become simple machines. I will be inconsolable. . . . Good-bye then to all industry, all skills, all kinds of intelligence. And if you cannot get them out of this difficulty who can?" Professor Cramer reminded him that, though the polyps did seem "to deal a heavy blow to the System of soul in animals," it was simply too early to draw conclusions. "Let me breathe a little," he replied. "You are overwhelming us with marvels."

In the summer of 1741, Bonnet, his sight increasingly blurred, began to look for other regenerating animals. Unable to find polyps in the ponds around Geneva, he began to experiment on aquatic worms. Extraordinarily, he discovered that they, too, reproduced themselves when cut in half. Réaumur, who was delighted by Bonnet's findings, urged him to look further. How far does this pattern extend in nature? What about sea nettles and starfish? he asked. If you cut them, will they regenerate? Within months Bonnet produced a worm with a tail where its head should be. It was too much for the young naturalist, who was still only twenty-one. He needed clear theological explanations.

The professors at Geneva were also confounded. In November 1741, Bonnet wrote to Réaumur himself asking him why God had given such miraculous regenerative powers to insects and not to his greatest work, Man. "*For what end?*" he asked. Réaumur did not know how to answer. He proposed a bland explanation: perhaps God had given animals that were designed as mass food for other animals the ability to reproduce the part that had not been eaten. Bonnet was not satisfied.

Abraham Trembley was not concerned with metaphysics; he wanted only to make the polyp regeneration visible to as many people as possible. If European philosophers were not prepared to believe the word of a low-born Genevan tutor living in Holland or to accept the endorsements of the members of the Academy of Sciences, he told the Bentinck boys, more people must see the experiments with their own eyes. The polyps must speak for themselves. Trembley began sending out packets of live polyps in glass jars by mail to addresses across Europe—universities, academies, salons—with meticulous instructions on how to undertake the experiments. Two years later he had 140 jars in his study, complaining in July 1743: "I am entirely taken up with dispatching polyps to one place or another."

Using the postal system, Trembley began to create witnesses in every city in Europe. It was a brilliant strategy at a time when all natural philosophers, influenced by the writings of Francis Bacon, stressed empirical observation and experimentation above speculation. Despite the disruptions to the postal networks caused by the War of the Austrian Succession,* the polyp experiments were performed and repeated like circus shows in public in Rome, Siena, and Germany; in Sweden in 1746, Abraham Beeck wrote, "Apart from electricity, naturalists did not deal with anything this year other than the polyp." Each one of those witnesses was part of a network of speculation and talk and correspondence across many different languages; each of those networks was part of a larger one. Everywhere the polyps converted the *incrédules*. Ambassadors carried news of them from court to court. Everywhere the philosophical talk spread.

*The war (1740–48) began as a challenge to the succession of Maria Theresa of Austria to the Habsburg thrones of her father, but in reality it was a struggle for power between France and Prussia on one side and the rest of Europe on the other.

British naturalists, however, remained resolutely skeptical; they quickly turned the polyp into a joke. Although the famous French naturalist Georges-Louis Leclerc de Buffon wrote to the president of the Royal Society in Britain, Martin Folkes, about the polyp experiments as early as 1741, and William Bentinck's brother Charles and Jan Frederick Gronovius had published excited reports in the journal *Philosophical Transactions of the Royal Society,* Folkes doubted Trembley's credentials and wrote to William Bentinck, Trembley's employer, for verification. Trembley was, after all, only a tutor, working with two young boys as his witnesses. Intrigued, Folkes also wrote to the famous French *salonnière* Madame Geoffrin for her assessment of the polyp talk in Paris. She gave him a long report by letter describing the experiments she had seen at Professor Réaumur's house that summer; philosophical opinion was divided among the Parisian savants, she told him, but she could certainly confirm that the polyp was preoccupying the finest minds in Paris.

Finally, in March 1743, at Folkes's invitation, Trembley dispatched a container of polyps to England. It arrived the following day. Folkes invited twenty fellows of the Royal Society to his home to witness the experiments. The polyps regenerated within hours. On March 17, he repeated the experiments at a meeting of the Royal Society, where in the course of a week more than 150 people saw them. It took Folkes some time to apologize to Trembley, but when he did so, he did it gracefully: "We are no less sensible of your great candour, and the Readiness you have shown not only to transmit to us faithful abstracts of your own experiments," he wrote, "but also to send us the Insects themselves, whereby we have been enabled to examine by our selves, and see with our own Eyes the Truth of the astonishing Facts, you had before made us acquainted with."

One of the witnesses in Folkes's house in March 1743 was the journalist, poet, and naturalist Henry Baker, son-in-law of Daniel Defoe and author of *The Microscope Made Easy.* Fascinated by what he had seen, he persuaded Folkes to give him three of the polyps and to lend him Trembley's letters. All through April he cut and multiplied the polyps in his London house until he had hundreds; he watched them under his microscopes; he sent out corked jars full of live specimens to everyone he could think of in Oxford and Cambridge. Friends brought him English polyps from the ponds of Hackney and Essex. As the spring came and the air warmed, the polyps, both Dutch and British, multiplied even faster as they fed on small worms that Baker had his servants dig out of the black mud of the Thames. Determined to publicize the discovery and frustrated that nothing was yet in print, Baker published a two-hundred-page account of his repetitions of Trembley's experiments with woodcut illustrations as *An Attempt Towards a Natural History of the Polype* in November 1743 before Trembley had published his own memoir. The polyp's regenerating skills were extraordinary, he wrote, but no one should waste any time on philosophical speculation. Any such hypothesizing, he told his readers, would be "a kind of Madness," a "Cobweb of the Brain."

British naturalists were now enthralled, the editor of the French journal *Library of Britain* (*Bibliothèque Britannique*) reported in the autumn of 1743: "The marvelous properties of the new Polyp . . . have become the object of such a curiosity and research of some of the members of the Royal Society, that Mr. Cromwell Mortimer, Secretary of this illustrious assembly, has but given in the no. 467 of the Transactions pieces which only relate to it." But although the polyp story had appeared in print in reports of the various scientific societies and in letters, in Baker's book, and finally in Abraham Trembley's long-awaited memoir, published in Leiden in 1744 and immediately pirated by a Parisian publisher, it largely traveled by word of mouth. In salons from Paris to Leiden to Oslo to Rome and Siena, the polyp, silently miraculous, was the subject of long, animated conversations. It had become a sensation.

Literary satirists immediately plunged in. Henry Fielding published a satirical pamphlet on the self-generating powers of money called "Chrusippus, Gold-foot or Guinea" in 1743; in 1751 the Scottish poet Tobias George Smollett, contemptuous of any philosophical subject that had no economic use, vented his spleen in *Peregrine Pickle* at the "muck-worm philosophers" who counted angels on a pinhead; in 1752 Vincent Miller published *Man-Plant, or a Scheme for Increasing and Improving the British Breed*. Within months of the Royal Society experiments of 1743, Charles Hanbury Williams satirized the philosophical pretensions of the aristocracy in a narrative poem called "Isabella, or the Morning." "Pray, Mr. Stanhope, what's the news in town?" the duchess asks her visitor.

"Madam, I know of none; but I'm just come
From seeing a curiosity at home:
'Twas sent to Martin Folkes, as being rare,
And he and Desaguliers brought it there:
It's call'd a Polypus."—"What's that?"—"A creature,
The wonderful'st of all the works of nature:
Hither it came from Holland, where 'twas caught
(I should not say it came, for it was brought);
To-morrow we're to have it at Crane-court,
And 'tis a reptile of so strange a sort,
That if 'tis cut in two, it is not dead;
Its head shoots out a tail, its tail a head;
Take out its middle, and observe its ends,
Here a head rises, there a tail descends;
Or cut off any part that you desire,
That part extends, and makes itself entire:
But what it feeds on still remains a doubt,
Or how it generates, is not found out:
But at our Board, to-morrow, 'twill appear,
And then 'twill be consider'd and made clear,
For all the learned body will be there."
"Lord, I must see it, or I'm undone,"
The Duchess cry'd, "pray can't you get me one?
I never heard of such a thing before,
I long to cut it and make fifty more;
I'd have a cage made up in taste for mine,
And, Dicky—you shall give me a design."

But what did it all mean? How would Dicky have explained the philosophical implications to the duchess if she had asked? Trembley himself, influenced by his mentor Réaumur's warnings against rash philosophizing, held back on offering meanings or philosophical interpretation. His memoir contains no philosophical conjecture, no theory, just detailed facts and descriptions of the hundreds of experiments he had made.

Bonnet, on the other hand, continued to find himself entangled in this philosophical "Cobweb of the Brain." All through the 1750s, as his eyesight weakened, he wrestled with the perplexing questions the polyp presented. Finally, late in that decade, he came to a position that seemed to reconcile all the polyp- and aphid-generated metaphysics that kept him awake at night and disturbed his religious beliefs. The polyp appeared to be the point of passage between the animal and vegetable kingdoms, which suggested either that the borders were permeable or that they did not exist at all. You could cut the polyp a hundred times and it would regenerate from the smallest part. An aphid could reproduce without sex. The answer was simple: God had originally created a multitude of germs on a graduated scale, each with the power of self-development. Over enormous lengths of time, the germs modified and improved. If there was progress, he argued, it took place within the species. Species boundaries were inviolable and designed by God. They were fixed. "Nature is assuredly admirable in the conservation of individuals," he wrote, "but she is especially so in the conservation of species. . . . No changes, no alteration, perfect identity. Species maintain themselves vigorously over the elements, over time, over death, and the term of their duration is unknown." The regeneration of the polyp proved not that there was no God but that species were indestructible, immortal. Man was immortal.

Trembley was shocked and surprised by his nephew's speculative claims about his polyps; close observation of nature and not philosophical speculation had been the object of his work. But it was too late. Now that the polyps' regenerating abilities had been witnessed by hundreds across Europe, their meanings were out of his hands. He wrote to his employer William Bentinck in January 1750 complaining that naturalists everywhere were being too quick to jump to metaphysical conclusions, including the great Comte de Buffon in his newly published first volume of *Natural History* (*Histoire naturelle, générale et particulière*): "Mr. de Buffon claims to explain nearly everything about generation, but I admit that I can only consider his system as a dangerous hypothesis. He tries to prove too much with the facts on which it is built. He seems to let himself be carried away by the imagination. If his work is very popular, I am afraid that he will do harm to Natural History by bringing back the taste for hypotheses."

Now that so many men and women had seen the polyp regenerate, the European imagination could not be so easily reined in. In Paris and

London, Berlin and Rome, microscope and microscope guidebook sales increased dramatically; across the world, intellectuals installed experimental apparatus in their studies; naturalists turned away from their botanical or beetle collections and instead scoured ponds and rock pools searching for more incomprehensible tentacled and budding creatures or peered down the eyeglasses of ever more powerful microscopes to watch the looping, oozing, and spawning of corals or infusoria. Ponds, seabeds, and rock pools, naturalists declared, were terra incognita teeming with minute, incomprehensible, inconceivable bodies that would expand the range of nature's laws beyond anything previously believed possible. For those inclined, like Charles Bonnet, to ponder, the polyp profoundly challenged contemporary beliefs in intelligent design: If man had dominion, if he was valued by God above all other creatures, why had he not been granted powers of regeneration like the polyp? If nature was really arranged on a scale with man at its pinnacle, why had such a simple creature been granted such complexity? And if nature's workings included laws as bizarre as animal regeneration, what other laws might yet be discovered?

In the four vignettes commissioned for Trembley's *Memoir,* he and the Bentinck boys stand in various parts of the estate collecting polyps or other water creatures, carrying nets and jars or fishing from the ornamental ponds. They are tiny figures dwarfed by a spectacularly groomed and ordered landscape in which each tree has been planted exactly equidistant from the next in its row. Apart from the occasional dog or flock of peacocks, Trembley and the boys are portrayed as entirely alone, wandering around large empty spaces. Where are the servants, the gardeners, the hundreds of European travelers who came every year to walk in those world-famous gardens? Where is the count? Or the count's brother Charles, who was always in and out of the house? Or the code breakers and spies the Bentinck brothers employed to shore up their political power and influence in Europe?

Why the emptiness?

Count Bentinck commissioned the vignettes to ensure that the pictures portrayed his beautiful estate, his tutor, and his two sons in the best possible light. But that does not explain why the landscape is so

Trembley and the Bentinck boys fishing for polyps
in the ponds of the Sorgvliet estate.
Abraham Trembley, Memoires *(1744)*

devoid of people. Trembley did not rattle around with the boys in a vast empty house. He was almost never alone there. Bentinck was one of the most important men in the Netherlands and the grandson of one of the most important aristocrats of England. He and his brother Charles were the two youngest children of the elder Willem Bentinck, who had been adviser to William III, king of England and stadtholder of the Netherlands. They were involved in Anglo-Dutch political decision making at every level. Their friends and relations and foreign ambassadors came to stay at his country estate all the time, bringing with them their own servants and friends.

Despite the emptiness portrayed in the picture and Trembley's insistence that he had merely been lucky, the great discovery of the polyp did not come out of an empty house or an empty head. And it did not happen just anywhere. It happened in The Hague, birthplace of the microscope and of experimental optics. It happened not just in The Hague but specifically on a country estate owned by William Bentinck. Trembley was part of an intellectual community that had multiple centers: The Hague, Paris, Geneva, and London. The linchpin to all of these communities was his employer, William Bentinck. There were scores of invisible people behind Trembley in those pictures—not just servants and gardeners, but also all the other people who passed in and out of the great house: microscope and instrument makers, enthusiastic amateur experimenters, encyclopedists, freemasons, freethinkers, booksellers, library curators, publishers, and illustrators. Trembley was working at a powerful crossing point; he was a node at the center of a series of political, trading, and scientific networks: The Hague.

At the time of Trembley's discovery, the Bentinck brothers, Charles and William, aged thirty-three and thirty-seven, were republicans; later they would be enthusiastic supporters of Rousseau, and they may have been leaders of a group of freemasons called the Knights of Jubilation based in The Hague, some of whom were pantheists, reformers, freethinkers, and Whigs. They had strong links with radical publishing networks in The Hague and read the writings of the French *philosophes.* The Lodge of the Knights of Jubilation had its own literary salon and magazine, the *Literary Journal* (*Journal Littéraire*), which promoted science as the means to reform or challenge certain established social institutions.

The freethinkers of The Hague who passed through the house and with whom Trembley shared conversations in the literary salons or coffeehouses of the city on his visits there were in the main first- or second-generation refugees who had come to the Netherlands by the boatload from France after the St. Bartholomew's Day Massacre of 1572 and then again after the revocation of the Edict of Nantes in 1685, which exiled 400,000 Protestants from France, forcing them into Russia, England, Prussia, Switzerland, and the Dutch Republic. These émigrés told horrifying stories of persecution at the hands of the Catholic Church: children detained by the French authorities and forced to convert, separated families, inquisition, imprisonment, expulsion, and systematic torture. At the same time, the increasing circulation of clandestine anticlerical books, the movement of travelers across the world, and the new familiarity with the diversity of world religions and beliefs further undermined the authority of the Catholic Church among these groups. Furthermore, Trembley's work was certainly spawned within a remarkably freethinking culture in The Hague, but the European and particularly French communities of savants to which he addressed it were considerably more radical. Ultimately, they would determine its meanings.

—

Trembley's discovery did not lead directly and immediately to a new flowering of ideas about the origins and transformation of species, but it did shake some of the premises on which all natural philosophy had been founded for at least a century. Seventeenth-century science had been characterized by a narrow and rigid empiricism and an aversion to large-scale theorizing, shaped as it was by Francis Bacon's opposition to hasty hypothesizing; Trembley's discovery spawned a new age of natural philosophical speculation. Trembley had created witnesses across Europe, men and women who had seen the polyp repeatedly cut and regenerated with their own eyes and who had been astonished and confounded by the philosophical implications of that regeneration. Increasingly powerful microscopes made the regeneration visible; a growing community of savants in Paris in particular talked it into a philosophical conundrum.

Trembley's polyp memoir was not the only bestselling science book in the bookshops of Paris in the late 1740s. Many Parisian intellectuals ordered both Trembley's memoir and a book that was as bold, outrageous, and fantastic as Trembley's was deferential, fact-obsessed, and myopic. This work, published in The Hague in French, appeared in the Paris bookshops in 1748 with the strange title *Telliamed, or Conversations Between an Indian Philosopher and a French Missionary* (*Telliamed, ou, entretiens d'un philosophe indien avec un missionnaire françois*). It proposed that the earth was billions of years old, that all animals had transformed through vast tracts of time from primitive aquatic creatures, and that the earth's crust had been formed by the gradual diminishment of the sea. It caused a scandal. For a hundred years it formed the backbone of all discussions about the origins of the earth and the beginnings of time, and it was still being discussed in radical circles on the eve of the French Revolution. It was this book, and the heretical speculations about the formation of the earth and the past and continuing mutation of species and landscape, that catapulted Trembley's polyp to even greater philosophical fame.

It came from The Hague, via Cairo.

6

The Consul of Cairo

CAIRO, 1708

As the call to prayer began across the city, as the air cooled and the moths gathered, the French consul, Benoît de Maillet, dressed in a white turban, silk robe, and embroidered slippers, sat writing at a table on the roof terrace of his house. Through the arch of the elaborate reed-woven pavilion he had designed to frame the view, beyond the roofs, domes, palm trees, and glittering minarets of the city, he could see the Nile, a line of silver-blue, now stained pink by the setting sun. From the city's harbor in the distance, trading boats set sail for Africa, Europe, and Asia carrying wheat, rice, and vegetables, cotton and linen cloth, leather, spices, coffee, and sherbet. Barques and feluccas ferrying goods from the upper Nile moored alongside them. Beyond the wide river, great pyramids jagged the desert skyline.

Immediately below Maillet's roof terrace, the streets of Cairo's French quarter branched out from a narrow, foul-smelling alleyway. It housed the forty-five or so French traders and their families and servants whose interests Maillet was here to protect; they were making their for-

Benoît de Maillet dressed for formal consul duties.
Benoît de Maillet, Description de l'Égypte *(1735)*

tunes shipping coffee and other exotic goods to the coffee shops and aristocratic tables of Europe. It was Consul de Maillet's job to maintain order among the French community, to ensure that taxes were paid, and to send reports back to the secretary of state for the navy, his employer, Jérôme Phélypeaux, Comte de Pontchartrain: detailed descriptions of commercial, trade, and taxation matters as well as local disputes, rumor, and gossip. He was, in Cairo, the eyes and ears of the French king. No one was more important in this small French community than Benoît de Maillet, though he struggled to make his compatriots acknowledge that.

There was very little sense of camaraderie in the French quarter, Maillet complained to his patron in Paris. The young French traders were unruly, greedy, and openly hostile to their consul; they accused him of being in the pocket of the Turkish pasha and were suspicious of his friendships with the high-ranking Arab and Turkish intellectuals and of his numerous trips into the desert. They fought and schemed among themselves and with the local traders; though French protocol dictated that they should attend church, few actually did; they resented paying taxes; many had taken Muslim women as mistresses and fathered half-Arab children.

Cairo, once a great capital, had been swallowed up by the Turks and was now little more than a provincial trading post of the Ottoman Empire, albeit a wealthy one. Maillet spent his days sorting out taxation disputes, trying to enforce regulations, and doing his best to control the unruly militia. Patient, firm, and canny by nature, over the sixteen years of his consulship he had become carefully attuned and attentive to the cultures, beliefs, and etiquette of the different groups who depended upon one another for their daily well-being and trade: the French, Turkish, Indian, and Jewish traders, the pashas, the Turkish militiamen, the missionaries, the ships' captains and sailors who passed through the port. He knew exactly when and how to flatter or threaten, bribe or stand his ground.

The consul was never at rest; in Cairo he was the spymaster; he had spies everywhere, men and women who were his eyes and ears. Cairo was a tinderbox; his power and that of the French ministers running the Paris Chamber of Commerce was continually contested. In 1708 he had finally come to the end of his consulship. When in 1704 eighty French traders and their relatives mutinied against consular authority, a royal inspector had been sent from Paris. In the ensuing investigation, Maillet had been required to answer the French traders' list of grievances, but he had also managed to establish thirty-one new articles of conduct and to bring about some expulsions. In return he had negotiated a transfer to the prestigious consulship of Leghorn, an important northern Italian port city on the Ligurian Sea.* Now that another, less controversial consul had arrived in Cairo to succeed him, Maillet, in preparation for departure, had begun to hand over his papers and files and auction off his furniture and effects. But the French traders, still furious about the new regulations, had begun making trouble; they chanted under his window at night. There were even rumors of assassination plots.

For seventeen years, at night, when the air was cool and the merchants quiet, Consul de Maillet had retreated to his roof terrace to write. His book, he told himself, would raise him forever above the petty gossip and machinations of the ignorant French traders and would couple his name across the world with those of Galileo and Copernicus. He believed he had discovered the origin of the earth and of species. When the time was right, when he had gathered enough proof—more measurements, more facts, more documentation—he would publish.

Since he had arrived in Egypt in 1692, Maillet had been gathering information about the country, collecting detailed geological observations about rocks, strata, and fossils, about sea levels and species. His

* Leghorn is the English name for Livorno's old form, Legorno.

mountain of facts, copied out carefully on paper and stored in so many carved chests with his precious books and manuscripts, had become an enormous cabinet of curiosities. But how much proof would he need? What weight of facts would be required to persuade the savants of Europe to overturn the beliefs of millennia? When would he ever have enough?

In 1697, sitting on his roof terrace in Cairo, Benoît de Maillet had begun an extraordinary book written in French with a very long subtitle called *Telliamed, or Conversations Between an Indian Philosopher and a French Missionary on the Diminution of the Sea, the Formation of the Earth, and the Origin of Men and Animals.* It was the first sustained attempt to prove that species had mutated, the first directly transformist account of earth history. Nearly forty years in the making, it was added to, edited, censored, and finally published ten years after its author was buried in a small cemetery in Marseilles.

Maillet wrote *Telliamed* on the border between East and West, at the crossing point of two cultures. Cairo's population in the late seventeenth and early eighteenth century was not characterized by the easy cosmopolitan alliances of Jahiz's ninth-century Basra. The city was much more volatile, but for someone as curious and imaginative as Maillet, it was an extraordinarily fruitful place in which to speculate about the origin of the earth and of species. The signs of prehistoric time were all around him, in the pyramids and their strange hieroglyphs, in the shifting coastlines and delta of the Nile.

Egypt wove its spell around Maillet from the moment he stepped off the boat that moored in Alexandria in 1692. The architecture of the domed mosques left him breathless. "One cannot but admire the beauty of these domes, their grace, their proportions, their boldness," he wrote. He became an Egyptologist by degrees. His consular duties in Egypt included collecting and shipping back to the Chamber of Commerce archaeological curiosities and reports about the Egyptian people's trade, customs, language, and manners, but it was the search for the ancient origins of Egypt that captured his imagination.

On board the boat that sailed from Marseilles to Alexandria, Maillet began his investigations into Egyptian history by rereading Herodotus' famous description of Egypt written in the fifth century BC. Under the swinging cabin lamp, he studied a map of the country, tracing the great Roman geographer's route up the Nile from Alexandria, marking out the places he had visited. In *Histories,* Maillet read Herodotus' excited observations of the different soils found in Egypt, his description of the omnipresence of salt and the layers of shells in the mountain rocks around inland Memphis, and he was persuaded by Herodotus' conviction that Egypt was "the gift of the Nile," that the land had been formed by the silting up of the great river and that there had once been a sea here. Herodotus described iron rings he had found in Memphis that were set into rock, rings that had been used to moor ships hundreds of years before, in the seventh century BC. Astonishingly, by the fifth century BC, when Herodotus stood contemplating those vestiges of an ancient harbor, there was no sea or river to be seen from even the highest roof terraces of Memphis.

Two thousand years after Herodotus had stared at the ruins of Memphis, contemplating the Egyptian sea that had disappeared from what had once been a port city, Benoît de Maillet set out to find Memphis for himself and to discover if the seashells and the iron rings that Herodotus had described were still there. But finding Memphis was not so easy. When he and his men finally arrived, the great ancient city that had once been the center of its own empire was no more than a looted ruin, a quarry of old stones in the desert, some twenty miles south of Cairo and now "25 Leagues [some 75 miles] from the Sea." Dusty from the desert wanderings and disappointed that there was so little of the great city left, Maillet was nonetheless elated to find the ruins and the rings that Herodotus had touched and the scatterings of seashells in the boulders on the plain.

In search of that lost sea, and for further evidence of shifting coastlines and falling sea levels, Maillet read Seneca, Plato, and Pliny, each of whom had been to Egypt thousands of years before. The ancient historians all noted that ships took a day and a night to reach Alexandria from the island of Pharos, whereas in Maillet's time the two landmasses were connected by a simple bridge. The sea had evidently been much higher and broader two thousand years earlier. It followed that sea levels were falling. As they fell, and as rivers moved more sediment around, Maillet was now convinced, new landscapes were opening up, imperceptibly slowly. The idea of the earth being in a constant state of flow fascinated him. He called it, borrowing a phrase from Bernard le Bovier de Fontenelle's *Conversations on the Plurality of Worlds* (*Entretiens sur la pluralité des mondes*), "this fame Flux and Reflux." What he could see happening in the Nile delta was happening all across the surface of the earth, and it was repeated in the star systems that Descartes described. Waxing and waning. Flux and reflux. Growth and decay.

As Maillet traveled up and down the Nile in his wooden consular barge accompanied by servants and bodyguards, or made expeditions by camel out across the desert to see the pyramids or to ancient white stone Coptic monasteries barely holding their own against sandstorms, he witnessed and noted down the marks of immense tracts of time inscribed upon the landscape. He saw the tracings left by the Nile as it flooded and contracted over thousands, perhaps millions, of years. And everywhere he looked for vestiges of a retreating sea, searching for seashells among the hieroglyphed ruins.

With the help of consular translators, Maillet taught himself Arabic so as to be able to read Arabic historical manuscripts and talk to Turkish and Arabic philosophers, historians, and naturalists. He befriended Christian savants, entering into elaborate correspondence with the Coptic and Greek patriarchs, the abbot of Sinai, and missionaries throughout the country, gaining access to specialized libraries, collections of rocks, archaeological finds, and mummies in remote monasteries and sending the carefully wrapped and crated objects back to his employer, Pontchartrain, in Paris. Although he was disliked by the traders in Egypt whom he represented, his fine manners and consular status endeared him to the pashas and their sages and curators, as well as to the commanding officers of the Turkish militia, some of whom also had libraries filled with rare books and manuscripts. Arabic, Latin, and Greek manuscripts took him back to the origins of Egypt, but when he reached the edge of those histories, he knew that the great expanse of time before man appeared on the earth could be read only in the rocks themselves. He taught himself to read them as if they were written in Arabic or Coptic.

The Great Sphinx of Giza as Maillet would have seen it—up to its neck in sand.
Description de l'Égypte *(1809)*

By 1697 Maillet had gathered enough information about Egypt past and present to begin a book that would not be completed until 1735, when it was published in Paris as a series of letters called *Description of Egypt* (*Description de l'Égypte*). Describing the origins of Egypt had forced him to expand his idea of the age of the earth, for the ancient historical accounts he read recorded events that had happened in this landscape thousands of years earlier, chronicling a stretch of time much greater than that of any Western historical account he had encountered.

Now that Maillet had reached that far back in time, there was no stopping his curiosity. He wanted to go further, beyond the formation of cities, beyond people, back into prehistory, back into deep time.

So in the last few years of the seventeenth century, while he was still gathering material for his book on Egypt, Maillet began to write a more ambitious and speculative book about the earth, the sea, and time itself. This was dangerous work for a public servant, however well connected. If he was going to retain his position and his salary, he knew, he would have to publish anonymously or under a pseudonym.

Perhaps it was by watching the way in which the Egyptians wrote their manuscripts from right to left that Maillet conceived of the idea of reversing the letters of his own name, much as Leonardo da Vinci had done with his mirror writing. Under the consul's pen, "de Maillet" once reversed became "Telliamed," and out of this strange reversal of his identity he shaped an alter ego, the mysterious Indian savant of his heretical book.* It gave him a fictional device, a seductive form of storytelling. It lent a frisson to his tale.

In the opening pages of *Telliamed,* the French missionary-narrator explains to the reader that an Indian sage had recently appeared in Cairo and announced that he was ready to pass on his secret discovery. Over several days, during which the missionary had been struck almost mute by the Indian philosopher's strange revelations, Telliamed had expounded a theory of the earth based on a hundred years of geological investigations, seabed mapping, diving, and fieldwork begun by his grandfather and continued by his family on a remote peninsula on the coast of India. The time had come, Telliamed had told the missionary, for the truth to be passed to the West. Then he had disappeared again, back into the Nile.

Consul de Maillet was at once the wise Indian philosopher Telliamed and the awestruck, unnamed French missionary, as well as the ancient sage, the Aeiul, Telliamed's grandfather. While it was of course Maillet who had gathered all the evidence and formulated his theory of the earth, in *Telliamed* it was the Indian philosopher who did the speaking. It was Telliamed, not the consul, who claimed that the earth was billions of years old, that it was shaped only by chance, and that all animals, including humans, had transmuted from primitive sea creatures. If the consul of Cairo had been asked, he would have said that it was Telliamed, not Maillet, who was the heretic.

The elaborate conceits that shaped Maillet's book, the fictionalized conversations, the anagrammed name, the mysterious Indian sage who issued from the Nile and disappeared back into it—all of this provided Maillet with a mask to protect his identity. But in using conversational form he was also imitating the shape of the great bestselling French science book of two decades earlier, Bernard le Bovier de Fontenelle's *Con-*

* Voltaire formed his pen name from an anagram of his original name in 1718. Given that a manuscript of *Telliamed* was circulating in Paris at this time, the two anagrammed names might have a connection.

versations on the Plurality of Worlds (1686), a book written as a series of conversations about astronomy held through a series of moonlit, star-studded nights between a fictionalized astronomer and a marquise.

Conversations was the first popular science book written in the vernacular; in it, Fontenelle, a poet and a playwright inspired by the new sciences and particularly by the work of René Descartes, had found a way of turning new Cartesian scientific ideas into an act of sustained persuasion that was also daring, entertaining, and seductive. Surveying the night sky, for instance, the astronomer and the marquise considered the possibility of travelers from the moon. "What if they were clever enough to navigate the surface of our atmosphere and, from a curiosity to examine us, should be tempted to draw us up like fishes; would that please you?" asks the philosopher. "Why not?" the marquise replies, laughing. "I would voluntarily put myself in their nets, just for the pleasure of seeing the fishers."

The ideas that Fontenelle's elegant and heretical book explored—ideas about alternative worlds, unimaginable races, great tracts of time, stars that waxed and waned through thousands of years—dominated the talk of the Paris salons for more than a century and enlarged ideas and imaginations: "Were the sky only a blue arch to which the stars are fixed, the universe would seem narrow and confined; there would not be room to breathe . . . [but now]," the marquise enthuses, "I seem more at liberty; to live in a freer air; and nature appears with astonishingly increased magnificence."

More deeply unchristian and speculative than *Conversations, Telliamed* made no attempt to apologize for its heretical claims; Maillet did not try to square his theory with the Bible or find a way of bringing God into the picture. He largely bypassed both. There was no God in Telliamed's system of the earth. There was no maker. There were no supernatural explanations or interventions. Instead, Telliamed claimed, everything on the earth, all the landscapes, all the trees, flowers, animals, and people that live on it, have been made through the operations of *le hasard,* chance. The earth was not thousands of years old as the Church claimed, but *billions* of years old. The sea was slowly shrinking; falling sea levels shaped—and continued to shape—the landscape. Most shocking of all, the Indian philosopher insisted in its pages that all life on earth had evolved from primitive sea creatures and that men had evolved from sea versions of themselves. Those remnants of an earlier sea race of humans were still alive in the world, he declared, living as exiles, hiding their sea tails.

The thirty-six-year-old Maillet had arrived in Egypt in 1692 with a set of questions shaped by his time in France, and particularly in Marseilles. Many of these questions, which pressed not only upon Maillet but upon a generation of European intellectuals, were stimulated by his reading of Fontenelle's *Conversations* six years earlier. The questions at the core of Fontenelle's book—where does time begin, how did the stars come into being, where does matter begin and end, how many living worlds are out there?—were expanded in the conversations Maillet had in the salons of Marseilles and in the circle of his patron Pontchartrain in Paris. Fontenelle's book and Claude Gadrois' treatise *Conversation on the Influence of the Stars* (*Discours sur les influences des astres*) (1675) introduced Maillet to René Descartes' cosmology for the first time—the theory that stars had been born from a vortex and continued to mutate through dark and luminous phases, shrinking and cooling, heating and swelling by turns. Maillet took a copy of *Conversations* with him to Cairo and searched out Fontenelle's other scientific papers; in one of these he read of Bernard Palissy's *Admirable Discourses* and gave an approving description of "the simple potter" in his account of fossil formation in the Second Conversation of *Telliamed,* chronicling the "striking proofs" Palissy had collected in his Paris cabinet and lectures. Palissy's arguments had convinced Maillet not only that the waters of the earth were infused with generative elements, as the potter argued, but that both sea and air were filled with minute seeds and that new species were being produced constantly at all the transmigration points.

After leaving Egypt in 1708 and settling in Leghorn, Maillet continued to add to his great work, collecting geological facts from savants, traders, sailors, and ships' captains; he visited more cabinets of curiosities, continued to scour the natural history pages of the European journals, and traveled the length and breadth of Europe measuring riverbeds, searching out shells at the tops of mountains and at ancient ruined ports, looking for further evidence of a fallen and falling sea and for reports of sea people. In Leghorn he discussed his ideas with his great friend the Constable Marie Mancini Colonna, an elderly divorcee who had been the first love of the king of France, who now lived an itinerant life dependent on patrons, exiled from several countries for her affairs and scandals and constantly under surveillance. It was Marie Mancini Colonna who provided Maillet with the stories of sexual and moral improprieties that he was required to send back to Pontchartrain for the king's ministers to use in their diplomatic and political negotiations.

His field researches were often aided by luck. In 1714, for instance, the engineers digging a ditch from the new infirmary of Leghorn to the old infirmary hit a layer of mud under several rock strata. Here they found a hollow log about twenty feet long embedded two or three feet in mud flecked with seashells, pinecones, animal horns, bones, and teeth. They sent for Maillet—he was well known to the miners, engineers, and road makers in the area—who confirmed that it was a pump from a wrecked ship that had once sailed on seas long retreated. With new evidence like this surfacing all the time, Maillet could not bring his book to a conclusion. There were always more facts to find, more libraries, more books, more reports and stories. His chests now contained hundreds and hundreds of pages.

Like Jahiz's *Book of Living Beings, Telliamed* was a melting pot, a synthesis of ideas, "facts," reports, and stories collected from across the world and from different traditions and cultures over many centuries, all of them convinced in different ways of the earth's formation through a retreating sea. Like the land of Egypt itself, crafted and carved by the Nile, *Telliamed* constantly changed its shape as new materials were brought into it. Though Maillet's questions and his search for origins may have been engendered in Marseilles and Paris, they were sharpened and remade by the Arabic and Persian authors he read, such as Ahmed al-Makrisi, a fifth-century Cairo geographer and historian who wrote *An Historical and Topographical Description of Egypt* and *A History of the Ayoubite Sultans and Memluks;* Maillet brought Arabic historians and geographers onto the stage of his book alongside Herodotus, Seneca, and Pliny as fellow heretics, and he allowed them to speak for themselves. Most important of these was Omar Khayyám, whom Maillet referred to as

Omar el Aalem, who "taught philosophy at Samarkand about 800 years ago." In *Telliamed,* Maillet devoted more than two pages to Khayyám's ideas about the diminution of the sea, stressing the intellectual risks Khayyám took and his collision with the authorities that resulted in his exile from Samarkand for heresy. As consul, Maillet had access to Khayyám's geographical and geological work, pages that have been lost to the West. Maillet was acutely conscious that with *Telliamed* he, too, was on extremely dangerous ground.

In 1717, Maillet returned from Leghorn, where he had been first consul and then inspector of French establishments in the Levant and on the Barbary Coast,* to Marseilles. The French port city, which received goods from the East, silk from Spain, rice and wheat from the Levant, Yemen mocha from the Barbary Coast, Arabic coffee from Egypt, and new American coffee beans from Martinique, was another crossing point, teeming with foreign traders and sailors. French ships set sail with flour, wine, eau-de-vie, marble mantels, chalk brick, soap, shoes, and French textiles. Maillet set up house in the rue de Rome near the old port where he could keep an eye on trading activities from his upper windows.

However, as inspector, it was difficult for him to focus on his book; he spent most of the following two years sailing between Algiers, Alexandria, Syria, and Cyprus, sometimes arriving incognito in an unmarked boat in order to undertake investigations or to broker treaties or inspect accounts. On his retirement in 1719, Maillet returned to his Marseilles house on a modest royal annuity, working on his manuscripts, ordering his library, and continuing to write reports. The new secretary of state for the navy, the third of the Phélypeaux dynasty, Jean Phélypeaux, no longer required him to provide reports of sexual scandal, but now wanted instead to mine the depth of his mercantile knowledge. After forty years, Maillet was still the eyes and ears of the French king.

When plague forced him to leave Marseilles in 1720, Maillet headed for Paris, determined to find a publisher. He now had working drafts of four books but had lost control of all of them: his book on Egypt, a short

*Maillet held this position for four years, from 1715 to 1719; for the first two years he was based in Leghorn, for the second two in Marseilles.

treatise on Ethiopia, a longer set of memoirs on Ethiopia, and *Telliamed*. The manuscripts had all become unwieldy. The Egypt book badly needed a structure; the arguments of *Telliamed* had been lost under all the evidence. He was now in his sixties, and time was running out.

A clandestine book trade had burgeoned in Paris in the first decade of the eighteenth century. A community of loosely networked savants and translators were busy identifying, translating, and reissuing controversial, materialist, and radical books that challenged religious, intellectual, and political orthodoxies. These books, which would often include new prefaces and commentary, were sometimes published in Paris but more often appeared in Amsterdam or The Hague, where the trade in clandestine books was less closely policed. The French authorities had established the first inspectorate of the book trade in order to control the growth of this seditious material; there were forty-one royal censors checking books in the early 1730s, and punishments included book burnings and imprisonment.

In Paris, Maillet hoped to find an intellectual patron who would have the courage to collaborate on *Telliamed* as well as to edit and restructure his Egypt book. First he approached his friend the Abbé Granet, a humanist, who was busy editing the many works of the great French dramatist Corneille. Granet turned the Egypt manuscript down. Next he approached the geographer and mapmaker Jean Baptiste Liebaux, who promised he would begin work on it when he could and took the manuscript away to read. Meanwhile, a Parisian publisher agreed to publish Maillet's treatise on Ethiopia as one of several additions to a reprint of a translation of the account of the voyages of a Jesuit Portuguese priest written in the previous century. Things were looking up.

But what was he to do with his most dangerous book? The manuscript of *Telliamed* had been circulating in Paris and Marseilles for several years, but though it was a controversial book about the history of the earth, it did not yet include Maillet's radical theories about species change and human origins. In 1726, relieved that his treatise on Ethiopia was now in press and the Egypt book in the hands of a prestigious editor, Maillet sent a copy of the manuscript of *Telliamed* to the sixty-nine-year-old Fontenelle, who was now perpetual secretary of the Academy of Science, asking for his help in editing it. He knew he was not the first to have developed a theory on the origin of the earth, he explained, but he believed he was the first to have discovered the way it had changed and been able to prove it to be true beyond doubt. He asked for Fontenelle's advice: although he had gathered a great deal of proof to support his theories about the origins of species, he considered them to be still too controversial for publication.

Fontenelle urged him to expand the book. He reminded him that the eminent German natural philosopher Gottfried Leibniz had speculated on extinct species in an essay on fossils that had been reported by the Academy in 1706. He assured him that Descartes' ideas were no longer so controversial in Paris and that the question of the origin of species was the most burning philosophical problem of the day. It was on everyone's lips. If Maillet had evidence and proof relating to the origins of species, he must disclose it.

So in Marseilles in 1722, in the aftermath of the great plague, Maillet began to sort through the materials he had gathered for the most controversial conversation of all: Conversation 3, "On the Origin of Species." He was still revising it fourteen years later in 1736, redrafting, rephrasing, putting the conversation in and taking it back out. He wrote to the Marquis de Caumont in 1737 in the last year of his life: "Telliamed has all the trouble in the world out of this conversation about which he is not happy." In the narrative, even the usually unperturbed Indian philosopher hesitates before he discloses these most dangerous of ideas to the French missionary. He also ensures that his boat is in the harbor, sails up, ready for him to effect a quick disappearance.

Maillet was quite secure in his conclusions now; he had gathered a mountain of evidence, scores of multiply witnessed sightings of sea people from across the world, from different centuries, and from many different sources and authorities. He could only hope that the weight of evidence would be enough to persuade some of his readers. But the words on the page trembled under his ventriloquist's pen.

Telliamed's species theory was startling, extraordinary, and bizarre: billions of years ago, he claimed, a great sea had covered the earth. All life came into being in that ocean, evolving from tiny, invisible seeds; some species, including a form of sea people, had transmigrated from the sea to the land. Beaks and claws and necks and limbs slowly changed their shapes as species adapted to new environments. Some intermediate forms of sea-human species still swam in the sea, he claimed; some walked on land. Many hundreds of these sea people—with webbed feet, scales, or tails—had been sighted and reported by people in positions of authority. He had even seen them with his own eyes.

Metamorphosis was at the heart of biological life, Telliamed claimed. All species were in a state of flux and reflux, just like the planets in the heavens, moving from life to death, through states of extraordinary change. It was the way of things. "The Transformation of a Silk-worm or a Caterpillar into a Butterfly," he wrote,

> would be a thousand Times more hard to believe than that of a Fish into a Bird if this Metamorphosis was not daily made before our Eyes. Are there not Ants which become winged at a certain Time? What would be more incredible to us than these natural Prodigies if Experience did not render them familiar to us? How easy is it to conceive the Change of a winged Fish flying in the Water, sometimes even in the Air, into a Bird always flying in the Air in the Manner I have explained?

After describing sea calves, sea dogs, and sea bears, Telliamed begins his lengthy catalog of sightings of sea people with the appearance of a sea man on the banks of the Nile. "Your Histories read," begins the Indian philosopher,

> that in the Year 592 of your Era, on the 18th of March, an Officer of one of the Towns of the Delta or the Lower Egypt, walking one Evening with some of his Friends on the Banks of the *Nile,* saw very near to the Shore, a Sea-man, followed by his Female, the Male raising himself often above the Water as far as his secret Parts and the Female only to the Navel. The Man had a fierce Air and a terrible Aspect, his Hair was red and somewhat bristly and his Skin of a brownish Colour. He was like to us in all the Parts which were seen. On the contrary the Air of the Woman's Countenance was sweet and mild, her Hair black, long, and floating on her Shoulders, her Body white, and her Breasts prominent. These

two Monsters remained near two Hours in the Sight of this Officer, his Friends and those of the Neighbourhood, who had come to see so extraordinary a Fact. An Attestation of it was drawn up, signed by the Officer and many other Witnesses, and sent to the Emperor Maurice who then reigned.

Maillet had been gathering these records of sea people sightings for decades, combing the pages of Arabic as well as European manuscripts, interviewing people, asking for information, compiling it. They included an account of a living girl rescued from the belly of a fish caught in the Caspian Sea in the ninth century who "fetched deep Sighs, and lived for but a few Moments"; a sea girl found in Holland in 1430 half buried in mud at the mouth of the river Tye near Edam after a great flood had receded, who was taught to spin and make the sign of the cross but could never speak; a French officer's report of seeing off the coast of Martinique in 1671 a "Sea-Monster of a human Form from the Middle upwards, and terminating below like a Fish"; a sea man caught at Sestri in 1682 who "survived for some Days weeping and uttering lamentable Cries"; another who was shot by a sentry while stranded on the shore by the receding tide at Boulogne; a mother and daughter captured from a herd of sea people spotted off the Indian coast by Portuguese sailors, who when caught and taken to King Don Emanuel "were so extremely melancholy that nothing could comfort them" but who when taken to the sea lagoon remained three hours under water "without coming above its Surface to respire"; and a man captured off the coast of Greenland who was "shap'd like us, with a Beard and Hair pretty long, but from the Middle downward, his Body was all covered with Scales." Maillet's chronological list of sightings came up to his present, ending with a report taken of a sea man seen swimming around a French ship moored on the bank of Newfoundland as recently as 1720. All these reports share the same features: the sea people were mute, all were irredeemably sad when taken from the water, and all died within days or months of capture.

In *Telliamed,* Maillet described genitals, breasts, fur, scales glimpsed or imagined just beneath the waves, a sea woman made to sit on a chair, male and female sea people playing together in the sea. Each one, when out of the water, he portrayed as an infinitely sad and fragile exile. His sea people of the land, the webbed or tailed humans, were similarly both melancholy and sexually irresistible, like the French officer seduced on a boat by a fifteen-year-old courtesan. "Being at Pisa, in 1710," Telliamed relates, recounting his most erotic story of all,

> I was informed that there was a Courtesan who boasted of having known a Stranger who had been there three Years before, and who was one of the Species of Men with Tails. This inspired me with a curiosity to see her, and examine her with respect to the Fact. She was at that Time no more than eighteen Years old and very beautiful. She told me, that in returning from *Livorno* to *Pisa* in a Passage-Boat in 1702, she met three *French* Officers, one of whom fell in Love with her. Her Gallant was large, well-made, and about thirty-five Years of Age; he was of a very fair Complexion, his Beard was black and thick, and his Eye-brows were long and shaggy. He slept all Night with her, and came very near that Labour for which *Hercules* is no less famous in the Fable than for his other Exploits. He was so shaggy that Bears themselves are hardly more so. The Hair with which he was covered, was near half a Foot long. As the Courtesan never met a Man of this Kind, Curiosity led her to handle him all over, and putting her Hands to his Buttocks, she felt a Tail as large as one's finger, a half a Foot long, and shaggy like the rest of the Body, which she grasped, upon which she asked him what it was. He reply'd with a harsh and angry Tone that it was a Piece of Flesh he had had from his Infancy, in Consequence of his Mother's longing for a Tail of Mutton when she was big with him. From that Moment, the Courtesan observed that he no more testified the same Affection for her; the Smell of his Sweat was so strong and particular, and smelt, as she said, so much of the Savage, she could not get quit of it for a month afterwards.

Maillet, now in his early seventies, drafted his "Origin of Species" conversation during what he described as "troublous times." When his treatise on Ethiopia finally came out in the edition of Lobo's travels in

1728, it was unrecognizable. The editor had taken enormous liberties with his manuscript; indeed, he had practically rewritten it. Maillet was furious. And despite frequent letters, he had heard nothing from Monsieur Liebaux, who still had the manuscript of his *Description of Egypt;* the geographer was ill, people said. Maillet was also short of money after a series of expensive lawsuits and was now forced to try to sell some of his furniture and art. He offered his eighty-volume oriental library to the Abbé Bignon, the royal librarian of the Bibliothèque du Roi, a library that included one of the very first Arabic editions of *A Thousand and One Nights,* a Coptic version of the Book of Ezekiel, the lives of Nouredin and Saladin, and his beloved three-volume edition of al-Makrisi's *History of Egypt.* The librarian declined his offer.

When one of Maillet's patrons in 1733 finally recommended a new editor for the Egypt book, the Abbé Jean Baptiste le Mascrier, Maillet traveled to Paris to meet him. The French capital overwhelmed his imagination, he wrote to the Marquis de Caumont. Brilliant gold and azure coaches seemed to him like flying chariots guided by celestial hands; from inside the coach people looked like "flying arrows transfigured into memory paths of passing flight." He bought a five-foot-tall Egyptian queen from a Paris dealer as a present for another patron, the governor of Marseilles, Pierre Cardan Lebret. His "belle Egyptienne," carved from striated green marble with black touchstone head and feet and removed from a pyramid, weighed more than a hundred pounds. He also sent the governor a crate of clocks that broke on the riverboat journey to Avignon.

The Abbé Jean Baptiste le Mascrier was a Jesuit-trained Catholic priest and poet living on the edge of a Parisian philosophical coterie, supporting himself financially by hackwork: editing, ghostwriting, translating, or prefacing controversial and sometimes anonymously or posthumously authored texts for Parisian publishers. He was thirty-three years old, and the exasperated and eccentrically dressed ex-consul who arrived carrying bundles and files of tattered papers was seventy-seven. Mascrier warned his visitor that he might not have time to turn his hand to the Egypt book for a while. He was writing a verse preface to a play by Montfleury that was being staged at the Comédie Française and co-translating from Latin sixteen volumes of Jacques-Auguste de Thou's history of the French wars of religion. But leafing through the descriptions of pyramids, ancient cities, and Egyptian customs and manners in Maillet's papers, Mascrier changed his mind. Everyone in Paris, he reflected, was enthralled by the mummies, carvings, and hieroglyphs that had arrived back in Paris from Egypt, and yet no one had written in detail about the country since Herodotus. This crazy pile of papers and descriptions could be a bestseller if it were handled right.

But Mascrier had not realized how much time the old man from Marseilles would demand from him. Letters arrived almost daily, sometimes several. In each one Maillet had more to say, more to restructure, more complaints and readjustments and corrections, even more ideas. The book was impossible to finish. As soon as Mascrier had knocked the Egypt book into shape, structured it as a series of letters, found a publisher, completed the proofreading, received the first batch of copies, and dispatched one to Maillet in the country, Maillet immediately sent back a list of errata and omissions and complained that Mascrier had put his own name on the title page as author and not as editor.

Nonetheless, Maillet respected Mascrier's editorial skills enough now to send him his five notebooks on Ethiopia and Coptic Christianity, instructing him to include an edited version in the second edition of the *Description of Egypt*. More important, he forwarded to him the extraordinary pile of papers that was *Telliamed*. Maillet knew he would not live much longer; his health was deteriorating. This was his last chance to see his life's work appear in print. Mascrier was a priest, albeit one whose faith had taken some dents as a result of long years spent reading and editing seditious books and working with atheist publishers. What would he make of this seditious book? If Mascrier had gained a reputation with publishers as an editor who defused heretical books, he was now doing that work with rapidly declining conviction. He had recently been commissioned to coedit a new pirated edition of a bestselling heretical book written by two Dutchmen called *Religious Ceremonies of the World* (*Cérémonies et coutumes religieuses de tous les peuples du monde*) that for its disrespectful representation of the Catholic Church as just one religion among many had been put on the papal list of forbidden books. Would Mascrier dare to take on the incendiary conversations of *Telliamed*?

As Maillet lay dying in his house in Marseilles, knowing that Mascrier was working his way through the editing process, he became increasingly anxious about the fate of his great book. He was still revising the last pages about the origin of man until a month before he died, propped up in bed under his Egyptian mosquito net, eating figs and sweet pears, drinking camomile tea, and writing by the light of a lamp in which he burned pure olive oil because, he told the marquis, olive oil rendered the light brighter and more constant. For three years he had sent Mascrier letters that were high-handed or anxious by turns, instructing the abbé to insert new material or to contact someone in Amsterdam or to rewrite yet another version of the preface. In his last months he became preoccupied with death and immortality. "Time is eternal," he wrote. "In nature nothing dies, but everything is enfolded back into the earth to be remade; soft to hard, hard to soft, the law of the earth, of the planets and of all bodies, is that of perpetual remaking." "Even after centuries of petrification," he wrote on January 8, 1738, "certain marine organisms could spring back into life in the womb of the earth." He had felt sure of this for some time, it seems, having made the Constable Colonna promise that she would make contact with him after her death. She had, he recorded, kept that promise.

Then in January 1738 the letters from Maillet to Mascrier stopped. Maillet had died of pneumonia in the midst of writing a letter that assured the Marquis de Caumont that a bound copy of *Telliamed* would arrive at any moment. But that bound copy was still a long way from the printer. Now that Maillet was no longer in a position to pay for Mascrier's time, the overworked editor, relieved, put the *Telliamed* manuscript aside and returned to coediting *Religious Ceremonies of the World.* The Parisian publishers of *Ceremonies* had grown impatient with the delays, and however heretical that work might be, Mascrier must have reasoned with himself, *Telliamed* was much more subversive. But as Maillet's own copies of *Telliamed* continued to circulate in Paris, the book was gaining a reputation. It was a dangerous book. It was also a book that Mascrier knew would sell.

Telliamed at last found its way into print in 1748 in Holland, ten years after its author's death. Mascrier realized that he would never get the book past the Paris book censors; Amsterdam was his only option. Fearful of his own safety now that he was under police surveillance as a subversive, he published the manuscript under the name of an obscure Dutch lawyer, Jean-Antoine Guer. He had done what was necessary to sanitize Maillet's last book. While the original manuscript papers of *Telliamed* claimed that the earth was billions of years old, Mascrier changed this to "thousands" or used vague phrases such as "a great number of years." Everywhere he did his best to square Telliamed's theory with the biblical version of creation and to introduce a sense of divine purpose behind Telliamed's description of blind forces at work in the world. He systematically deleted Maillet's references to *le hasard,* or chance, and all of Maillet's powerful refutations of the biblical flood. To imitate Fontenelle's *Conversations,* he divided *Telliamed* into six days. Finally, he dedicated the book to Cyrano de Bergerac in order to persuade readers that it was a semifictional work.

Mascrier's editorial acrobatics and neutralizations did not prevent *Telliamed* from provoking a scandal. Outraged reviewers and naturalists wrote angry or mocking repudiations in scientific journals and newspapers. The naturalist and collector Dezallier d'Argenville denounced the book in the third edition of his *Natural History,* published in 1757: "What a folly in this author to substitute Telliamed for Moses, to bring man out of the depths of the sea, and, for fear that he should descend from Adam, to give us marine monsters for ancestors! Only a kind of godlessness could invent such dreams." Twenty years later, when *Telliamed* had become one of the bestselling books of the century, with copies to be found in the majority of great French libraries, Voltaire was still outraged by its claims. "This consul Maillet was one of those charlatans who wanted to imitate God, and create a world with words," he wrote. "It was he who, abusing the story of some upheaval that arrived in the world, claims that the seas had formed the mountains, and that fish have turned into men."

By the time Maillet had met Mascrier, the abbé was already under surveillance by a senior police agent, the new inspector of books in Paris, Joseph d'Hémery. Despite the large number of people he was watching, d'Hémery was especially perceptive, even intuitive, in his assessment of the slow radicalization of this particular Jesuit priest. He composed his portrait of Mascrier from fragments of interviews with friends, colleagues and acquaintances, booksellers and publishers, and from overheard conversations in cafés. "He was a Jesuit for a long time," d'Hémery wrote.

> He edited *Telliamed* and various other publications for the booksellers. He contributed to *Religious Ceremonies of the World* and worked over Maillet's *Description of Egypt,* which does great honour to him by its style. He turns poems very nicely, as is clear from a prologue to a play that was performed some years ago. . . . The Benedictines, where he worked, agree that he is a man of talent. Too bad he isn't more creative. He has published an excellent work of poetry, a book that is useful to every true Christian, but the people who know him most intimately think that the need to produce copy is making him gradually shift to different sentiments.

What file would Joseph d'Hémery have kept on Benoît de Maillet? Alienated from his own community, judgmental, inflexible, and high-handed, Maillet was connected to a global community of savants that stretched from Egypt across Europe. He was an outsider, enabled by his consular position and by his command of languages and cultures to access libraries and collections across the world and to synthesize different scientific positions, beliefs, and ideas. He was imaginative. He could stand in a landscape and propel himself back in time to conjure the great forces that had been at work to carve out mountain ranges and riverbeds. He was a traveler whose experience of diverse beliefs, peoples, landscapes, and cultures had enabled him to imagine a world in flux and reflux and to embrace a version of a world that was for him, as it was for Aristotle and Epicurus, timeless and eternal, pulsing its way through darkness and light, growth and decay. He spent an entire fortune on his obsession, funding his own research centers, paying for measurements, proofs, stories, and translations, and building a library of materials that he did not live to see published, and always to the end, like Jahiz, dependent on the goodwill of enlightened and wealthy men.

A hundred and twenty-two years after Maillet's death, in February 1860, Charles Darwin added the name Demaillet to his list of nineteen predecessors and posted that list, now entitled "A Historical Sketch," to the botanist Asa Gray in New York asking him to include it in the first authorized American edition of *On the Origin of Species by Natural Selection.* On those pages Darwin paired Maillet with the great naturalist the Comte de Buffon, making them the only two people on Darwin's list who stood in that great void of time between Aristotle and Lamarck—that void of time that frightened Darwin because he knew that his grasp of history had always been weak.

Two months later, in April 1860, Darwin opened the pages of the *Edinburgh Review* to find Richard Owen's poisonous notice of *Origin* in which the writer sneered at him as another fantasist as deluded as the merman believer Benoît de Maillet. Darwin was both embarrassed and enraged. Without reading *Telliamed,* he now erased Maillet's name from the new version of the "Historical Sketch" that he was preparing for the third English edition of *Origin.* When the Scottish horticulturalist Isaac Anderson-Henry offered to send him his own copy of *Telliamed* in 1867, Darwin wrote, no doubt with a frisson of remembered embarrassment, "I am bound to read it as my former friend and present bitter enemy Owen generally ranks me and Maillet as a pair of equal fools."

That copy of *Telliamed,* printed in London in 1750, ordered by Anderson-Henry, and posted from Edinburgh to Kent in 1867, disappeared deep into a box stored in the family attic at Down House; it was not retrieved until 1993. The archivist noted that in Maillet's yellowed pages, Darwin had placed a single line in the margin to mark up paragraphs of special interest to him: passages on fish transforming into birds, sea men, ape-human breeding, and finally Telliamed's statement that he was absolutely sure that all species might have come into being through adaptation and transmigration without any divine intervention. Darwin also wrote the words "Men with Tails" beside the passage in *Telliamed* that Owen had used to ridicule him in that terrible review. Understandably, the book failed to find a place on Darwin's study bookshelves; it was relegated to the box in the attic instead.

7

The Hotel of the Philosophers

PARIS, 1749

At 7:30 on the morning of July 24, 1749, two police agents carrying search warrants climbed the stairs above the upholsterer's shop at 3, rue de l'Estrapade in the Latin Quarter of Paris.* They had come to interrogate the man living with his wife and young son on the third floor. As a writer of books deemed contrary to religion, the state, and the king, the suspect, Denis Diderot, had been under police surveillance for six months. His last book, *Philosophical Thoughts* (*Pensées philosophiques*), which placed Catholicism on the same level as all other religions and implied that none had any exclusive claim to truth, had been banned by the Parliament of Paris and burned in the public marketplace. But Denis Diderot was well connected and had influence in high places. The inspector of books, Joseph d'Hémery, and his men would have to keep their wits about them.

To the police agents' frustration, when they reached the third floor,

*The street is named after a form of torture used on heretics in a nearby square. The word means "bucking," a term used to describe the movements of the victims suspended by their arms.

the man with large dark eyes and flowing hair who appeared at the doorway wearing only his dressing gown seemed to have been expecting them. The rooms had been tidied since their last visit. On the family dining table, books and handwritten papers had been arranged neatly; natural history books were to be seen everywhere, but there was not a single illegal or banned book. They found a microscope and twenty-one cartons containing hundreds of manuscripts stacked in piles on the floor. He was editing an important work, Diderot told them by way of explanation, his eyes flashing provocatively, a book in many volumes that would bring knowledge to the people and dispel superstition and ignorance. But Joseph d'Hémery was not interested in the *Encyclopedia* (*Encyclopédie*). He was looking for proof that Diderot had, despite police warnings, published a still more dangerous book than his last, an anonymous work called *Letter on the Blind* (*Lettre sur les aveugles*) now circulating in the cafés and salons of Paris. They found two copies of that manuscript on the desk. "In the presence of the said Diderot," Commissioner Rochebrune recorded in the police files, "we continued our search in the other rooms, and having opened the wardrobes and chests of drawers, found no papers therein." Finally d'Hémery produced his arrest papers, the notorious *lettres de cachet,* and explained to Diderot's furious wife, Nanette, as they bundled her husband into a carriage, that Monsieur Diderot was under arrest. No, he was not going to the Bastille. The Bastille was full. Madame Diderot would find him at Vincennes, a medieval fortress and former royal residence six miles east of Paris.

Denis Diderot, aged around fifty (an engraving from a painting, ca. 1767).
Getty Images

Denis Diderot was only one of hundreds of Parisian subversives in d'Hémery's files that year.

It was 1749; there was a dangerous spirit brimming among the savants of the city. The citizens of Paris were hungry and the king and his government ministers unpopular. Louis XV had won the War of the Austrian Succession but had thrown the victory away by negotiating a weak treaty. Now he had imposed a punishing new tax on his people to pay for a war that had achieved very little. Government ministers, fearing revolution, had appointed hundreds of new police agents and charged them to round up and imprison anyone who threatened the status quo by writing, publishing, translating, or selling subversive propaganda or satirical poetry, ballads, or books. Across Paris, poets, intellectuals, actors, and booksellers passed illegal manuscripts or antiroyalist propaganda from one to another, in garrets and cafés and public gardens. To control this rising tide of sedition, d'Hémery and his agents opened new files every day, gathered lists, followed paper trails, recorded overheard conversations, interviewed witnesses, friends, priests, and concierges, recruited more spies, and interrogated suspects, and back in their offices they studied the books and manuscripts they had impounded to prepare heresy and sedition cases for the courts.

In Paris in 1749, everyone was watching everyone else.

Diderot was not only clever, d'Hémery reflected a few days later as he leafed through the philosopher's file while preparing to take a carriage ride out to Vincennes to interrogate him, he was also potentially one of the most dangerous men in Paris, a silver-tongued and persistently impious man, a provocateur.

The book burning in 1746 did not stop "the Boy," as d'Hémery called Diderot. It added luster to his reputation among the intellectuals of Paris and increased his book's sales on the black market. Two years later a parish priest reported Diderot to the authorities, claiming that he was writing yet another subversive book. The priest's report, filed away by the police, was uncompromising: "M. Diderot is a young man who passed his early years in debauchery," it began.

> The remarks that Diderot sometimes makes in the household clearly prove that he is a deist, if not worse. He utters blasphemies

> against Jesus Christ and the Holy Virgin that I would not venture to put into writing. . . . It is true that I have never spoken to this young man and do not know him personally, but I am told that he had a great deal of wit and that his conversation is very amusing. In one of his conversations he admitted to being the author of one of the two works condemned by the Parlement and burned about two years ago. I have been informed that for more than a year he has been working on another book still more dangerous to religion.

In January 1749, in the hope of locating a network of *philosophes* as well as getting hold of this book that was "still more dangerous," d'Hémery had found time to interrogate Diderot himself. He impounded the manuscript he found in the philosopher's atelier, a book called *The Skeptic's Walk* (*Promenade du sceptique*) that made a number of disparaging remarks about the Church. Because it was veiled in allegory and written as a meandering conversation, it was almost impossible to pin the author down to particular claims. Diderot, in his dressing gown in midwinter, had been a charming, though frustratingly evasive, suspect; d'Hémery had found himself quickly ensnared in the philosopher's headache-inducing wit and intellectual parrying, his words ink clouds as opaque as the book under investigation. He had, to d'Hémery's relief, finally responded to threats of prison by promising not to publish the book.

Now here they were again six months later. Different neighborhood, different spy network, same dangerous philosopher.

In his headquarters, the pages of *Letter on the Blind* laid out in front of him, Joseph d'Hémery found the new work as impossible to pin down as *The Skeptic's Walk*. Ostensibly a series of philosophical speculations based on an eye operation that Professor Réaumur had performed on a blind girl, it was all twists and turns of logic and rhetoric, a speculation on the psychology of the blind, on the relativity of morality and of ideas about God, and on the question of how anyone could know anything beyond the tangible physical world. The priest had been right. Despite its tricks and ink clouds, there was no doubting that *Letter on the Blind* was a more dangerous book than the last. *Philosophical Thoughts* had been the work of a deist, but *Letter on the Blind* had been written by an atheist. And there was a new scientific turn in the writing, too, d'Hémery noted. Hidden away among all the philosophical panache and fireworks, among all the conversational speculation on blindness and sight, d'Hémery found a series of extraordinary propositions about the earth and nature. The speaker—whoever he was supposed to be—claimed that all higher animals including man had transformed since the birth of time from "a multitude of shapeless things . . . some with no stomach and others no intestines"; that all these monsters had died out gradually, leaving only organisms that "could exist by themselves and perpetuate themselves" and that the world was "subject to revolutions that all prove a continual tendency to destruction"; that through unimaginable stretches of time "a rapid succession of beings . . . follow each other, thrust one another aside and disappear," giving life on the earth only an appearance of "a passing symmetry, a momentary order." There was no God here. There was no Maker. According to Diderot, species had not been made at the birth of time by a benevolent God with Man's best interests at heart; they had mutated from shapeless monsters across millions of years.

D'Hémery respected cleverness; he admired new ideas. He was well read. But he had people to report to, and his job was to say one way or another whether the men and women in his files were heretics. He had to assess degrees of dangerous thinking. And with Diderot there was no question. But he was also one of the cleverest heretics in Paris, probably in the world. You had to admire him—his imagination, his vision. But a heretic was a heretic, and it was d'Hémery's job to bring them in.

And what of the Boy? How had this confident, dangerous atheism come into being in the mind of the young philosopher of the rue de l'Estrapade? After all, Diderot had a wife and a young child. Was he really prepared to risk his own safety and theirs to articulate a vision of a godless earth? To what end?

Growing up in Langres, a stifling and conservative town in northeastern France, Diderot had been under the constant surveillance of his father, of his pious and supercilious elder brother who was training for the priesthood, and of the local priests. Everything in Langres revolved around the church. All truth was their truth. There was no questioning them. So young Diderot had run with the pack. He had done what was expected of him in the hope of one day getting away. He was inducted into minor orders at the age of thirteen and experienced a period of intense religious belief under the influence of his Jesuit teachers, but when he turned sixteen he moved to Paris to attend school and to learn to think for himself.

Paris, with its freethinkers, libraries, and cafés, its philosophical and metaphysical conversations, worked its spell on him. Living on very little money, he put together a substantial library scavenged from booksellers in the Marais, attended lectures and the theater when he could afford to, read voraciously, and worked as a tutor giving mathematics lessons to private students. For a brief period he considered first becoming a priest and then practicing law; he even wrote the occasional sermon for money. But most of his time he devoured books and conversation. He married a beautiful young seamstress and settled down to a life dedicated to learning. He could, he told his father, make a living in Paris.

Then in the early 1740s the Paris savants began discussing Trembley's polyp, the pond creature that could regenerate itself. Until the mid-1740s, Diderot, then in his early thirties, would argue at these gatherings in salons, garrets, and coffeehouses that though you could not prove definitively that God existed, you had only to look at the complexity of design in a butterfly's wing or in the eye of a mite to see a designer at work. But the polyp took him in a new philosophical direction. He bought natural history books or borrowed them from friends or from the Royal Library or from the library of the Parisian botanical garden, the Jardin du Roi. He reread Lucretius, Empedocles, and Epicurus, trying to understand their answers to these same questions about life. He read works by the Dutch theologian and physicist Bernard Nieuwentijt, by the French physicist and clergyman Jean-Antoine Nollet, and by the English clergyman and natural philosopher William Derham, and John Needham's book on the microscope. He read illegal books, too, books that pursued metaphysical questions through the natural sciences,* and that asked questions that were contrary to the teachings of the Church.

First there was the dangerous *Vénus physique,* part erotica, part natural philosophical treatise, written by the French mathematician and philosopher Pierre-Louis Maupertuis and published in 1745. For Maupertuis, the polyp proved that nature had self-organizing and self-patterning (but blind) powers based on motion and gravitational attraction. Then there was the physician-philosopher Julien Offray de La Mettrie, who came to the most radical conclusions of all in his *Man Machine* (*L'Homme machine*) in 1747, arguing not only that all matter contains within itself the power that produces its own activity and organization but that there was no spiritual or supernatural presence in the universe. All talk of souls and spirits could now be swept away, La Mettrie declared.

By the middle of the 1740s, Diderot's curiosity about the natural sciences had gained a political edge. Everywhere in the coffeehouses and salons of Paris, young men talked of the corruption of the Church, condemning Catholicism's recent history of torture and intimidation, the Edict of Nantes, censorship. They said that the priests kept the people deliberately in the dark to control them, that Christianity was no more than a set of myths and rituals of no greater truth or validity than any other world religion. This was heady talk, subversive and liberating. For Diderot and for many of his new friends, politics, the natural sciences, metaphysics, and theology were all intimately connected. Their questions multiplied, multiheaded, like the polyp itself. They could not be contained, branching and forking into anatomy, philosophy, microscopy, physics, mineralogy, mathematics, and optics. Diderot began to hatch a plan for an encyclopedia, an attempt to bring knowledge directly to the people, bypassing the priests, pollinated in this world of the salons and the freethinking of the cafés. The light—the new knowledge—was not in the pulpit, he declared to anyone who would listen; instead, the answers were in the shaft of a microscope, in experiment, in what you could touch, in the branches of knowledge.

For a philosopher fascinated by questions derived from the natural sciences, there was no shortage of new books to read or people to de-

* The term "biology" did not come into common usage until the nineteenth century.

bate with. In 1749 an acquaintance of Diderot's and a man he much admired, Georges-Louis Leclerc, the Comte de Buffon and eminent director of the Jardin du Roi, the botanical garden at the heart of Paris, had just brought out the first volumes of his ambitiously comprehensive history of animal species. He had set out to map, define, and describe the entire animal world, not as a tedious, myopic work of taxonomy or classification, but as a kind of literary encyclopedia of natural history and as an attempt to understand the connections between all species. Although Buffon believed that species were fixed and that there was no evidence to support the transformist ideas of Maillet or the materialism of Maupertuis or La Mettrie, he kept circling around transformist theories, returning to them, rebutting them, redefining them, opening them up and closing them down. But even in 1749 he was describing the "imperceptible shadings" he saw between one species and another and was prepared to speculate that all species had been made from a common plan or mold. Diderot also read Benoît de Maillet's *Telliamed* and his epic description of men with tails and time stretching back millions of years. Polyps, men with tails, eons of time, chance, a cosmos in a continual state of growth and decay, flux and reflux: all of these ideas were taking shape in his mind and provoking yet more questions.

A new set of questions now vexed the philosopher, waking him in the night and usurping the theological questions that had fascinated him as a boy: How did life begin? What nature of a being is man? Where is the soul if an animal can regenerate itself from any fragment of its own body? What is the relationship between man and other animals? Are species fixed or variable? How does inert matter come into life? How long has it all taken? For forty years Diderot would go on searching out answers to this same set of metaphysical questions derived from the natural sciences in every library, salon, conversation, and book in France. He was still deep in those questions when he wrote *Letter on the Blind,* and when he was arrested and taken by carriage to the prison of Vincennes.

On July 31, 1749, d'Hémery and his superior, Nicolas-René Berryer, the lieutenant general of the police, interrogated the still defiant philosopher of the rue de l'Estrapade. He refused to admit to anything. Diderot even declared under oath that he had not written *Letter on the Blind* or any of the other heretical books that had been attributed to him. The following day, Berryer called Diderot's publisher, Laurent Durand, in for questioning. Durand, probably under duress, confirmed under oath that Diderot was the author of *Philosophical Thoughts, Letter on the Blind,* and the other books under investigation. Instead of acting, Berryer decided to bide his time. The philosopher was suffering. He was desperate. It was only a matter of time before he would confess.

Although his prison room in Vincennes was large and airy enough, Diderot had no books with him other than the first three volumes of Buffon's *Natural History,* which he studied intensively, taking copious notes, and copies of Milton and Plato; he had no candles, no contact with his philosopher friends, no Rousseau or d'Alembert to talk with. His wife, whose mother had died that year, wrote distraught letters complaining that she was struggling alone with their small son, François, who was sickly, and that the publishers had canceled the stipend they paid him for his editorial work. Durand, who had invested so much money in the *Encyclopedia* project, complained to Diderot about being harassed; the police were watching him; he was losing money. Now he had to issue an explanation to the subscribers. "The detention of M. Diderot, the only man of letters we know of capable of so vast an enterprise and who alone possesses the key of this whole operation," he warned, "can bring about our ruin." The police agents threatened to tell Nanette about Diderot's mistress if he did not cooperate with their investigations. After another week of sleepless nights, anxiety, and nightmares, Diderot confessed to everything.

When Diderot was released from prison in November 1749, he had become a celebrity not only in Paris but all over Europe. All copies of his *Letter on the Blind* had sold out. As he returned to his endlessly postponed and arduous work on the *Encyclopedia,* picking up the reins of this complex and ambitious project, his friend Jean-Jacques Rousseau introduced him to new admirers. In the long entry "Animal" written for the first volume of the *Encyclopedia,* Diderot wove together extracts from the first two chapters of Buffon's *Natural History,* adding to and contradicting his account and describing intermediary creations that bridged kingdoms. In the pages of the article he spliced together Maillet, Maupertuis,

Buffon, and Trembley: "Nature advances by nuanced and often imperceptible degrees," the article declared unapologetically. Hidden away in the pages of the *Encyclopedia,* Diderot had taken Buffon across the line into terrain where the conservative philosopher was no longer prepared to go—into transgression, into heresy.

While Diderot was writing this entry, Buffon's *Natural History,* now a bestseller across Europe, had run into its own theological tangle. The Jansenists declared it a work of heresy, despite the fact that it had been published by the royal press; it clearly contradicted Genesis, it was a scandal. They demanded that it be censored or burned. In the autumn of 1750, the theologians of the Sorbonne, fearful of public disorder, called Buffon in for interrogation. Buffon agreed to sign a letter retracting fourteen "reprehensible statements" and to publish the retraction in all further editions of the book. He never did so, but he now resolved to be especially careful with the claims he made in future editions. *"Sur la scène du monde, je m'avance masqué,"* he wrote: "I advance on the world's stage as a masked man."

In his entry, Diderot was taking considerable risks again, only shortly after having been released from prison. But now he was hoping that the sheer volume of the material covered by his new book might serve to shield the ideas from the eyes of the censor and the book police. He thought he could hide anything in there.

Nanette was pregnant again. Tragically, their four-year-old son, François, died in June 1750 from a violent fever, and the new baby boy, born a few months later, did not survive the year. But still Diderot could not stop, driven now by his philosophical quest. He worked night and day, writing his own entries, checking the proofs of others, asking for revisions, refining and honing. In November 1750, Durand printed eight thousand copies of the prospectus of the *Encyclopedia,* and in 1751 the first volume appeared. In April 1751, the work had 1,002 subscribers. By the end of the year the number had risen to 2,619. The figure eventually rose to around four thousand.

Soon after his release from prison, Diderot met a man who would become the most important friend in his life—a patron, fellow philosopher, polymath, translator, and interlocutor, Paul Thiry, Baron d'Holbach. There is no record of their first meeting, but it may have been in the publishing house of Durand on the rue Saint-Jacques—for Durand published both writers—or at a dinner party arranged by Rousseau or in one of the many salons. D'Holbach had arrived in Paris in 1749 from the University of Leiden. Half German, half French, newly married, clever, very rich, and well connected, he lived in an elegant and spacious six-story house in the rue Royale in the middle of a tangle of streets where he threw lavish Thursday and Sunday evening dinners to which he invited some of the most interesting intellectuals of Paris. He was in his early thirties when they met; Diderot was in his early forties.

ENCYCLOPÉDIE,
OU
DICTIONNAIRE RAISONNÉ
DES SCIENCES,
DES ARTS ET DES MÉTIERS,
PAR UNE SOCIÉTÉ DE GENS DE LETTRES.
Mis en ordre & publié par M. DIDEROT, de l'Académie Royale des Sciences & des Belles-Lettres de Prusse; & quant à la PARTIE MATHÉMATIQUE, par M. D'ALEMBERT, de l'Académie Royale des Sciences de Paris, de celle de Prusse, & de la Société Royale de Londres.
TROISIÉME ÉDITION ENRICHIE DE PLUSIEURS NOTES,
DÉDIÉE
À SON ALTESSE ROYALE
L' ARCHIDUC
PIERRE LÉOPOLD
PRINCE ROYAL DE HONGRIE ET DE BOHEME, ARCHIDUC D'AUTRICHE,
GRAND-DUC DE TOSCANE &c. &c. &c.
TOME PREMIER.
À LIVOURNE
DE L'IMPRIMERIE DES ÉDITEURS
M. DCC. LXX.
AVEC APPROBATION.

Title page of the first volume of Diderot and d'Alembert's *Encyclopedia,* issued in 1751.

The two men quickly discovered a mutual interest in science, particularly in natural history; the baron had already been signed up by Durand to translate German scientific works. Diderot recruited d'Holbach as one of the contributors for the second volume of the *Encyclopedia.* Soon Diderot was virtually living at the rue Royale, borrowing books from the baron's enormous library, rummaging through his sprawling and rare natural history collection, or borrowing his microscopes, telescopes, and other optical instruments.

When Diderot and d'Holbach met, the baron was on a mission: he wanted to put his own knowledge and fortune to good use. Fascinated by the application of science to industrial processes and frustrated by the lack of good translations of some of the most important new books, he had determined to translate German copies in his own library into French. He began with an important seventeenth-century book on glassmaking that he had managed to sell to Durand and publish in 1752. Next he began translations of two German editions of books on mineralogy and hydrology by the Swedish professor of chemistry J. G. Wallerius. Durand gave him more and more work. Over the next fifteen years, he translated a dozen scientific volumes for Durand and other publishers.

Diderot was also on a mission to bring knowledge directly to the people. He was electrifying—not just clever, but passionate, dogged, disciplined, charming, well read, and rhetorically acrobatic. He contributed more than four hundred articles to the *Encyclopedia* between 1751 and 1765, long essays on fossils, glaciers, the sea, mountains, stones, strata, earthquakes, volcanoes, mines, and metallurgy, as well as thirty articles on the constitution of the early Roman Empire. He wrote about Iceland; he wrote travel books. And he kept buying books to add to his vast library.

The baron's house, often nicknamed the Hotel of the Philosophers or, as Diderot preferred, the Synagogue, was by the early 1750s not only the nerve center of the Paris intelligentsia but also a center of production for the *Encyclopedia*. It had in addition become a virtual translation factory characterized by a similar intellectual industry to that of the Abbasid Empire at its peak, except that in the Hotel of the Philosophers the emphasis was on the dissemination of clandestine knowledge to the people as broadly as possible rather than amassing libraries of manuscripts for the use of an intellectual elite. Diderot was its engine room. "Let us hasten," he wrote, "to make philosophy popular. If we want the philosophers to march on, let us bring the people up to the point where the philosophers are now."

Unlike Diderot's lodgings, where spies—concierges, priests, servants—might be listening everywhere, d'Holbach's house was closed to the police and their enormous network of agents. The baron chose his servants very carefully and ensured that none was present during the dinners that took place every Thursday and Sunday. No one in Paris in the 1750s seemed to be safe from accusation and counteraccusation, not even

Buffon himself. The French Enlightenment depended on there being at least one place of absolute intellectual safety and freedom in the capital. D'Holbach guaranteed this.

If Diderot was the engine room of the d'Holbach coterie, d'Holbach was its choreographer. The baron was a collector of people and a conversation maker above everything else. Rousseau described him as a fervent recruiter of freethinkers. He had a great deal to offer his guests: not only did he employ one of the best cooks in Paris and maintain an extensive wine cellar, he had a library of three thousand books, often with several volumes each in French, German, English, Italian, Latin, Greek, and Hebrew, beautiful pieces of art, a breathtaking natural history cabinet, and a country house in Grandval, just outside Paris, to which he took his friends for shooting, fishing, and walking. Once those conversations began in the rue Royale, with Diderot guaranteeing spectacular intellectual pyrotechnics night after night, it was difficult for the intelligentsia of Paris to stay away.

D'Holbach was a radical when he met Diderot, and almost certainly a deist. A member of the salon claimed that Diderot converted d'Holbach to atheism while d'Holbach was trying to convince him of the existence of God:

> [D'Holbach] pursued the incredulity of Diderot even into those workshops where the editor of the encyclopaedia, surrounded by machines and workers, was taking sketches of all the manual arts. . . . [D'Holbach] asked him if he could doubt that they had been conceived and built by an intelligence. The application was a striking one, but it did not, however, strike either the mind or heart of Diderot. Diderot's friend, bursting into tears, fell at his feet . . . he who fell on his knees a deist, got up an atheist.

But d'Holbach's atheism was almost certainly also reinforced by the death of his young and pious wife in 1754: she is said to have died in moral agony in fear for her soul. D'Holbach was devastated. Though he married again a year later—his new bride was his wife's sister Charlotte—and the marriage was a happy one, blessed with several children, d'Holbach's inability to rescue his first wife from that deathbed terror of hellfire haunted him for the rest of his life and may well have contributed to the furious evangelism of his atheism.

Certainly Diderot was himself an unrepentant and unwavering atheist by the time he and the baron met. He enjoyed challenging religious belief where he found it, taking on some of the young priests he knew and running rings around their theological claims. But his new atheism did not bring him easy answers to the metaphysical questions about life, its origins, the nature of man, or the variability of species, which continued to fascinate and plague him. In 1753, the year Nanette gave birth to a daughter, Angélique, who was to be Diderot's only surviving child, he published a further attempt to answer those questions, a book called *Thoughts on the Interpretation of Nature* (*Pensées sur l'interprétation de la nature*). It was part of a new intellectual exchange with La Mettrie's *Man Machine* and with Buffon, but he had to be careful. He was still under surveillance. He would not risk being taken back to prison, which would certainly be the breaking of him, so he developed new rhetorical and literary strategies. He had learned a good deal from English satirical writers like Jonathan Swift and Laurence Sterne and from La Mettrie himself, who often used evasive phrasing or rhetorical questions to introduce controversial propositions, such as *"Ne pourrait-on pas dire que"* (Is it not possible to say that . . . ?). If Diderot wanted to bring La Mettrie's materialism to a wider audience, he had only to appear to be rebutting such ideas while actually pushing them as far as they would go.

With prison memories sharp in his mind, Diderot now devised still more acrobatic rhetorical strategies, using questions and making radical propositions before veering away from them or refuting them, leaving their color and texture and brilliance still hanging in midair: "May it not be," he wrote,

> that, just as an individual organism in the animal or vegetable kingdom comes into being, grows, reaches maturity, perishes and disappears from view, so whole species may pass through similar stages? If the faith had not taught us that the animals came from the hands of the Creator just as they are now, and if it were permissible to have the least uncertainty about their beginning and their end, might not the philosopher, left to his own conjectures,

> suspect that the animal world has from eternity had its separate elements confusedly scattered through the mass of matter; that it finally came about that these elements united—simply because it was possible for them to unite . . . that millions of years have elapsed between each of these developments; that there are perhaps still new developments to take place which are as yet unknown to us . . . and that [man] will finally disappear from nature forever, or rather, will continue to exist, but in a form and with faculties wholly unlike those which characterize him in this moment of time?—But religion spares us many wanderings and much labour. If it had not enlightened us on the origin of the world and the universal system of beings, how many different hypotheses would we not have been tempted to take for nature's secret?

May it not be . . . If the faith had not taught us otherwise? It is tempting to imagine poor Joseph d'Hémery in his office, his pen poised over the report he was duty-bound to prepare for the censor, tearing his hair out at such phrases and passages. It was like trying to catch a fish with one's bare hands. But unfortunately for Diderot, while the rhetorical strategies may have kept the police agents at bay, most reviewers complained that *Thoughts on Nature* was opaque, incomprehensible, and at best obscure.

Diderot felt the perpetual presence of d'Hémery, the police agents, and the Paris censors. He sensed them on street corners. No doubt, after his early run-in with d'Hémery, he could now anticipate the police agent's questions. "He who resolves to apply himself to the study of philosophy," Diderot wrote resignedly in *Thoughts on the Interpretation of Nature,* perhaps the most guarded and carefully worded of his books,

> may expect not only the philosophical obstacles that are in the nature of his subject, but also the multitude of moral obstacles that will present themselves, as they have done to all the philosophers preceding him. When, then, it shall come about that he is frustrated, misunderstood, calumniated, compromised, and torn to

> pieces, let him learn to say to himself, "Is it in my century only, am I the only one against whom there are men filled with ignorance and rancour, souls eaten by envy, heads troubled by superstition?" I am then, certain to obtain, some day, the only applause by which I set any store, if I have been fortunate enough to merit it.

Buffon chose a more conservative path. By the time the sixth volume of *Natural History* came out in 1756, the self-censored silences of his work were audible to Diderot and the members of d'Holbach's coterie. Friedrich Melchior, the Baron von Grimm, Diderot's coeditor on the *Encyclopedia,* reviewed it in the *Literary Correspondence* (*Correspondance Littéraire*), complaining that this was a book that showed patent signs of censorship: it had been produced, he declared, "in the middle of the persecution incited against philosophy. It was not achieved without frequently sacrificing the liberty and the boldness that speaking the truth demands." Buffon was, Grimm suggested, either too timid or too clever to speak the truth.

In 1759 Diderot was still circling around the polyps. After nearly twenty years, the regenerations of the pond creatures no longer vexed but had begun to inspire him. Though he now agreed with La Mettrie's conclusions, that all living matter drove itself, he could not agree that man was simply a machine. It seemed too reductive, too impoverished a way of seeing. There was something more to be understood from the polyps, something both grandly immortal and communal about their ability to remake themselves, to transcend both the individual body and death itself.

Diderot met a new interlocutor, Sophie Volland, a brilliant, philosophically minded woman, in 1755. Sophie's mother, concerned about scandal, made it difficult for them to see each other, so Diderot wrote to her instead, long, delightful letters describing his books and the conversations he was having at Grandval or in the rue Royale. The polyps were on his mind again in a letter to her in 1759. "Tell me," he wrote from Grandval on a windy, storm-threatened night in October, recalling part of a conversation he had had the previous night with members of the salon,

> have you ever thought seriously about what living means? . . . The only difference I know between life and death is that now you live in the mass whereas in twenty years' time you will live in fragments, dispersed and scattered in molecules. Twenty years is a long time! . . . Perhaps those who have loved one another in life and have themselves buried side by side are not as mad as we think. Perhaps their ashes come together, mingle and unite. Who knows, perhaps they have not lost all feeling or all memory of their former state? Perhaps they still have the remains of warmth and life, which they can enjoy in their own way in the confines of their cold urn. In judging whether elements have life or not, we are guided by what we know of the life of large masses. Perhaps the two things are quite different. People think there is only one sort of polyp, but why shouldn't all of nature be like the polyp? When it is split into a hundred thousand fragments the original parent polyp no longer exists, but all its elements continue to live. . . . Do not take this fancy away from me; it is dear to me, for it would give me the certainty of living eternally in you and with you.

By 1759, when Diderot was forty-six, life had become more dangerous within the d'Holbach circle. The baron, now a zealous atheist, had turned from translating books of science to publishing anti-Christian or deistic works, many by English writers, and he had now determined to import into France and publish openly atheist books, translating them when necessary. Despite the protection of his fortune and his connections, his reliable networks of people who could smuggle manuscripts across the border into Amsterdam to be published, his use of made-up authors' names or none at all on the title pages, and his ability to pay the substantial costs of the colporteurs, or peddlers, and publishers, the baron was taking substantial risks. The first of the anti-Christian books, *Christianity Unveiled* (*Le Christianisme dévoilé*), was published secretly in Nancy in 1761 and then republished at considerable risk and expense in Paris in 1767.

Copies of the book were dangerous to anyone who bought or sold them. In October 1768, Diderot wrote to tell Sophie about a distressing incident. When a student-apprentice sold his apothecary-tutor one of his two copies of *Christianity Unveiled,* the master denounced the student to the police as an act of revenge. "The pedlar, the pedlar's wife, and the apprentice," he wrote, "were all arrested and they have just been pilloried, whipped and branded, the apprentice condemned to nine years in the galleys, the pedlar to five and the wife to life imprisonment."

The police case rattled Diderot. Writing and publishing these books endangered the whole network of people who had come into contact with them. While he might be able to hide his own dangerous ideas in ink clouds, d'Holbach's prose was uncompromising, bald, and aggressive. It was like gunfire. Once he had started, there was no stopping him. All through the 1760s, d'Holbach produced book after book, each more incendiary than the last.

"It is raining bombs in the house of the Lord," Diderot wrote to Sophie in 1768;

> I am in fear and trembling lest one of the intrepid bombers should be hurt by the recoil. We have had the *Philosophical Letters,* translated or supposedly translated from the English of Toland, the *Letter to Eugénie, The Sacred Contagion, The Life of David, or the Man After God's Own Heart;* it is like a thousand devils running riot. Ah! Madame de Blacy [Sophie's moral younger sister], I very much fear the Son of God is at the door, the coming of Elijah is nigh, and the reign of the Anti-Christ is upon us. Every morning when I get up I look out of my window to see if the great whore of Babylon is walking the streets with her cup in her hand and whether there are signs in the sky. What are you doing at Isle [her family's country château in Isle-sur-Marne]? Hurry back here, so that we can be present together at the general resurrection of the dead. If you wait for the sun to be extinguished, how will you get back to Paris? It is impossible to travel when you can't see the back of your hand.

Diderot was making his own bomb. During the arrests and the trials and d'Holbach's full-on assault on the establishment and on the Church, during the long, hot, late summer of 1769, he was incubating his most heretical book yet, the extraordinary *D'Alembert's Dream*. He was virtually alone in Paris. Nanette and Angélique, now aged fifteen and the apple of her father's eye, were away at Sèvres, staying at the country house of a family friend; Sophie and her sisters were at their country estate; Grimm was traveling in Germany; d'Holbach was in a temper at Grandval. Diderot was shackled to his desk, in sole editorial charge of the journal *Literary Correspondence* in his coeditor Grimm's absence and preparing two volumes of illustrations for the *Encyclopedia* at once. "So you can see me," he wrote to Sophie, "surrounded by engravings from head to foot." He added, "I do not believe I have worked so hard in my life."

For twenty-five years or so Diderot had been keeping up with scientific advances in anatomy, microscopy, physiology, and the natural sciences, speculating about the nature of life itself, and the origins of time and of species. The previous summer he had agreed to help the tutor of d'Holbach's children, Nicolas La Grange, with his translation of Lucretius, and now Lucretius' ideas about atoms were echoing in his head, too, alongside ideas drawn from Spinoza, Descartes, Hobbes, Toland, Buffon, Réaumur, Trembley, Robinet, Bonnet, Needham, La Mettrie, and Maupertuis. The challenge now was to choreograph and synthesize his own vision of nature, whereby "everything is bound up with everything else."

He had tried out his new ideas on the company at Grandval earlier that summer. Diderot had declared that not only species but also planets can become extinguished. When asked what would happen to life, Diderot told the company with absolute certainty that though everything would disappear, the whole cycle of life would begin again just as if the sun had reignited. Would man also reappear? someone asked. Yes, he answered, man, "but not as he is. At first I don't know what; and then at the end of several hundreds of millions of years and of I-don't-know-whats, the biped animal who carries the name man."

Diderot's ideas were always formed in conversation, in dialogue with himself and with men and women living and dead. When he came to put his vision of nature's complex interconnectedness into words, it came in the shape of a play for voices in three acts, a book written at white-hot heat in a few days in that sweltering summer. It was an expression of the complexity and multivoicedness of his own inner world and the world of d'Holbach's salon. It also resembled the looping and digressive philosophical conversations that characterized Laurence Sterne's anarchic *Tristram Shandy,* a book that Diderot had read and greatly admired a few years earlier.

D'Alembert's Dream was a rehearsal space for the improvising philosopher. He gathered the speakers he needed from his immediate circle: Jean le Rond d'Alembert, the mathematician and philosopher who had been his collaborator for many years; d'Alembert's younger mistress, the brilliant and articulate Julie de l'Espinasse, who ran a salon in the rue de Bellechasse that had come to rival d'Holbach's; and Théophile de Bordeu, a distinguished doctor with a special interest in natural philosophy who was a collaborator on the *Encyclopedia.*

In August 1769, Diderot wrote to Sophie to tell her about the new work. "It is the height of extravagance," he wrote, "but at the same time it is the most profound philosophy. It is quite cunning to have put my ideas into the mouth of a dreamer. You often have to dress up wisdom as folly to gain admittance for it. I had rather they said 'But this isn't as mad as you might think,' than 'Listen, here are some great truths.' "

Through the sleeping d'Alembert, Diderot described a self-sufficient, perpetually renewing and reorganizing universe in which "the whole is constantly changing. . . . All creatures are involved in the lives of others, consequently every species . . . [and] all nature is in a perpetual state of flux. Every animal is more or less a human being, every mineral more or less a plant, every plant more or less an animal. . . . There is nothing clearly defined in nature." All living forms have developed from earlier different forms: "If the question of the priority of the egg over the chicken or of the chicken over the egg embarrasses you," declares d'Alembert, "it is because you suppose that animals were at first what they are now. What madness! We no more know what they were than what they will become. The imperceptible earthworm that wriggles in the mud is perhaps on his way to becoming a large animal."

Every living thing on earth is made up of germ cells organized in different ways, d'Alembert explains. Just as individual organs are submerged in the life of the body to which they belong, so are organisms submerged in the life of the collective. Nothing stays the same. Forms

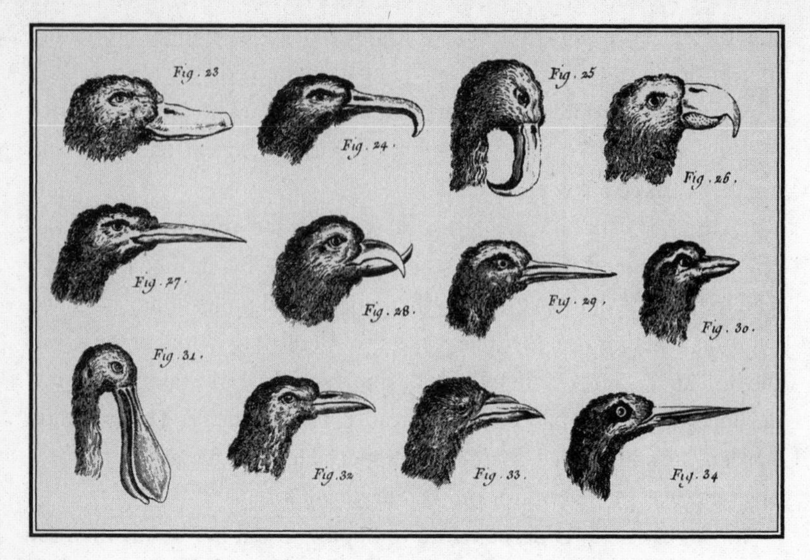

Details from a plate illustrating the variations in birds' beaks from volume 6 of the *Encyclopedia*.

change not in response to a changing environment, but through an internal dynamism:

> Let us assume a long succession of armless generations, and at the same time unremitting efforts, and you would see the two members of these pincers get longer and longer, cross each other at the back and come round to the front again, possibly develop fingers at the extremities and so make new arms and hands. The original shape of a creature degenerates or perfects itself through necessity and habitual functioning. We walk so little, work so little but think so much that I wouldn't rule out that man might end by being nothing but a head.

Diderot's vision of universal weblike connectedness was reflected in the organization of his encyclopedia, in which he arranged knowledge in branched patterns rather than distinct categories. It also gave him an organic explanation for the relationship he believed existed between the individual and the people. "Stop thinking about individuals and answer me this," d'Alembert continues in his dream.

> Is there in nature any one atom similar to another? No . . . Don't you agree that in nature everything is bound up with everything else and that there cannot be a gap in the chain? Then what are you talking about with your individuals? There is no such thing; no, no such thing. There is but one great individual, and that is the whole. In this whole, as in any machine or animal, there is a part which you may call such and such, and when you apply the term individual to this part of a whole you are employing as false a concept as though you applied the term individual to a bird's wing or to a single feather of that wing. You poor philosophers, and you talk about essences! Drop your idea of essences.

The webs in *D'Alembert's Dream* differ from those Jahiz imagined in the deserts of the Abbasid Empire. These were not communities of interdependent organisms particular to the desert or to the mountain or the city, but living webs of connectedness in which every organism was joined to, and part of, everything else. D'Alembert illustrates his point by asking Julie to imagine trying to cut a swarm of bees apart from one another with a pair of scissors. He even suggests that polyplike humans might exist on other planets, splitting and regenerating constantly like Trembley's tiny pond creatures. Imagine, d'Alembert tells her: in such a world, no one would be reluctant to die. In the *Dream,* flux is at the heart of d'Alembert's vision: *"Tout est en un flux perpetual,"* he declares. "If everything is in general flux, as the spectacle of the universe shows me everywhere, what wouldn't the duration and vicissitudes of a few million centuries bring about?" Even Diderot's prose reflects his vision of nature in a permanent state of flux, spilling out and over its own edges, with conversations ebbing and flowing from idea to idea.

When Grimm returned to Paris in mid-October 1769, Diderot read the dialogues to him. The two agreed that, not only because of the deeply heretical nature of the material but also because it had been put into the mouths of d'Alembert, Mademoiselle de l'Espinasse, and Bordeu, publication was out of the question for the moment. Nonetheless Grimm took the manuscript to get it copied. The dialogues were read aloud in small circles in Paris, to gasps and applause. Certainly d'Holbach would have seen them or heard them "performed." When the secret inevitably leaked out, Mademoiselle de l'Espinasse was horrified, both on her own behalf and on behalf of her lover. D'Alembert insisted that the manuscript be destroyed. Diderot claimed later that he had indeed destroyed it, but at least one copy had already been made and was now in Grimm's house, safely locked away.

Angélique was now sixteen and in love. This was not the time for her father to be throwing bombs among his friends; there must be no scandal. Vexed by the problem of finding a dowry for Angélique and wanting to involve his family in the negotiations, he tried to engineer a reconciliation with his brother, the Abbé Diderot, who had many years before washed his hands of his reprobate younger brother. Diderot traveled to Langres in the late spring of that year, but though he waited there six weeks, his brother still refused to see him. He would meet with his heretic brother, the abbé told family members, only if the heretic promised to write no more against religion. Diderot refused. The negotiation, Angélique later recorded, thus went to the devil.

Then, just a few months after the engagement had been agreed between the two lovers and their families, d'Holbach, increasingly politicized by new atheist friends such as the philosopher Jacques-André Naigeon, set off his own incendiary device, the ticking bomb that had been sitting on his desk in the rue Royale for several years, an atheist tract called *A System of Nature: or, the Laws of the Moral and Physical World* (*Système de la nature, ou des lois du monde physique et du monde moral*), which he had smuggled across the border into Amsterdam to be published. The baron had shown a draft of the work to Diderot, who complained privately that there was no wit, no shadow, no nuance, no multisidedness, no art, no pleasure or erotic play to be found in its pages. It denied God, arguing that all religions were created out of fear, ignorance, and anthropomorphism; that the mind was no more than the workings of the brain; that souls did not outlive the body; that the world was determined by strict laws. The baron underscored his case with references to the new sciences, but his real interest was in destroying the power of religion, not in exploring science and metaphysics. Worst of all, Diderot admitted to friends, the baron had made atheism dull. His book was an act of war, not of argumentation. It would set back the cause rather than advance it. But Diderot kept his own counsel. He was loyal to his old friend despite the differences of their temperament and their methods. They were, he reminded himself, on the same side.

Diderot, knowing that the authorship of the anonymous book would be attributed to him, left Paris on August 10, 1770, in case a warrant was issued against him. On the eighteenth, *System of Nature* was condemned to be burned along with Voltaire's *God and Men* and d'Holbach's *Discourse on Miracles, The Sacred Contagion,* and *Christianity Unveiled.* D'Hémery and his men meanwhile moved against the colporteurs and the middlemen of the trade and spent little time investigating the identities of the authors of such books. They wanted to contain the scandal and silence the books, not make martyrs out of famous intellectuals. When Diderot returned to Paris in October, he kept to himself, excusing himself with poor health, the weight of work, and his daughter's illness. "[I have] acquired a taste for solitude which makes me shun company," he wrote to Sophie in November. "I live here in my study, working, dreaming, and writing, not happy, but happier than I should be elsewhere it seems, since I don't have to take off my dressing-gown."

D'Holbach's book caused a great stir among the Paris savants, effectively dividing the deists from the atheists. Voltaire, committed to deism and long impatient with d'Holbach's declamatory ways, called it "a chaos, a great moral sickness, a work of darkness, a sin against nature, a system of folly and ignorance." He wrote to the philosopher Jean-Baptiste-Claude Delisle de Sales: "I think that nothing has debased our century more than this enormous stupidity."

Time passed. The scandal died down. Angélique married. Diderot fell in love again. In 1773, he traveled to Russia to meet with his great patron Catherine II and to The Hague, where he met the now very aged Bentinck brothers, who had employed the great Abraham Trembley. "With their solemn manner and their serious, austere way of speaking, I really felt as if I were with a Fabius and a Regulus," he wrote, referring to two distinguished Roman politicians. And he often went back to gaze at the North Sea to listen to its rise and fall, thinking perhaps of Maillet's description of "this fame Flux and Reflux" of time and motion and thinking, too, about his own impending death. He was an old man now. "I hardly ever go out," he wrote. "When I do, I always go to the sea, which I have not yet seen either calm or rough. The vast monotonous expanse and the murmuring noise make you dream; I have good dreams there."

Buffon was aging, too, and prepared to take more risks. For thirty years he had been philosophically guarded in return for protection from the members of the Sorbonne's Paris Faculty of Theology, and for the continuation of his important public position at the Jardin du Roi. For his efforts he had achieved public acclaim, power, and royal patronage. Now he was seventy-nine years old. What had he to lose by breaking his contract? In 1778 he published his most controversial book yet, *The Epochs of Nature* (*Époques de la nature*). The book was seriously at odds with the beliefs of the Church and with Genesis. Buffon discussed the origins of the solar system; he proposed that the planets had been created by comets colliding with the sun and that the earth was much older than the Church claimed, having been in existence for seventy-five thousand years (he calculated the planet's age in a series of experiments with cooling iron balls); he declared that Noah's flood had never occurred; he also argued that some animals retained parts that were vestigial and no longer useful, which implied that they had evolved rather than been created. He described scenes from the beginning of time: torrents of molten matter pouring from the sun to form planets, a barely cooled earth teeming from the waters that rained from its atmosphere and quickly boiled away in thick vapors, volcanoes erupting, and gigantic earthquakes caused by the collapse of underground caverns. Although Buffon had consistently rejected evolutionary conclusions, through his bestselling and widely influential volumes of *Natural History,* he brought the idea of species mutability into public discussion along with a much expanded conception of the earth's longevity. "My old age does not leave me the time to examine sufficiently to draw the conclusions that I glimpse," he wrote. "Others will come after me. . . . They will weigh. . . . They will see."

Buffon's *Epochs of Nature* was immediately attacked by leading religious figures, and the Sorbonne had no choice but to denounce the book. Buffon offered a new retraction, very similar to the one he had signed in 1750. He promised to print it in the next edition, but when the time came he refused to do so. "The people need a religion," he wrote scornfully to the young magistrate Marie-Jean Hérault de Séchelles in

1785. "When the Sorbonne picked petty quarrels with me, I had no difficulty giving it all the satisfaction that it could desire: it was only a mockery, but men were foolish enough to be contented with it."

In 1782, Grimm's successor at *Literary Correspondence* suggested to Diderot that the three dialogues of *D'Alembert's Dream* be copied and circulated to subscribers. Mademoiselle de l'Espinasse had died in 1777 at the age of forty-four. D'Alembert, grieving, had retired into isolation. There were few people left to offend. Diderot, himself old and tired, gave permission, and a version of the text appeared in a limited edition of four successive numbers from August to November 1782. The full manuscript would not be printed until 1831, and then only in England.

When Diderot died in 1784 with no deathbed conversion, his body was buried in the Church of Saint-Roch, where Maupertuis was also buried, only a few streets away from the Hotel of the Philosophers, his coffin accompanied by fifty priests hired by his daughter. He had long intended this, explaining to d'Hémery forty years earlier that when he died he would allow the usual religious rituals to take place for the sake of his family but that he would refuse the sacraments. He was not the only man to compromise his beliefs. A year later, Buffon told Hérault de Séchelles, who visited him in Montbard, "When I become ill and feel my end approaching, I will not hesitate to send for the sacraments. One owes it to the public cult." Diderot's funeral rites, wrote the *Mercure*, "were of a splendour rarely accorded to power, opulence, dignity. A numerous gathering of distinguished people, academicians, and men of letters accompanied the funeral procession . . . such was the influence of his name that twenty thousand spectators waited in this sad procession, in the streets, in the windows, and almost on the rooftops, with the curiosity that the people reserve for princes."

There was no such grand procession for the Baron d'Holbach's funeral cortège when he died four years later in January 1789 at the age of sixty-six. He was buried quietly in the presence of his family alongside Diderot in the same crypt of the Church of Saint-Roch. Four months later, the first wave of the French Revolution began with the convocation of the Estates-General; in July, thousands of insurgents stormed the

Bastille; in August, the new assembly published the Declaration of the Rights of Man and of the Citizen and a group of women led the march on Versailles to force the royal court back to Paris under the protection of the royal guards. In November, only ten months after d'Holbach had been buried, the Assembly declared that the property of the Church was "at the disposal of the nation." Four years later, in July 1793, during the Reign of Terror, priests were defrocked, sent into exile, or executed; the Church of Saint-Roch, where Diderot and d'Holbach lay entombed, was desecrated and stripped of its paintings, carvings, statues, stucco, marble, and votive objects. In October 1795, its cloisters became the refuge of freedom fighters and insurgents during a period of fierce street fighting; its façade is still riddled with bullet holes. Masons digging an air vent into the crypt a hundred years later found no bodies there.

The house on the rue Royale, the most important center of Enlightenment Paris, carries no plaque. The present occupants—a dental surgeon and a software company—have never heard of the Baron d'Holbach or his salon. "Who knows what breeds of animals will succeed ours?" Diderot asked in *D'Alembert's Dream.* "Everything changes, everything passes, only the whole remains."

8

Erasmus Underground

DERBYSHIRE, 1767

In late June 1767, three hundred feet under the rounded hills and grit-stone escarpments of Castleton, in Derbyshire, four men led by miners carrying lamps threaded their way through the narrow tunnels of a lead mine called Tray Cliff. They were the two mine agents, the brothers Anthony and George Tissington; John Whitehurst, a Derby clock and instrument maker who had investments in the mines and was writing a book about the caves and mountains of Derbyshire; and Erasmus Darwin, a doctor in his midthirties from Lichfield. Darwin's friend Whitehurst had brought him here to show him the natural caverns the miners had only recently tunneled their way into. Tall, overweight, and ebullient, with pockmarked skin and a stutter, the doctor never stopped talking and asking questions. But when the miners raised their lamps to reveal the inverted forest of white stalactites in the first cavern, gleaming and dripping against the darkness, Dr. Darwin stopped talking. He was, like everyone who visited the caves, stunned into silence.

Under the flickering light of the lamps, the wet cave walls and the shards of white stalactites rippled like the sea. The seam of the rare fluo-

Tourists being shown the interior of the Gailenreuth cave, near Muggendorf in Bavaria, accompanied by guides with torches in or around 1816.
William Buckland, An Assemblage of Fossil Teeth and Bones *(1822)*

rite rock called Blue John, found only in this mountain, forked through the limestone like veins, banded in yellow and purple. Tilting their lamps, the miners pointed out the shapes of marine creatures and plants standing out from the dripping rock here and there, as if carved in relief, excavated from softer rock by swollen underground rivers that had passed through these rocks for unimaginable tracts of time. Some men of a religious disposition had gone mad down here, the miners told the visitors, convinced that the gaping shaft openings that disappear into ink-black pools were the entrance gates to hell itself. But the rock shapes of Tray Cliff Cavern and the nearby Peak Cavern made Erasmus Darwin think of the capillaries and arteries and veins of the human body, of Greek and Roman creation myths, of Ovid's *Metamorphoses,* of Orpheus and Eurydice. To him, Tray Cliff Cavern was not the gateway to hell but a temple of mysteries, the altar of the goddess Nature; it seemed to him to be lined with symbols—hieroglyphs and mystic scrolls; it was a place of revelation, of magical transformations and secrets. It held the secrets of time itself.

Over two days the four friends visited other caves in the Castleton area. They met rope makers who lived and worked in the enormous cave mouth of the Devil's Arse Cavern and were led down into tunnels where they were ferried over underground rivers into spectacular cathedral-like caverns. The miners showed the doctor the fault lines and the sinking shafts, the strata twisting and snaking like petrified rivers, and the tiny shrimplike creatures that twitched and contracted in the pools that gathered at the base of stalagmites. Most of the strata contained only one kind of shell or marine organism, Whitehurst pointed out, as if each layer had been laid down separately with long intervals of time in between.

For decades Derbyshire miners had been chiseling shells and plant shapes out of the limestone, hoping to get a good price for them from the mineral and souvenir sellers in Castleton or the agents of the aristocratic collectors such as the Duke of Devonshire or Sir Ashton Lever. They called them petrifactions. Some people claimed they were magical, others that they were the wanton sportings of nature, nature's freaks, rejected designs, like two-headed calves or seven-toed cats. Others declared them to be the remains of the deluge, God's signs of his covenant, his promise to man.

But John Whitehurst had other ideas. He had read in the Comte de Buffon's *Natural History* the extraordinary list of places in which shell fossils had been found: in mountains, quarries, and mine workings in every corner of the known world, hundreds of miles from the sea. The strata all around them proved that minute marine animals, entombed here in the lowest bands of rock, were the very first organisms on the earth, formed long before the first land had surfaced. He knew that fossils held the secret of the history of the earth. Erasmus Darwin's head crowded with questions about time and the origins of life on the planet. Could life, he wondered, have started long ago, in the underground pools of an earth only recently lifted from the sea?

Living and working in the fissured, cave-laced landscape of Derbyshire, Erasmus Darwin had long been fascinated by fossils. Collections of fossils and colored, veined, and glittering rocks filled the display cabinets in

Erasmus Darwin, aged around forty (an engraving from an oil portrait by Joseph Wright, ca. 1770).
Getty Images

the drawing rooms of the country houses of his patients and friends. Tiny shells flecked the stone walls of the Castleton and Derby shops. Fossilized shells, fish, and ammonites, and sometimes the bones of larger swimming animals with fins or jaws like lizards, surfaced in the workings of canals and roads, in the foundations of buildings, on the dripping walls of mines, in the foundations of churches. A rare fossil "crocodile" had washed up nearby in Nottingham in 1712 when workmen were rebuilding the well of a rectory opposite Erasmus's father's house. Robert Darwin sent the rock to William Stukeley, the antiquary, who wrote a paper about it published in *Philosophical Transactions* and put it in the Royal Society museum, calling it a marine reptile, perhaps even a survivor of the flood, "a rarity, the like whereof has not been observ'd before in this Island."

A slab of limestone with part of an "encinite," or fossil sea lily, as illustrated in a popular eighteenth-century book on fossils.
Barthélemy Faujas de Saint-Fond, Histoire naturelle de la Montagne Saint-Pierre de Maastricht *(1799)*

Erasmus had forged a number of close friendships with like-minded local men such as John Whitehurst—printers, industrialists, manufacturers—who shared his interest in astronomy, geology, electricity, and meteorology. They had begun to meet informally at one another's houses to discuss scientific matters, calling themselves fellow schemers or the Birmingham Philosophers; later, when their gatherings formalized into regular meetings held on the Sunday nearest the full moon, they came to call themselves the Lunar Society.

A few weeks before he visited the Derbyshire caves, Erasmus Darwin's friend and fellow member of the group, the industrial potter Josiah Wedgwood, had sent him a box of giant fossilized bones and tree trunks and rocks. His navvies, he explained, had dug them out of the workings of the Harecastle Tunnel as they carved out the Trent and Mersey Canal. He hoped that, as Erasmus knew something about anatomy, he might be able to identify the bones. "These various strata," Wedgwood wrote to another friend at the same time,

> seem from various circumstances to have been in a Liquid state, & to have travel'd along what was then the surface of the Earth; something like the Lava from Mount Vesuvius. They wind & turn about, like a Serpentine River. . . . [But] I have got beyond my depth—These wonderful works of Nature are too vast for my narrow microscopic comprehension. I must bid adieu to them for the present, & attend to what suits my capacity. The forming of a Jug or Teapot.

Erasmus Darwin could not identify the bones Wedgwood sent him. He was baffled and fascinated. To be able to make an accurate identification, he told his friend, he had to see the bones and fossils deep in the earth's rock, as they surfaced, not in some dusty box or display cabinet. He needed Whitehurst to arrange a trip into the Derbyshire caves.

Erasmus wrote to Wedgwood excitedly when he returned to Lichfield after the two-day journey. "I have lately travel'd two days into the bowels of the earth, with three more able philosophers," he wrote, "and have seen the Goddess of Minerals naked, as she lay in her inmost bowers, and have made such drawings and measurements of her Divinity-ship, as would much *amuse,* I had liked to have said, *inform* you." A few weeks later he wrote to another member of the group, Matthew Boulton: "I want to see you and Dr Small much if you will fix a Day. . . . I have been into the Bowels of old Mother Earth, and seen wonders and learnt much curious knowledge in the Regions of Darkness. . . . And am going to make innumerable Experiments on aqueous, sulphurous, metallic and saline Vapours. Food for Fire-Engines!"

Erasmus Darwin's friends in the Lunar Society were engaged in a series of collective investigations and experiments. They exchanged information, facts, questions, evidence, inventions, and objects by mail and at their meetings. They tested their knowledge on one another, experimented, took notes, mixed, distilled, and transformed. Many of their questions were practical and forward-looking, concerned with reform and progress: How do we fix this? How can we make that? How do we improve upon this? But some of the men, like Erasmus, were driven by questions about the past rather than the future, by theoretical and speculative questions as well as practical ones. How and when did life begin? How did species come to be?

Erasmus Darwin was a busy man. Every year he traveled thousands of miles across the potholed and sometimes nearly impassable roads that stretched across this craggy and spectacular landscape to visit patients in remote farms and country estates. To make his life easier he employed a carpenter to fit out his carriage like a small study with bookshelves, writing materials, and notebooks and spent long hours trying to invent suspension techniques that would protect travelers from the endless jolts, falls, and accidents.

In July 1768 he was immobilized for several weeks by a carriage accident. In the weeks of forced convalescence that followed, he appears to have begun work on a book that drew together his various developing theories on the relationship between mind and body, a book of medicine and medical theory that would also contain his emerging ideas on life itself, its generation and its origins. These ideas had arisen from questions formed in the Tray Cliff mine workings.

Immersed in the freethinking debates of the Lunar Society discussions, in which no line of questioning or proposition was prohibited, Erasmus could no longer quite tell whether his new book, now tentatively entitled *Zoonomia, or the Laws of Organic Life,* would be regarded as heretical in the wider world. To gauge this, in August 1768, he sent a copy of the most potentially controversial opening sections to his friend and theological interlocutor, the eloquent, philosophically inclined, and well-read vicar of Duffield, Richard Gifford, who had a keen interest in the relationships between the mind, the soul, and the body. Gifford's reply seems to have been full and frank. He told Erasmus that it was "not pious" to inquire into the "living Principle of Life." Though Erasmus protested in reply that surely "the Lord" would want his subjects to inquire into "the Wonders of his Works," he also assured Gifford: "I do not mean to attack the Christian Religion; and that I do not mean to go up to the first Cause of any Thing, but endeavour to trace it one Step higher than others have done." Gifford warned him of the dangers of being linked to the materialists and modern skeptics. "I believe I shall never publish these Papers," Erasmus told Gifford with a degree of weariness. "I would not certainly if I could see any bad effect they might have on the Morals of Men."

If certain beliefs could not be expressed outright for fear of offending the Church, Erasmus reflected, perhaps they might be implied instead. Three scallop shells adorned the Darwin family crest. The emblem of Saint James, worn by pilgrims traveling long distances to the great pilgrimage site of Santiago de Compostela in Spain, the scallop shell symbolized piety. In classical mythology it also symbolized fertility—Venus was supposed to have been born from a scallop shell; so were Castor and Pollux. Erasmus now added a new dangerous motto to the family crest: *E Conchis Omnia*—"Everything from Shells." Though the motto was short and not exactly explicit in its meaning, once it was painted on Erasmus's carriage door to be seen everywhere across Derbyshire, it seemed to some in Lichfield to be a flagrant declaration of his unorthodox views. Erasmus also had a seal made so that every letter he sent carried the motto on its envelope.*

It did not take Erasmus's neighbor, the watchful Canon Seward, long to see what he was up to. Rather than confront him directly, Seward composed and circulated an anonymous satirical poem entitled "Omnis

*Even the mourning seals Erasmus used on a letter sent in November 1770 made of black wax to mark his wife's death contained the motto. He was not too self-conscious to use it even in late 1770.

e Conchis" in which he denounced Erasmus's "hodge podge of iniquity"; he wrote that, like Epicurus,

He too renounces his Creator,
And forms all sense from senseless matter;
Makes men start up from dead fish bones,
As old Deucalion did from stones;
Great wizard! he, by magic spells
Can build a world of cockle shells;
And all things frame, while eye-lid twinkles
From lobsters, crabs and periwinkles;
O Doctor! change thy foolish motto,
Or keep it for some lady's grotto;
Else thy poor patients well may quake
If thou no more canst mend than make.

Erasmus understood the veiled threat in this satire penned by the influential canon of Lichfield: the doctor had better be careful, Seward implied, for if he lost the trust of his patients he would soon lose his practice and his livelihood.

It was a difficult time for Erasmus. His wife, Mary, had been suffering from an undiagnosed illness for years; multiple births, stillbirths, and addiction to opium had weakened her further. By the early summer of 1770 she began to have paranoid delusions, convinced that someone was going to kill her surviving children. Erasmus described how she would beseech the phantom repeatedly: "Don't kill them all, leave me one, pray, leave me one." She died in June that year.

Sometime in the months after his wife's death, Erasmus—concerned for the reputation of his practice and family—had the motto removed from the carriage door and from his seal and went back to his less controversial inventions and canal-building schemes and to the care of his three children, Charles, Erasmus, and Robert, then only eleven, ten, and four years old.

Though Erasmus Darwin was wary enough of local gossip to remove his materialist motto from his carriage door and seal, he was not afraid to risk other kinds of gossip. Seventeen-year-old Mary Parker, nursemaid to his boys, who had joined the household a few months after his wife died, became his lover and was soon carrying his child. Everyone in Lichfield must have known of the liaison as Mary made her way around the town with the older Darwin boys. A year after her first child, a daughter, Susannah, was born in 1772, she was pregnant again with a second daughter, who was born in 1774 and christened Mary, after her mother. Mary Parker continued to live with Erasmus, to raise their daughters and raise and teach Erasmus's sons for a short time; when the relationship came to an end around 1775, with no apparent animosity, Mary left the household, moved to Birmingham, married a merchant in 1782, and began another family. Erasmus's two daughters continued to live with him, raised and educated openly alongside his boys.

Every natural philosopher that Erasmus Darwin knew who had big ideas about the origins of the earth seemed hesitant to publish. The Scottish geologist James Hutton came to stay with Erasmus in June 1774, using Erasmus's house as a base for his geological expeditions. Like Erasmus, Hutton was a deist; while he did probably believe in a Christian God, it was a "caretaker" God who did not interfere in the operations of the cosmos. He did not feel a need to square his scientific theories with a literal interpretation of the Bible as his friend Whitehurst was trying to do. Hutton, too, believed that the earth was much older than anyone would admit and that the landscape was a complex composition still shifting constantly and imperceptibly from continent to continent. He did not, however, believe in species change. He had a book, a grand theory of the earth, which he had been writing for at least fifteen years, but, he told Erasmus, he had much more proof to gather before he would risk putting his ideas into print.

There are glimpses of Erasmus's frustrations in his letters. In November 1775 a patient, Joseph Cradock, sent him a copy of his book *Village Memoirs*. Erasmus wrote to thank him:

> What shall I send you in return for these? I who have twenty years neglected the Muses, and cultivated medicine alone with all my industry! Medical Dissertations I have several finished for the press, but dare not publish them, well knowing the reception a

> living writer in medicine is sure to meet with from those who wish to raise their own reputation on the ruin of their antagonists. Faults may be found or invented; or at least ridicule may cast blots on a book were it written with a pen from the wings of the angel Gabriel.

If he could not run the risk of publishing his ideas in direct prose, he told Cradock, he would write poetry instead. "I lately interceded with a Derbyshire lady to desist from lopping a grove of trees, which has occasioned me . . . to try again the long-neglected art of verse-making, which I shall inclose [*sic*] to amuse you, promising, at the same time, never to write another verse as long as I live, but to apply my time to finishing a work on some branches of medicine, which I intend for a posthumous publication."

Erasmus had not given up on *Zoonomia*, but he could see no way of publishing it before he died. An optimistic man who believed passionately in progress, perhaps he hoped that if he waited long enough the scientific world might become more tolerant. He resolved to bide his time.

Meanwhile in 1775 he was falling in love with a beautiful married woman, Elizabeth Pole, a neighbor and patient, the Derbyshire lady of the poem. Married to a retired military man, she was raising her children in a large country house four miles east of Derby. Mrs. Pole shared his interest in botany, in gardens, and in the raising and education of children. Erasmus sent her unsigned love poems, expressing his feelings indirectly, playfully, under cover of classical myth and allusion, even disguising himself as a wood nymph.

In 1776 Erasmus bought a few acres of mossy land in a valley a mile from his house in order to turn it into an experimental garden—"tangled and sequestered," as his friend the poet Anna Seward, daughter of Canon Seward, described it. Wet with springs and lined with rare aquatic plants, it was also the site of an old bathhouse. He employed local men to widen the brook into small lakes and "taught it to wind between shrubby margins." He planted a variety of trees and plants, "uniting the Linnaean science with the charm of landscape." Perhaps in this charmingly entangled bank full of drosera and bog plants and insects he and

Mrs. Pole could meet without exciting gossip.* It was also here that he began to think about putting his heretical ideas about sex, nature, and the origins of life into poetry. "The Linnaean system," he told Anna Seward that year, referring to the new system of plant classification introduced by the Swedish botanist Carl Linnaeus, "is unexplored poetic ground, and a happy subject for the muse. It affords fine scope for poetic landscape; it suggests metamorphoses of the Ovidian kind, though reversed. Ovid made men and women into flowers, plants, and trees. You should make flowers, plants, and trees into men and women. I will write the notes, which must be scientific, and you shall write the verse." It would be a way of popularizing botany and promoting Linnaeus's ideas about plant sentience and sexuality. But, Anna replied, it would not be proper for a female poet to write about botany and plants' sexual parts. Erasmus must write the work himself.

Each project perpetually branched into others in Erasmus's writing life. He tried to find ways of holding it all together, refusing to compartmentalize his ideas, understanding the ways in which they were interdependent. Soon after he discussed the botanical Linnaean poem with Anna Seward in 1779, he began writing it, trying to put the scientific prose notes together at the same time as the poetry, but there were obstacles to overcome. When he failed to find an English translation of Linnaeus, he formed the Lichfield Botanical Society in the hope of getting some translations commissioned, and when the contributors failed to produce anything very useful, he began translating Linnaeus's Latin phrases into English himself. Although the translation work was laborious and slow, he found it counterpointed and complemented the poetry in surprising ways.

For four years Erasmus slipped between the two manuscripts piled on his desk, the embryonic poem now called *The Loves of the Plants* and its prose notes and the translation project, describing plant couplings and birthings, coining new words to describe the sexual parts of plants (bristle-pointed, end-hollowed, scollop'd, thread-formed, lance-prickled, vein-hollowed), amusing himself with shaping elaborate sexual innuendoes and courting Mrs. Pole. *Zoonomia,* direct, claim-making, plain-

* It was perhaps here, too, that she comforted him after the death of his eldest son, William, who had succumbed to an infection contracted at the dissecting tables of Edinburgh.

speaking, heretical *Zoonomia,* would have to wait in the wings a little longer. It was still not the time to publish controversial books. France had signed a treaty with America against Britain. In June 1779, Spain, supported by the French, declared war.

In matters of philosophy and science, Erasmus Darwin was a Francophile. In the late eighteenth century, as relations between Britain and France deteriorated, the British grew first scornful and then afraid of all things French. France represented everything Catholic, autocratic, and volatile. French natural philosophers, their British counterparts would say, wrote speculative science with grand theories; they built beautifully crafted castles in the air, in the manner of the Comte de Buffon. The British, by contrast, excelled by using plain facts. Erasmus, a grand-theory man and a Buffon enthusiast, must have felt that philosophical conservatism keenly.

When John Whitehurst finally published his revealingly entitled *Inquiry into the Original State and Formation of the Earth; Deduced from Facts and the Laws of Nature* in 1778, it was a mass of contradictions. Though the evidence in the rocks and shells he had studied and collected for twenty years all seemed to undermine both the creation story and the Flood, he had worked hard to force a square peg into a round hole, to reconcile his science with the Bible. In the end he simply divided his book in half, starting with a long attempt to resolve his findings with the biblical account and putting all the facts he had collected over twenty years into an appendix that seemed to say something quite different. He was a practicing Christian, more observant than the other Lunar men, but the contradictions in his book were probably as much the result of an attempt to distance himself from French theorizing as the result of a struggle with his own belief.

Josiah Wedgwood and Erasmus Darwin were astonished at how straitlaced Whitehurst's book had become in comparison to the earlier drafts they had seen. Wedgwood complained to Thomas Bentley, his partner in the porcelain factory, that Whitehurst's "manuscript has undergone as many alterations since its first formation . . . as his world has suffer'd by earthquakes, & inundations. . . . I own myself astonish'd beyond measure at the labour'd & repeated efforts to bring in & justify the mosaic account beyond all rhime [*sic*] or reason." "I should like to tumble a little of his world about his ears," he continued in another letter, "but I shall forbear, for I love the man." Was Whitehurst struggling with his religious beliefs, or had he simply decided to play it safe? After all, he had a business to run. He could afford neither to alienate himself from the scientific or industrial communities in Derbyshire or London nor to be ostracized as an unbeliever.

When Elizabeth Pole's husband died, on November 29, 1780, leaving her a rich widow, it was Erasmus Darwin she married four months later rather than any of her younger suitors. They already had eight children between them from their previous relationships; they would have another six of their own over the next eight years, the first born in January 1782. Erasmus's new wife shared his interest in science. In the summer of 1782 he took Elizabeth and her daughters on a geological expedition up into the peaks around Derby to show them first the copper mine at Acton and then Thor's Cave. He was still thinking through his earlier ideas about life having emerged from aquatic filaments nourished in underground caves and great subterraneous lakes.

Although *The Loves of the Plants* was quickly taking shape, Erasmus was deeply ambivalent about becoming a published poet. It was one thing to play about with private love poems, but going public with a poem about the sexual lives of plants, a poem freighted with scientific material, some of it controversial, was something else. On the one hand, he told Elizabeth, a published poem might bring them much-needed extra income—his friend Anna Seward was making a good deal from her poetry now; on the other hand, it would put him at the mercy of scornful reviewers. He would seek opinions, he assured her. In the spring of 1784 he sent the manuscript to Joseph Johnson, a radical publisher in London, explaining, "I would not have my name affix'd to the work on any account, as I think it would be injurious to me in my medical practice, as it has been to all other physicians who have published poetry."

As he prepared *The Loves of the Plants* for anonymous publication, Erasmus began a new long poem called *The Economy of Vegetation,* which he had decided to publish paired with *The Loves of the Plants,* as two volumes called *The Botanical Garden.* He was now writing poetry at an extraordinary pace in any spare time that opened up: indoors on a writing board supported by a cushion, taking refuge from the children in the summerhouse or on the road in his carriage. The young daughter of one of his patients observed the famous doctor climbing out of his mud-splattered carriage around this time. He was, she remembered,

> vast and massive, his head almost buried in his shoulders, and he wore a scratch-wig, as it was then called, tied up in a little bob-tail behind. A habit of stammering made the closest attention necessary, in order to understand what he said. Meanwhile, amidst all this, the Doctor's eye was deeply sagacious, the most so I think of any eye I remember to have seen; and I can conceive that no patient consulted who was not inspired with confidence on beholding him; his observation was most keen; he constantly detected disease, from his observation of symptoms so slight as to be unobserved by other doctors.

James Hutton's long-awaited *Theory of the Earth* came out in 1788; Erasmus devoured it, admiring Hutton's bold account of a regulated earth cycling on through time and space, continents shifting and remaking themselves infinitely slowly, with "no vestige of a beginning and no sign of an end." The earth increasingly seemed to Erasmus, too, to have its own internal economy, rhythms and cyclings replicated by the human bodies he tended and nursed every day. Hutton's book confirmed him in his view that his several unfinished manuscripts were all in dialogue with one another, all part of the same investigation of nature's very simple but elusive laws. Inspired by Hutton's book, Erasmus worked hard to extend his geological knowledge and began to put together an ambitious fossil collection.

The Loves of the Plants was finally published in 1789, just as the Revolution broke out in France and in the same month that Elizabeth gave birth to their sixth baby. The critical response to the poem was rapturous. This was his attempt, Erasmus explained in the introduction, "to enlist imagination under the banner of science" and to interest readers in botany. The book was well suited to the atmosphere of 1789—it sketched a Rabelaisian vision of a world in a state of progressive change, but it did so tentatively. "Perhaps all the products of nature are in their progress to greater perfection?" Erasmus wrote, attaching a question mark and a "perhaps" to soften it and then tucking the idea deep into a footnote, where it might escape notice.

Erasmus buried scientific speculation in his footnotes. Perhaps he thought an enlightened reader would dig these thoughts out, someone who knew what to look for. His friend the fellow Lunar Society member and industrial chemist James Keir noticed. "You are such an infidel in religion that you cannot believe in transubstantiation," he wrote, teasing Erasmus about the materialism of the poem, "yet you can believe that apples and pears, hay and oats, bread and wine, sugar, oil, and vinegar are nothing but water and charcoal, and that it is a great improvement in language to call all these things by one word, oxyde hydro-carbonneux." It was an astute comment. What Keir saw was that Erasmus Darwin was replacing the mystery of the Catholic belief in transubstantiation (the bread being transformed into Christ's body in the sacrament, the wine into his blood) with his own materialist chemical transubstantiation. It was a dangerous kind of heresy, but so disguised as to have gone almost unnoticed.

In January 1790, six months after the storming of the Bastille, three months after seven thousand women had marched on Versailles, two months after the National Assembly had declared all Church lands and buildings in France the property of the people, Erasmus Darwin resolved to publish *Zoonomia*. He wrote to fellow Lunar Society member James Watt, making light of his decision: "I have some medico-philosophical works in MS which I think to print sometime, but fear they may engage me in controversy (which I should not much mind) and that they will not pay so well (which I mind more)."

It was a dangerous time, certainly, but for many radically minded deistic or atheist intellectuals in the later eighteenth century it was also a hopeful time. The French Revolution proved that reform was possible, that tyrannies might fall, that human societies might evolve toward democratic or republican systems. As William Wordsworth famously wrote, recalling the atmosphere in Paris during the early days of the Revolution some years afterward, "Bliss was it in that dawn to be alive /

But to be young was very Heaven!" (*The Prelude,* X, 690–94). For reformists in England it now seemed possible to speak, to take risks. More important, it perhaps seemed imperative to do so. It was time to take a stand for reform alongside other liberals, dissenters, and radicals. "I know you will rejoice with me," Wedgwood wrote, describing the events in France as "a glorious revolution." Erasmus, determined to make his own stand, inserted a long celebration of the Revolution into *The Botanic Garden.* He portrayed the French people rising up like the giant Gulliver from the repressions of "Confessors and Kings" and Liberty as volcanic lava that, erupting, would form newly fertile landscapes.

But as the French Revolution shifted into a more violent phase over the following year, opinion in Britain turned against dissenters. Erasmus's friend the chemist and dissenting priest Joseph Priestley, one of the Lunar Men, continued to be among the most outspoken supporters of liberal reform, dubbed the "arch-priest of Pandaemonium liberty" by the *Gentleman's Magazine.* In 1790 his house was attacked. Priestley continued to give libertarian speeches, stressing that the reformers must hold their nerve and bear persecution, for "the world may bear down on particular men, but they cannot bear down a good cause." A Birmingham dinner arranged to celebrate the anniversary of the fall of the Bastille provoked a riot; the rioters marched on Priestley's New Meeting House and his Old Meeting House, destroying all the books in the library and torching the building, then marched on Priestley's house, where they burned all the furniture and books and manuscripts in the library, destroyed all the instruments in his laboratory, and razed the house. Priestley and his wife managed to escape to London. Over the following week the rioters and looters, daubing their slogans of "Church and King for ever" on walls and shutters, encouraged by Anglican clergy, local landlords, and even a justice of the peace, attacked four meeting houses and twenty-seven residences before dragoons were brought in from Nottingham to make arrests. Erasmus declared the Birmingham riots "a disgrace to mankind."

This was a riot against philosophers and science. The rioters had been persuaded that *philosophes* like Voltaire and Rousseau had been responsible for starting the revolution in France and that such men must be silenced in Britain if a similar revolution was to be prevented. One witness remembered that "the highroads for full half a mile of the house were strewn with books and that on entering the library there was not a dozen volumes on the shelves, while the floor was covered several inches deep with the torn manuscripts."

The Economy of Vegetation, one of the two volumes that made up Erasmus's poem *The Botanical Garden,* with its 2,440 lines of verse and 80,000 words of scientific notes, was a series of speculative essays in verse on geology, the economy of nature, and the atmosphere as well as a celebration of the French Revolution. It came out in June 1792, only a year after the Birmingham riots, and only three months before the September massacres began in Paris, in which thousands of aristocrats and priests were hunted down, imprisoned, and executed on the guillotine. The French would call it the Terror. No one had yet smashed the windows of Erasmus's Derbyshire house. His new poem was admired in literary circles for its originality, inventiveness, and range, and it became one of the most widely read books of the season, taking its place on bookshelves alongside Mary Wollstonecraft's *Vindication of the Rights of Women* and Tom Paine's *Rights of Man.*

With each new month of patriotic bombast and antiliberal rhetoric from the press and the failure of liberal ideals in France as the Revolution entered its most violent stage, Erasmus strengthened his resolve to publish both *Zoonomia* and the increasingly materialist new poem-in-the-making *The Temple of Nature* in his lifetime. "I am now too old and harden'd to fear a little abuse," he protested. Publishing the two books would be his greatest risk yet. Both manuscripts, in prose and poetry, contained undisguised evolutionary speculation. Erasmus added more scraps and ideas and evidence and case studies to them daily, through 1793 and 1794, and edited pages of others. In May 1794, twelve reformers were arrested and imprisoned in the Tower of London and brought to trial one by one. The government declared habeas corpus suspended for eight months. In October 1794, William Godwin wrote: "This is the most important crisis in the history of British liberty that the world ever saw."

Thus far Erasmus had escaped censure by dressing his reform-centered radical ideas in classical garb or disguising them as burlesque or masking them as mock heroic verses inspired by the work of Alexander

Pope. But he was tiring of disguises. The gamble he played was whether he could reach enough people, make some money for his family, propagate his ideas, and die before anyone put him on trial. And if they did, what charges would stick with the material he had written so far? *I was writing about plants, my lord. Trying to bring botany to the people.*

But with something of a witch hunt sweeping the country, by 1794 Erasmus was no longer getting away with it. In late 1794 an anonymous author wrote a parody of *The Botanic Garden* called *The Golden Age* written as a letter by Erasmus himself to the reforming physician Thomas Beddoes, making the implied prorevolutionary sentiments of *The Botanic Garden* absolutely explicit. Erasmus was so troubled by the accusation that he put a notice in the *Derby Mercury* to deny that he had written it.

The first edition of *Zoonomia, or The Laws of Organic Life,* 586 pages long and weighing four pounds, appeared in the bookshops of England in the early summer of 1794, a year after the revolutionaries had beheaded the French king and war had broken out between France and much of Europe. Though his enthusiastic followers would be eagerly watching for an advertisement from his publishers announcing a new volume of poetry, Erasmus had, they were surprised to find, published a specialist medical treatise, the fruit of a lifetime's study and treatment of the human body, an attempt to "reduce the facts" to a "theory of diseases," a classification of illness. The prose was bald and uncompromising. There were lists, charts, case studies, and classifications and absolutely no eroticized compound adjectives, rhymes, or poetic flights of fancy or comedy.

But there was a hand grenade. Buried deep inside the immensely detailed case histories and notes about specialized medical treatment, in which Erasmus Darwin described the human body as no more than a complex bundle of fibers and nerves, and families as passing down patterns of experience from one generation to the next, sat a fifty-five-page chapter called "Generation." Those fifty-five pages boiled down to one astonishing claim, the claim Erasmus had been avoiding uttering for twenty years: that species—the human species, indeed all living species—had descended from minute aquatic filaments swimming in a prehistoric sea. "Would it be too bold to imagine," he wrote,

> that in the great length of time, since the earth began to exist, perhaps millions of ages before the commencement of the history of mankind, would it be too bold to imagine, that all warm-blooded animals have arisen from one living filament, which THE GREAT FIRST CAUSE endued with animality, with the power of acquiring new parts, attended with new propensities, directed by irritations, sensations, volitions, and associations; and thus possessing the faculty of continuing to improve by its own inherent activity, and of delivering down those improvements by generation to its posterity, world without end!

Despite the apparent tact in this sentence, and the fact that he had couched his most daring claim as a rhetorical question, Erasmus Darwin's "Generation" chapter shot straight from the hip. Evolutionary speculation had finally risen from the footnote underground of his prose and into the main body of the work. The fifty-five pages did not apologize for themselves. Erasmus cited evidence to prove that species were mutable and had adapted to their environment: "Some birds have acquired harder beaks to crack nuts, as the parrot. Others have acquired beaks adapted to break the harder seeds, as sparrows. Others for the softer seeds of flowers, or the buds of trees, as the finches. . . . All which seem to have been gradually produced during many generations by the perpetual endeavour of the creatures to supply the want of food."* He recognized fossil discoveries as ways of tracing descent; he described sexual selection; he suggested that man was an animal among animals and thus not uniquely possessed of a soul; and he made absolutely no attempt to square his theories with the Bible.

Over the next few months Erasmus Darwin watched for reviews, looking out for the first signs of alarm and accusation.† In September 1794 the *Monthly Magazine* declared *Zoonomia* to be "one of the most important productions of the age"; but that came as no surprise, since

*This was not a Charles Darwin–type concept of adaptation produced through natural selection—it was closer to Lamarckian ideas of adaptation in which disused limbs gradually atrophy.

†It eventually went into five American and three Irish editions and was translated into German, Italian, French, and Portuguese.

the journal was radical and run by Joseph Johnson, Erasmus's publisher. The reviewer simply ignored the controversial chapter and praised the rest of the book. All the subsequent reviews followed suit.

Was Erasmus frustrated by the silence? It is impossible to know. He seems to have been living on tenterhooks. He was under surveillance; he knew that. Three years earlier, John Reeve, a judge in London, had set up the Association for Preserving Liberty and Property Against Republicans and Levellers, employing spies in every town who were instructed to watch local subversives. In March 1795, Erasmus complained to Richard Lovell Edgeworth, the politician, inventor, and fellow Lunar Society member, that there were spies in his street: "I have a profess'd spy shoulders us on the right, and another on the opposite side of the street, both attornies!" He was sure, he wrote, that both he and Wedgwood were on Reeve's list: "And I hear every name supposed to think different from the minister is put in alphabetical order in Mr Reeve's doomsday book, and that if the French should land these recorded gentlemen are to be all imprison'd to prevent them from committing crimes of a deeper dye. Poor Wedgwood told me he heard his name stood high on the list." Joseph Johnson was also on that list; everyone Erasmus knew seemed to be thinking of emigrating. "America is the only place of safety," he wrote to Edgeworth, keeping up the veneer of banter, "and what else does a man past 50 (I don't mean you) want? Potatoes and milk—nothing else. These may be had in America, untax'd by Kings and Priests." A single review might tip the balance. But the accusations still did not come. The conspiracy of silence held.

It took a year. In 1795, a reviewer in the *British Critic,* a right-wing periodical formed in 1793 to guard against the spread of radical ideas, condemned *Zoonomia* as subversive and urged the public not to read it. The situation worsened. In October 1796, Erasmus received a polite letter from Thomas Brown, an eighteen-year-old law student from Edinburgh, expressing his surprise that "no one had yet answered" the claims of *Zoonomia* and declaring that he had determined to be the one to do so, in print. He sent Erasmus a rather alarmingly detailed manuscript a few months later; Erasmus, furious, wrote to him twice that winter to tell him his book was both "hard" and "impertinent."

Just as Erasmus thought he had disposed of Brown, in April 1798,

George Canning, under secretary of state for foreign affairs, published a long satire of Erasmus's materialist ideas (and his poetic style) called *Loves of the Triangles* in his journal the *Anti-Jacobin or the Weekly Examiner,* effectively calling Darwin a revolutionary sympathizer. Serial publication allowed Canning to prolong the attack through April and May. Then in May Brown published the 560 pages of *Observations on the Zoonomia of Erasmus Darwin* denouncing Darwin's materialism, his evolutionary ideas, and his classification of diseases. By the end of the year, Erasmus had begun to appear in political cartoons as a subversive. Prison beckoned. In February 1799 his publisher, Joseph Johnson, who had been publishing seditious books for decades, was tried and imprisoned for six months, charged with being "a malicious, seditious, and ill-disposed person and being greatly disaffected to our said sovereign Lord the King."

In that burst of revolutionary idealism back in 1791, in his commitment to putting progress before his own reputation, Erasmus Darwin had resolved to publish his two most dangerous works before he died. He published only one of those two books. With his publisher in prison, certain that his own name had been put on a list of subversives, with his face and name beginning to appear in political cartoons, he seemed no longer in any great hurry to publish *The Origin of Society* in his lifetime. Perhaps he had lost the energy or resolve; perhaps he knew he was dying. He put the final touches to the poem and wrote the short preface in January 1802; he died four months later.

When Joseph Johnson published the poem, honoring his friendship with the late good doctor but protective of his own safety after his time in prison, he changed its title to the less provocative *The Temple of Nature, or The Origin of Society, a Poem with Philosophical Notes.* These were Erasmus Darwin's last bold words on the origin of species. There was a simpler, quieter beauty about the poetry; it lacked the overwrought, highly eroticized language of *The Botanic Garden.* The whole poem turns on a question, asked in the first two lines: How did life begin? How was it "kindled"?—a question asked of the Muse by a now-dead Derbyshire philosopher. Urania, Priestess of Nature and Muse of Astronomy, deep in the earth, deep in the Temple of Nature, gives her answer:

Organic Life began beneath the waves . . .
Hence without parent by spontaneous birth
Rise the first specks of animated earth.
First forms minute, unseen by spheric glass,
Move on the mud, or pierce the watery mass;
These, as successive generations bloom,
New powers acquire, and larger limbs assume;
Whence countless groups of vegetation spring,
And breathing realms of fin, and feet, and wing.
Canto 1, 295–302

Under the lyrical soprano of the poem, the bass voice of Erasmus's lengthy prose footnotes explains, glosses, extends, and makes connections, drawing out mystery after mystery, synthesizing, cross-referencing to his own earlier work, drawing on Egyptian, Roman, and Greek creation stories as if to say: *These things have always been known, but they have been hidden or repressed. I have excavated them for you.* The story he tells is a simple one with a happy ending, based in its essentials on Lucretius' *De rerum natura:* the universe was formed by "chemic dissolution," and as the earth's surface changed over millions of years, organic life developed beneath the sea, adapting constantly to survive ("One great Slaughter-House the warring World"), budding and breeding, constantly improving, transmigrating and transmuting through unimaginable tracts of time:

In countless swarms an insect-myriad moves
From sea-fan gardens, and from coral groves;
Leaves the cold caverns of the deep, and creeps
On shelving shores, or climbs on rocky steeps.
As in dry air the sea-born stranger roves,
Each muscle quickens, and each sense improves;
Cold gills aquatic form respiring lungs,
And sounds aerial flow from slimy tongues.
Canto 1, 327–34

Shout round the globe, how Reproduction strives
With vanquished Death—and Happiness survives;

How Life increasing peoples every clime,
And young renascent nature conquers Time.
Canto 4, 451–54

Every line of both the poetry and the footnotes in *The Temple of Nature* put forward the fullest version of Erasmus's evolutionary hypothesis, developed and modified over twenty-five years.

Erasmus Darwin had a sophisticated understanding of how species had evolved from a single-celled aquatic ancestor, and of how both rocks and species were the result of millions of years of adaptation; he also had some emerging sense of natural selection. But though he read widely and synthesized ideas across many different disciplines, he did not come to understand evolutionary processes through a detailed study of natural history. Instead, his evolution emerged from his medical knowledge. It was good knowledge, important knowledge, but it was not broad enough to enable him to plot and document natural selection in a way that would provide the all-important evidence to persuade enough people. Erasmus wanted to convince his readers that progress was nature's way, and thus that reform was to be embraced, not feared. His nature, unlike his grandson's, was a progressive one. It emerged in an age of reform.

There was not a single good review. *Horror, monstrosity, atheism, degradation,* cried reviewer after reviewer. The liberal *Monthly Review* derided "the tendency of Doctor Darwin's poetry to degrade the human species and exalt animals of an inferior nature." The poem, he added, was "in no way adapted to improve either the judgement or the morals of his readers." The reviewer in the *Anti-Jacobin Review and Magazine* warned its readers that the poem "teems" with heretical ideas; he was scandalized by the poem's "total denial of any interference of a Deity." The *Critical Review* attacked the poem for "trying to substitute the religion of nature for the religion of the Bible"; the *Gentleman's Magazine* declared the poem "glaringly atheistical"; in America, Joseph Priestley wrote, "If there be any such thing as atheism, this is certainly it"; and the reviewer for the *British Critic* simply declared: "We are full of horror and will write no more."

The poet Samuel Taylor Coleridge, fascinated by Erasmus Darwin's earlier works, was, he claimed, nauseated by *The Temple of Nature.* The idea of "Man's having progressed from an Ouran Outang state," he wrote to William Wordsworth, is "contrary to all History, to all Religion, nay, to all Possibility." He preferred, he wrote, "the History I find in my Bible . . . that Man first appeared with all his faculties perfect and in full growth." Can any believe, he wrote elsewhere, "that a male and female ounce [leopard] . . . would have produced, in course of generations, a cat, or a cat and a lion? This is Darwinizing with a vengeance." Still, the poem found at least one admirer. In London the young radical poet Percy Bysshe Shelley, who had been expelled from Oxford in 1811 for publishing a tract called *The Necessity for Atheism* and who, estranged from his father, had lived a migrant life since then, ordered and read and reread every Erasmus Darwin volume he could get his hands on.

There is no question that Erasmus Darwin was sure about his theory of the origin of species. The trouble was that, partly due to his own personality and taste, partly because he had stopped believing that any plainly argued theory of the earth might be believed or respected, and partly also because he knew that the consequences of such straight talking might be the loss of his practice and his position in his tight rural community that otherwise tolerated a great deal, he was never able to say any of this outright. So Dr. Darwin, a freethinking polymath ahead of his time, had disguised his infidel ideas in poetry, or buried them deep in footnotes or deep within a medical textbook. But though the poetry of Erasmus's vision may have seemed dark, heretical, and nauseating to some, it excited others who, like Erasmus himself, saw in his vision of human mutability a giant pulling himself warily to his feet.

It is June in the year 1816 and midnight in the Villa Diodati in Geneva. Three young English travelers, Mary Godwin, Percy Shelley, and John Polidori, have come to visit Lord Byron. Because the weather is so cold and wet, they have been holed up in his villa by the lake, discussing philosophy for several days. Polidori suggests they all write ghost stories, but as they think of ghosts and monsters, the conversation roams wildly. Percy Shelley, aged twenty-two, not long out of Oxford, has been reading about microscopy, the solar system, magnetism, and electricity. He tells them of a discovery made by Dr. Darwin—how in a paste of flour and water you can make tiny organisms increase in number and size, even without air, and they will come back to life again when dried out. That is how life began, he tells them. Not with a garden in Eden, but with tiny organisms in a pond. Might it not be possible, he speculates, to find a way of harnessing the vital principle that drives the minute water creatures back into life from death?

Mary Godwin, Shelley's brilliant and intellectually voracious lover, the daughter of William Godwin and Mary Wollstonecraft, is interested in theories of life for different reasons. Only eighteen years old and unmarried, she has been pregnant almost continuously since she and Shelley became lovers when she was sixteen. After eloping in 1814 and traveling across Europe with barely any money, she lost her first child, a daughter, at only two weeks old in February 1815; the loss devastated her. "Find my baby dead," she wrote. "A miserable day." Pregnant again only eight weeks later, she gave birth to her second child, a boy, William, in January that year. She is already pregnant again.

Mary—who became Mary Shelley on her marriage to Shelley later that year—described the late-night conversations at the Villa Diodati in her introduction to the revised single-volume edition of *Frankenstein* in 1831: "They talked of the experiments of Dr Darwin . . . who preserved a piece of vermicelli in a glass case till by some extraordinary means it began to move with a voluntary motion. Not thus, after all, would life be given. Perhaps a corpse would be reanimated; galvanism had given token of such things: perhaps the component parts of a creature might be manufactured, brought together, and endued with vital warmth."

The vermicelli is a misremembering on either her or Shelley's part. Darwin had actually written "vorticellae" in his notes on spontaneous generation in *The Temple of Nature.* He was describing a microscopic aquatic filament found in lead gutters that when dried out shows no sign of life but "being put into water, in the space of half an hour a languid motion begins, the globule turns itself about, lengthens itself by degrees and assumes the form of a lively maggot . . . swimming vigorously through the water in search of food."

Unable to sleep, her head full of these speculations about life, Mary Shelley dreamed of a "pale student of unhallowed arts kneeling beside the thing he had put together. I saw the hideous phantasm of a man stretched out and then, on the working of some powerful engine, show signs of life and stir with an uneasy half-viral motion."

Published in January 1818, *Frankenstein* became an instant bestseller. Now regarded as the origin of modern science fiction, it is the world's most famous horror story. "It was on a dreary night of November," Mary Shelley wrote, "that I beheld the accomplishment of my toils. With an anxiety that almost amounted to agony, I collected the instruments of life around me that I might infuse a spark of being into the lifeless thing that lay at my feet. It was already one in the morning; the rain pattered dismally against the panes, and my candle was nearly burnt out, when by the glimmer of the half-extinguished light, I saw the dull yellow eye of the creature open; it breathed hard, and a convulsive motion agitated its limbs."

It had been quickened by the evolutionary speculations of Erasmus Darwin.

9

The Jardin des Plantes

PARIS, 1800

On the left bank of the Seine, beyond the wrought-iron entrance gates, the botanical garden of the Jardin des Plantes stretched back from the edge of the water, all straight lines and symmetrized borders, to the classical façade of the Museum of Natural History at the far end. Originally named the Jardin du Roi, a garden of medicinal plants created for a king, in the eighteenth century it had housed a community of eminent French botanists who flourished under the leadership of the Comte de Buffon, who had improved and extended it, adding greenhouses, a labyrinth, and new collections of rare plants. With the overthrowing of the monarchy in 1793, the French revolutionary government renamed the Jardin du Roi the Jardin des Plantes, established a Museum of Natural History in the grounds, and appointed twelve new professors, all of equal rank and with equal pay, to study nature in all its forms and to extend the boundaries of the known world for the glory of the French people.

During the months leading up to the eruption of violence that came to be known as the Terror, while mobs rioted in the streets, during the beheadings and executions and the savagery, as aristocrats fled France or were rounded up into the prisons of Paris and marched to the guillotine, wagons carrying crated-up natural history collections confiscated from country palaces and elegant town houses rattled in through these gates. The twelve professors and their assistants unpacked bones, fossils, stuffed animals, minerals, bottles of preserved invertebrates, rare shells and corals. As money poured in from confiscated churches and palaces, masons and carpenters marked out the ground plans of new greenhouses, museums, laboratories, and lecture halls; horses and oxen hauled stones on wagons from the quarries of Montmartre; masons dug and drained, ditched and leveled, lifted stone onto stone, set glass into frames, tiles into roofs; carpenters dovetailed joints for new shelves and display cabinets and ranks of polished seating in libraries, lecture halls, and museums.

The collections of the Jardin swelled with the spoils of war as well as with the confiscated property of the executed and exiled aristocracy of France. As French armies marched across Europe capturing country after country in the last years of the eighteenth century, they appropriated the finest menageries and cabinets of natural history objects in the newly conquered countries and sent them back to the professors at the Jardin. Live animals, including two elephants from Holland, and stuffed animals and crates of labeled bones and fossils piled high on wagons made their way across the French roads and waterways and into Paris, together with the vast confiscated collections of the princes of Condé and the stadtholder of Holland. When Napoleon Bonaparte rose to power through the republican army—as an officer to begin with, then as a general, appointing himself first consul in 1798, and then emperor of France in 1804—he employed experts to identify the most famous collections of paintings, books, and natural history specimens in Europe and then to cherry-pick the finest pieces and oversee their packing and shipping back to Paris. When French scientist-explorers went to sea in search of new landscapes, they also shipped thousands of preserved species—peeled off rocks, dug out of holes, caught in jungle traps—back to their capital.

In 1800 the Jardin housed fifty-six employees and their dependents. In addition to the professors who lived in the grand houses that were built or renovated for them near the museum buildings in which they worked, there were naturalists' assistants, gallery superintendents, gardeners, animal keepers, custodians, taxidermists, carpenters, glaziers, and masons. They lived in apartments at the museum with their families, if they had families, and sometimes not just with wives and children but also with siblings, nieces, nephews, or parents. Some had housekeepers. Others kept small gardens and domestic animals. The Jardin des Plantes had become a world unto itself.

But despite the stories the French newspapers told about the utopia that the republic and then the emperor had made in the Jardin des Plantes, life was not entirely harmonious behind its high walls and iron gates. Three of the twelve professors were engaged in a battle of ideas here in the first decades of the nineteenth century. They were fighting about the nature and definition of life itself, and the idea, which by then was discussed widely in Europe, that existing species might have transformed from earlier species. In the early years of the nineteenth century, the French had no single word for species mutation, although they would eventually call it *transformisme.* Later, the English, looking across the Channel at what seemed to be an aggressive, hostile, and godless country, called the idea *transmutation,* using a word drawn from alchemy, and labeled it both inherently heretical and dangerously French. It struck at the heart of the Christian religion, clerics declared; it contradicted the biblical account of creation; it sullied the idea of a benevolent God who had designed the universe for the enjoyment of man; it belittled the position of man in the universe; it challenged belief in man's divinely granted dominion; it contradicted biblically based estimates that the earth was around six thousand years old. It was not to be supported.

Three professors; three different versions of nature. In these opening years of the nineteenth century, as Napoleon began his war with Europe, Professors Georges Cuvier, Jean-Baptiste Lamarck, and Étienne Geoffroy Saint-Hilaire first set out their battle positions in the Jardin des Plantes. The three men, united by a common drive to discover nature's laws, were profoundly different from one another.

Jean-Baptiste Lamarck, in his fifties as the century turned, was the most senior. The eleventh son of a minor provincial aristocrat, he had served in the army in the Seven Years' War and then scratched out a living as a botanist in Paris. Buffon, impressed by his work, had made him his protégé. With the reformation of the Jardin, Lamarck had been appointed professor of insects, worms, and microscopic animals, with special dominion over the enormous collection of invertebrates. Widowed twice, he now lived with his third wife and eight young children in an apartment on one floor of the large house Buffon had once occupied, adjacent to the Museum of Natural History.

Étienne Geoffroy Saint-Hilaire (known by his shortened surname, Geoffroy) and Georges Cuvier were both young men in 1800. Cuvier turned thirty-one that year; at twenty-eight, Geoffroy was the youngest professor in the Jardin. When Cuvier first arrived in Paris from Germany, he and Geoffroy had shared lodgings and collaborated on papers. It took Cuvier, who came from a German provincial town and had few high-powered patrons in Paris, four years to negotiate a position for himself in the Jardin. Ambitious by nature, he worked hard to get it. Soon after he took up his position, he appropriated the attics of a vast building adjacent to his lodgings that had been recently purchased for the Jardin. With plans to establish his own museum of comparative anatomy there, he purloined unused animal skeletons that he found "piled up like bundles of firewood" in the old galleries. He was certain that natural history could be turned into an exact science through analysis of the relationships between body structures. France would lead the way, he told friends, overturn old mistakes, old orthodoxies, remap the natural world; it was time to look *inside* animal bodies, to examine and analyze their internal structures rather than their external characteristics as earlier naturalists had done.

Soon Cuvier's Museum of Comparative Anatomy, carved out of the attics of an old hackney-carriage building and former flour depot, had spaces for dissection, a study, a library, and rooms filled with the skeleton and fossilized body parts Cuvier had assembled and pinned together to illustrate his new and radical understanding of the relationship of animal body parts to each other.

Cuvier had to be careful in those early years. He was not yet a professor himself but only a stand-in for an elderly professor of animal anatomy. In the library and corridors of the Museum of Natural History,

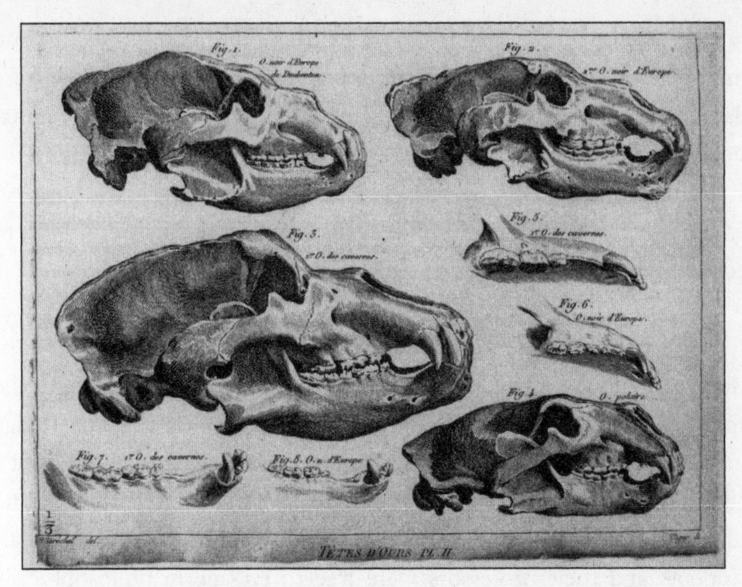

Cuvier's comparison of the skulls and teeth of living and fossil bears.
Georges Cuvier, Ossemens Fossiles *(1812)*

students and professors gossiped and whispered about one another, making and breaking reputations. To Cuvier, increasingly dedicated to measurement, fact, precision, and evidence, Lamarck, who roamed widely in his work, crossing the boundaries between meteorology, botany, chemistry, and physics, must have seemed an old-fashioned generalist, a man of science who belonged to the old century. Cuvier found Lamarck adversarial and obsessive, always pitting himself against experts in other disciplines. The older man rarely listened to new ideas; he seemed not to be interested in anything except his own theories and systems.

By contrast, Georges Cuvier was a plain facts man, a smartly dressed professional with immaculately formal manners. He was a defender of rules, hierarchies, law, and political order. Grand theories and speculation were all part of the last century, he would complain to his brother, Frédéric Cuvier, and his secretary, Charles Laudrillard; they were part of Buffon's century. The proliferation of "systems" in the previous century had muddled things, confused the picture. In the new natural philosophy, claims to truth must be based only on close observation and accurate recording. Buffon's wilder and later ideas, his notions about flux and mutability in species, were not just unprovable, he would say, they were dangerous; they challenged the natural order and imperiled the reputation of natural history.

Lamarck and Cuvier belonged to different traditions. Relations between them were, in early 1800, cordial enough, but they were about to come under strain.

By May 1800, the museums of the Jardin des Plantes, glutted with the spoils of war, had become the largest repositories of natural knowledge in the world. Hundreds of young lawyers, medical students, hospital administrators, and travelers signed up to attend the public lectures and to study alongside the professors. In the absence of censoring priests, students and their professors were free to engage in philosophically heady debates about the nature of life and creation, and their conversations spilled into the streets of Paris.

The lecture season in the Jardin des Plantes had just begun. Sixty-four young men, most under thirty and some as young as eighteen, sat in the ranked seats of the lecture theater waiting for the appearance of Jean-Baptiste Lamarck, the professor of insects, worms, and microscopic animals. They were mostly French, although there were also students from Italy, England, Ireland, Brazil, and Switzerland. They had come to study at the greatest center for natural history in the world, in the square mile where the great Buffon had lived and worked. None of them knew that this lecture would go down in history, that Lamarck, who had previously been a fixity-of-species man, had changed his mind about the origins of species. From years spent studying the relationship between living and fossil forms among the shells and invertebrates in his enormous collection, he had come to believe that all species had evolved from earlier forms and were continuing to mutate. And he was about to go public with that claim.

Lamarck began his lecture that day by declaring that natural history entailed a great deal more than the collection of facts. The naturalist had to be a philosopher who was prepared to take risks like the great Buffon, he insisted; he must always be searching for the larger picture.

Then—extraordinarily—he began to describe nature as being in a state of perpetual flux, constantly producing new species and organisms in inexhaustible ways. Diversity of habitat, temperatures, actions, and modes of life, he declared, all influenced the development of living beings. As a result of these various influences, he continued, the limbs and body shapes of animals developed and strengthened with use and then diversified under the effect of long-preserved habits. And imperceptibly, over thousands of years, the nature, shape, and state of parts as well as organs were preserved and propagated by generation.

Lamarck's new transformist ideas were hedged with perhapses and maybes, but his theory was simple and, no doubt for his students, astonishing: *nature was on the move.* He had used the expression *la marche de la nature* before, to mean the scale of nature, but now when he used it he meant something different; he meant it was was actually moving; it was *on the march,* progressing from the simplest forms toward the most complex. "Little by little," he wrote, "nature has reached the state in which we see it today."

Lamarck used birds—birds' legs elongating, birds' claws hooking, birds' feet webbing—to give those students sitting in the auditorium that day a vivid but bizarre picture of the timeless and still continuing process he described. He asked them to imagine a species of water bird stretching out its toes repeatedly to move through the water and producing—generation by generation—webbed feet. He asked them to imagine a bird perching on a tree; might it not inevitably acquire, over time, longer, clasping, hooked toes so as to hold on better? Imagine, he told them, a shore-dwelling bird wading constantly in the mud. Generation by generation, over enormous tracts of time, might it not eventually develop longer, featherless legs? The wording he used was vague and evasive: the tree-perching bird "inevitably acquired" claws, he claimed; the mud-wading bird "finds itself raised" on long bare legs. It was a picture that would lend itself easily to parody, but it was also one that contained no God or divine plan.

It would not have taken long for Cuvier to hear the students of the Jardin discussing Lamarck's new ideas. Cuvier was not a religious man; the fact that Lamarck's ideas were at odds with the biblical creation story would not have seemed to him to be a problem in itself. To Cuvier it was merely bad science. It was preposterous, he had declared since 1796, to think that all quadrupeds had descended from a single species; such a theory "would reduce all natural history to . . . variable forms and fugitive types." Convinced by years of painstaking comparisons between the physiological and anatomical internal structures of animals, he was certain that all organisms were functionally integrated wholes. Any change to that balance would make them unviable, unfit to survive. But though he found Lamarck's ideas ludicrous, he kept his counsel, believing that if he refused to respond to them at all, others would follow suit. Besides, it was part of the etiquette of French science to avoid public controversy. He would let Lamarck's ideas stand or fall by themselves.

For two years, between 1800 and 1802, the fifty-seven-year-old Lamarck worried that his poor health would not allow him to finish all the projects he had planned. He wrote frenetically, drafting and publishing several books simultaneously, taking the central tenets of his system in each subsequent book further and further across the border into speculative heresy. In *Natural History of Animals Without Backbones* (*Système des animaux sans vertèbres*) (1801), a systematic review of invertebrate animals accompanied by the text of his lecture of 1800 and a brief memoir on fossils, he claimed that fossils were traces of "the changes that living beings have themselves successively experienced" on the surface of the globe.

In the twinned books *Hydrogeology* (*Hydrogéologie*) and *Researches on the Organization of Living Bodies* (*Recherches sur l'organisation des corps vivants*), both published in 1802, Lamarck expanded and defended his transformist theories within a vast series of contemporary debates on anatomy, zoology, botany, mineralogy, geology, chemistry, and natural history. The earth's surface had been shaped and modified, he argued in *Hydrogeology,* by the regular and ongoing action of water moving above and below the surface, producing continual changes to the earth's crust. The same principles applied to the changing shapes of animal bodies: "It is not organs—that is, the nature and shape of an animal's body—that have given rise to the animal's habits and specific faculties," he wrote. "On the contrary it is the animal's habits, its mode of life, and the circumstances in which its individual forebears found themselves that, over time, have determined the form of its body, the number and state of its organs and lastly, the faculties with which it is endowed."

There was a melancholy poetry in these books and in much of Lamarck's subsequent work. His transformism was a story of erosion and invasion, of organisms and landscapes pitting themselves against the destructive forces of nature caught forever in a balance between antagonistic processes: "In this imperceptibly slow process, the sea is constantly breaking up, destroying, and invading the continental coasts it encounters on its path," he wrote; "meanwhile on the opposite coast, the sea is constantly falling, withdrawing from the land it has raised, and forming new continents behind itself, which one day it will return to destroy." These processes were both mighty and slow, so slow indeed that the earth's age, he wrote, "utterly transcended man's capacity to calculate." Lamarck told optimistic stories, too, descriptions of animals straining to survive and improve against all odds, including eventually the famous giraffe that his enemies would later use to mock his ideas. "To the examples I cited," he added as an afterthought to his *Researches on the Organization of Living Bodies,* "I could add that of the shape of the giraffe (*camelopardalis*), a herbivorous animal which, as it dwells in a place where the land is arid and grassless, is obliged to browse on tree leaves and to strain itself continuously to reach them."

As Lamarck's books appeared through 1801 and 1802, the numbers of students attending his lectures almost doubled, rising from 70 in 1801 to 131 in 1802. These were mostly young pharmacists, doctors, medical students, lawyers, and hospital administrators, such as the idealistic forty-five-year-old hospital administrator at the Paris Institute for Deaf-mutes, Joseph Alhoy, who was also writing a book-length poem about reforming the treatment and education of abandoned children, or the Irish doctor John Butler, or the fourteen-year-old navy recruit and son of an engineer Christophe Paulin Fréminville de la Poix, who had been attending Lamarck's spring lectures since he was twelve and had already determined to be an archaeologist and an adventurer and to dedicate himself to science. Although most of these young men would have attended Lamarck's lectures simply to learn about invertebrate zoology, for some of them his account of nature's process must have been radical and enthralling. Nothing was fixed in nature, it seemed. Everything was remaking itself, striving and stretching—like them and like Napoleon—toward a future in which human will would triumph and in which those with talent and adaptability, rather than money and an old name, would thrive.

In 1802 Lamarck was a rising star among these students even if his fellow professors ignored or dismissed his speculations. If reading Buffon's *Natural History* or *Epochs of Nature* had brought them to the Jardin to study nature, they found in Lamarck's books and in parts of his lectures on invertebrates similarly gripping ideas, and sensational descriptions of nature's processes working across unimaginable tracts of time and space. Many of those sixty-four students sitting in the auditorium during Lamarck's lecture in 1800 or those who read *Researches* after 1802 or *Zoological Philosophy* (*Philosophie zoologique*) after 1809 would later return to their work in the provincial hospitals or museums of France and across Europe; others went to sea on expeditions or as sailors in the French navy, as did Fréminville, and they took Lamarck's metamorphosing birds and astonishing time scales with them, scattering them into further conversations at provincial dinner tables. Following Lamarck, they claimed that the first life-forms on the earth had been infusorian or wormlike organisms; from these, over immense periods of time, nature had successively developed all other life-forms, all the way to the most complex animals walking the earth. The environment—changing temperatures, rising water levels, scarcity of food or water, or the spread of predators—caused animals to adopt new habits to survive. These new habits led infinitely slowly to the appearance of new structures through the inheritance of acquired characteristics: longer limbs for running, longer tongues for catching food, flat-topped teeth for chewing; other structures, if no longer needed or used, would gradually atrophy. It was a distinctive, simple, and radically new way of seeing the world. It was also, of course, deeply heretical to anyone of a religious disposition.

The idea of species change—whether it was called descent with modification, transformism, transmutation, or the development theory—could be heard discussed in the coffee shops, libraries, universities, and workingmen's clubs of Europe in the first decades of the nineteenth century. Lamarck was now its most brilliant and articulate spokesman.

—

By 1798, when he left for Egypt, Geoffroy Saint-Hilaire had been professor of vertebrates at the Jardin for five years. Like many serious young men who had been training for the priesthood, he had transferred his studies from theology to medicine after the Revolution. Discovering that he preferred mineralogy to medicine, he took lessons from a professor of mineralogy at the Collège of Navarre, René-Just Haüy, who was also a practicing Catholic priest. When Haüy was arrested and thrown into prison with several other priests during the street hunts of the Terror, Geoffroy, despite his revolutionary sympathies, disguised himself, borrowed a ladder, and tried to rescue his professor in the dead of night, resorting finally to alerting the professors at the Jardin, who secured his release. Terrified for his own safety, Geoffroy then fled Paris and returned to his parents' home in the provinces, where he collapsed with a "nervous fever" that kept him in bed for months. There were rewards for such reckless loyalty. On his return to Paris, Haüy recommended his "young liberator" for a vacancy as a keeper in the Jardin's cabinet and as assistant to the elderly professor of mineralogy Louis-Jean-Marie Daubenton. Eighteen months later, when the Jardin was transformed into the Museum of Natural History, Geoffroy found himself promoted to professor of zoology, a subject about which he knew very little.

After months spent arranging, assembling, identifying, and labeling hundreds of skeleton specimens in the new museum arriving from collections across Europe every day, Geoffroy was struck by the fact that so many of the animal skeletons he assembled—birds, lizards, apes, dolphins—seemed to be structured along the same architectural principles. He was twenty-four years old when he first articulated his big idea. "It seems that nature is confined within certain limits," he wrote, "and has formed living beings with only one single plan, essentially the same in principle, but that she has produced variation in a thousand ways in all her accessory parts."

In 1798, when Napoleon set sail to conquer Egypt in order to establish a base in the East and to disrupt British trade with India, Geoffroy volunteered to join the community of 167 of the most promising and prominent men in the French sciences whom Napoleon took with him to study Egypt's history, culture, zoology, and landscape. Luck turned against the savants from the start. The ship carrying their equipment—scalpels, microscopes, tweezers, jars, pins—sank, and they had to wait for new supplies. Once Napoleon had taken Alexandria, the entourage of scientists followed the soldiers down the course of the Nile, where they had their first views of the pyramids, and into Cairo, where they requisitioned the harem quarters of a mansion recently built and then abandoned by a wealthy Mameluk and his retinue. French engineers, zoologists, archaeologists, and astronomers set up a study center there, holding meetings and reading papers in airy, marbled, and pillared rooms while Napoleon and his army of twenty-five thousand men fought for control of the country. In 1799, when Napoleon deserted his army and returned to Paris to orchestrate a coup d'état, handing over command of Egypt to General Kléber, the situation in Cairo became increasingly precarious.

Geoffroy was overwhelmed by what he found in Egypt: birds, reptiles, and insects; fish and other aquatic animals that Europeans had no name for; images of ibises, dogs, beetles, monkeys, cats, and scorpions engraved deep into sacred temples and tombs crumbling in the desert. There were lion-headed, crocodile-headed, and bird-headed gods. Hieroglyphs everywhere. For Geoffroy, as for Maillet before him, in this landscape etched with signs of immensely ancient civilizations, time began to change its contours. What relation, he wondered, did these ancient tomb-carved animals, seemingly united by a common structural plan, bear to modern species?

Within a year the naturalists on the expedition had established a printing press, a zoo, workshops, and laboratories in the garden of the Mameluk palace in which they lived and worked. With all the beauty and diversity around them, it was impossible not to roam from subject to subject; in fact, they needed to do so in order to make sense of the scale of the new ideas and knowledge. They worked together in new intellectual collaborations, sharing skills and methodologies, giving and reading scientific papers to one another.

Collecting specimens was difficult and skilled work. Geoffroy recruited local people—fishermen, hunters, snake charmers—to bring him new species of animals from the deserts and market stalls of the fishing villages. In the cool hours of the morning and evening, he and his assistants dissected, preserved, stuffed, and classified the thousands of creatures that were brought to the harem rooms in the palace. He and the twenty-one-year-old botanist Marie Jules César Savigny made three expeditions to the Nile delta, Upper Egypt, and the Red Sea, where they found a new species of mongoose as well as hares, bats, foxes, rats, and hedgehogs. They sailed down the Nile to the delta town of Damietta and found the sea and the marshes alive with nesting, feeding, breeding, and migrating birds. "I have never seen such water birds," Savigny wrote. "Flamingoes, cormorants, ducks . . . at night one can't hear anything but the calls of all these birds. Serene temperatures. Soft air. I have never seen so many water birds and so many species. The water's surface is rippling with wings."

On a trip to the ruins of Memphis, which Benoît de Maillet had discovered a century earlier, local guides took Savigny and Geoffroy by rope down into the Well of Birds, where they were shown hundreds of pots of mummified ibises stacked up along the walls of subterranean cellars. At the Red Sea they paid the local divers to collect starfish, corals, sea urchins, and crustaceans for them. Within a year the Mameluk palace was piled high with boxes of labeled specimens, bones, jars of preserved sea creatures, and mummies of cattle, cats, crocodiles, and birds.

Geoffroy fell sick with a nervous illness: he grew thin and was unable to eat; he suffered fevers and skin eruptions and several bouts of ophthalmia that blinded him for weeks at a time. But he did not stop collecting. He could not sleep. The theory he had formed in Paris had become a kind of mania to him. Convinced that he could see a basic body plan under all this diversity of hair, skin, color, texture, and shape, he set out to compare the fish of the Red Sea and the Mediterranean and to describe all the fish of the Nile.

In the workshop in the garden of the Mameluk palace, Geoffroy and his assistants sorted through and preserved in wine spirits the fish brought to them every day by the fishermen and traders who worked the riverbanks: sharks, rays, puffer fish, lungfish, fish of every color and shape. The day after Napoleon abandoned Cairo, leaving it in the hands of General Kléber, Geoffroy wrote to Cuvier to describe a new species of freshwater fish he had found, *hétérebranche,* a previously unknown species of lungfish with multiple dorsal fins, and another he called the *Silurus anguillaris* that when opened up proved to have bronchioles closely resembling—indeed, almost identical to—the structure of the human lung and three hearts that resembled those of a cuttlefish. This was a revelation for the young zoologist. It was proof of the theory he had begun to explore back in Paris—that all animals were built to a common plan. He wrote in jest to Cuvier that he would now "demand no less than the throne of anatomy." But his fluctuating moods and enthusiasms troubled him, too. "I am so overwhelmed with business," he wrote to Cuvier, "I do not know what I am doing or saying."

By 1799, Geoffroy's companions were concerned about his agitated state of mind, but then everybody had become a little high-strung in Egypt, especially after August 1799, when Napoleon had abandoned them and returned to France. The soldiers of the French army were hungry, unpaid, and dangerous; they accused the savants of hoarding looted treasure. The Turks were confronting the French troops in violent clashes. The British were anchored off the coast. Their lives were in danger.

But in this atmosphere of political volatility, Geoffroy's mania for zoological discovery seemed only to increase. He proposed a series of experiments with six hundred fertilized eggs and an incubator; he began speculating on the nature of dreams and sleepwalking. But he also longed for Paris. In November 1800, he wrote to Cuvier to beg him to do everything he could to secure his return to the Jardin. That winter, the same year that Lamarck presented his first transformist ideas at the Jardin, Geoffroy gave an extraordinary number of papers to his fellow savants as they waited to be evacuated from Egypt: papers on the coexistence of male and female germs in all animals; on the formation of the egg; on the organs of respiration; on the sepulchres in the tombs at Memphis; on the crocodile of the Nile; and on the animals known to the ancient Egyptians. He was also dissecting two more new species of fish. "The bombing, the fires, the ambushes, and the plaintive cries of the victims," he wrote, all paled before "my problems of natural philosophy."

Back in Paris, Cuvier worried about Geoffroy and the direction his work was taking out in Egypt. Despite the new commitment in the work of the Jardin to facts and verifiability and the rejection of Buffon-like castles in the air, Geoffroy, like Lamarck, it seemed, was also allowing himself to be seduced by abstract, unprovable systems. It must have been the heat, Cuvier told himself, congratulating himself on his decision to stay in Paris and consolidate his networks and connections, turning back to the solid facts and details that thrilled and sustained him. In the three years that Geoffroy had been in Egypt, Cuvier had been appointed to an additional post at the Collège de France, had published two volumes of *Lessons on Comparative Anatomy* (*Leçons d'anatomie comparée*), and had risen to a high rank in the administration of public education. He was making headway not just in France but across Europe, convincing people that comparative anatomy, the close and systematic study of animal structures and internal characteristics, a branch of science that was becoming synonymous with his own name, was the key to deciphering nature's laws. Geoffroy's fevered declarations of yet another new theory or system must have filled him with dread.

Still, Cuvier told his assistants, Geoffroy had done France a great service. When the triumphant British generals had tried to get the savants to hand over all their natural history collections and notes, Geoffroy had told them that he would burn them all rather than relinquish them to the enemy; the British generals, convinced that this wild-eyed Frenchman might be capable of anything, relented, confiscating only the extraordinary Rosetta Stone.* French science, Cuvier knew, would be forever in Geoffroy's debt for that act of defiance.

On his departure from Egypt in September 1801, Geoffroy wrote to tell Cuvier that he would soon send him the manuscript of a "very vast theory" that would revolutionize science. "I hope," he added, "to re-enter France worthy of you and my illustrious colleagues." But as the tired savants sailed from Egypt to Marseilles, where they were stuck for weeks in tedious quarantine, the mathematician Joseph Fourier, permanent secretary of the Egyptian Institute, described Geoffroy's nervous

*This slab of stone, discovered by a French soldier working on strengthening a defensive wall, carried inscriptions in ancient Greek, demotic Greek, and Egyptian hieroglyphs. It enabled the later decoding of the hieroglyphs.

illnesses and crazy babblings; he passed Geoffroy off as a fool and his theories as the ravings of a madman. Geoffroy, now deeply anxious, wrote distraught letters to Cuvier, asking him to protect his reputation and to quell any rumors he heard. He finally arrived back in Paris in late January 1802.

On his return, Geoffroy was unable to convince a single one of his colleagues of the truth of his great theory. He was unable even to get them to listen for very long. Overcome by fear for his career and reputation in an environment that seemed in his absence to have turned away from all large-scale theorizing toward an exclusive obsession with facts, Geoffroy locked his manuscripts and his great theory away. He resolved to avoid controversy, at least for the moment. He unpacked his boxes, unwrapped his specimens, arranged his collections, and kept his ideas to himself.

Everywhere Cuvier went he praised not Geoffroy the philosopher or the theorist, but Geoffroy the collector. All knowledge of nature and nature's laws would now be advanced, he declared, by the extraordinary collection Geoffroy and his colleagues had brought back to Paris, which included not only hundreds of new living species but also mummified ancient species. Geoffroy's mummified animals, he declared, would unlock the truth about the supposed transformation of species. If Lamarck continued to propose—preposterously, in Cuvier's eyes—that birds had evolved their shapes from reaching for leaves or wading in mud or grasping trees, then the mummified ibises, wrapped and embalmed three thousand years earlier, would settle the problem once and for all. They had only to unwrap them.

Students and professors gathered around the musty-scented, cocoon-like gray parcels assembled on the laboratory tables of the Jardin to watch Cuvier free the birds and the other mummified animals from their swaddling cloths. Cuvier was excited. "One cannot master the transports of one's imagination when one sees again," he wrote, "conserved down to the slightest bones, least bit of hair, and perfectly recognisable, an animal that had once had priests and altars, two or three thousand years ago in Thebes or in Memphis."

As Cuvier had predicted, inside the cocoons, the three-thousand-year-old animals looked no different from modern species. On the dis-

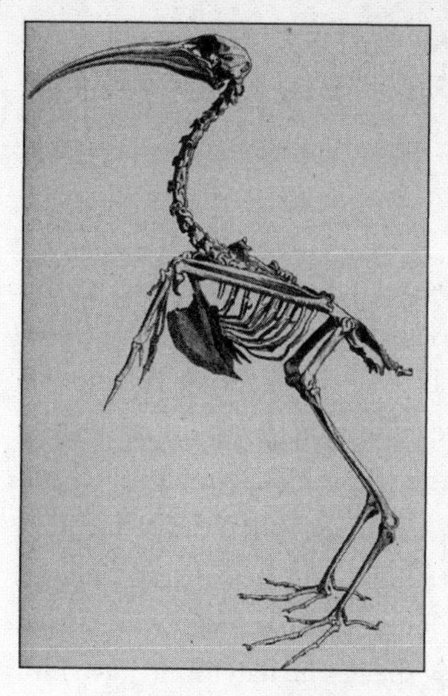

Two illustrations used by Cuvier to illustrate his claim that the mummified remains of the ibis of Egypt (left) were identical to the modern bird (right).
Georges Cuvier, "Ibis des Anciens Égyptiens" (1804)

section table, the mummified cat appeared to be just the same as the modern alley cats of Paris. The mummified ibises were no bigger or smaller than the living ibises that waded along the edge of the Nile. There was nothing strange, no extra or missing limbs, shorter necks, or additional fur or feathers. There was nothing out of the ordinary.

This was a good autumn for Cuvier. He had publicly challenged Lamarck's ideas about species transformation without embroiling himself in controversy or debate; he had made sure that the official report on the mummies, written by himself, was signed not only by his colleague Lacépède but also by Lamarck himself; and, a month after the report was published, he was finally appointed to a vacant professorship in the Jardin. He even managed to persuade his colleagues to change his title from professor of animal anatomy to professor of comparative anatomy. It was a small change but an important one.

Lamarck stayed quiet at first, telling his daughters and students in private that Cuvier's victory was a hollow one. Of course, he declared to a larger audience in his introductory discourse of 1803, the birds and cats had not transformed their shape. The climate of Egypt had remained stable for the last three thousand years, so they had had no need for change. It took much longer scales of time than that to effect biological change. Three thousand years was nothing in the age of the earth, Lamarck declared.

In 1803 all three men returned to their books, collections, research projects, and lectures. Geoffroy, now curator of mammals and bird collections and determined to keep out of trouble in order to further his career, began to prepare a catalog of the collection. It would be his first book, worthy and uncontroversial. Yet the whole classifying project still remained a contradiction for him. He was a philosopher, not a classifier, he complained; "true science should be sought on a broader and higher plane." He struggled. He fell ill again. Then he abandoned the whole project. Publishing papers about the new species he had found gave him pleasure, but he repeatedly failed to gain a seat at the National Institute of the Sciences and the Arts, the prestigious learned society founded in 1795 to group together the various academies. A new vacancy came up in 1803; Geoffroy, now engaged to be married, begged Cuvier to intervene on his behalf, but the position went to a physician of no great status who had higher-ranking and more influential friends.

Geoffroy married the daughter of an eminent Parisian politician in 1804; his wife gave birth to a son a year later, Isidore, who would become an eminent and important zoologist in his own right. In the same year, after several rejections from other, younger women, Cuvier married a widow whose first husband had been guillotined in 1794. She and her four children now moved into Cuvier's house in the Jardin des Plantes, where his first child, a son, christened Georges, was born in 1804. The baby lived only a few weeks. The bereaved Cuvier plunged back into his brilliant magnum opus, *Lessons on Comparative Anatomy,* a detailed five-volume study of the four classes of vertebrates and an explanation of his own radical methods of anatomical investigation. He also began to sketch out a new series of lectures on geology.

In the spring of 1805, before an audience of society people drawn to the public lectures delivered in the Royal Athenaeum of Paris by talk of giant fossil animals and by his theatrical and flamboyant lecturing style, which, people said, he modeled on the performances of the great Parisian actor François-Joseph Talma, Cuvier spoke about the new discoveries in geology. He described earth time as a sequence of epochs interspersed with revolutions or catastrophes and claimed that human history had begun only relatively recently, after the last such revolution. The epochs were, he added, remarkably close to the descriptions of the seven days of creation described in Genesis.

A twenty-six-year-old Italian nobleman, Giuseppe Marzari-Pencati, taking notes in the auditorium that day, was disappointed to see Cuvier, who was well known for his religious skepticism, making what seemed to him a show of piety; the professor was, he speculated in a letter written to a friend in Geneva, surely "eyeing a Cardinal's hat." The pope was in Paris, he pointed out, to witness Napoleon crown himself as self-styled emperor, an act intended to show the world that Catholicism was recognized once again as the majority religion in France. Marzari-Pencati was convinced that he was watching Cuvier compromise himself and the cause of science.

But the reality was more complicated. The revival of Catholic conservatism in France posed a problem for everyone in the Jardin, especially for geologists; books such as François-René Chateaubriand's recent bestselling *The Genius of Christianity* (*Génie du Christianisme*) declared science, and particularly geology, to be impious. Nature's laws were, Chateaubriand claimed, forever hidden from man; only the beauty and goodness of the Creator were proper objects of discovery. Cuvier, who had never before tried to square his scientific evidence with the biblical record, was now walking something of a rhetorical tightrope in these public lectures; he wanted to persuade his bourgeois audiences that geology need not be antireligious, and at the same time he wanted to use both the mummified ibises and the fossil record to refute what seemed to him to be the scientific heresy of transformism. The result required some considerable rhetorical gymnastics.

Lamarck was now in his sixties and, with a large family to support, burdened by debt. With no substantial salaried positions like Cuvier's, he was struggling to make ends meet while refusing to make compromises or concessions. Only seven auditors signed up for his lectures in 1805: two Italian brothers from Naples and their young professor of zoology and comparative anatomy, a German doctor, two French doctors, and a student. There had never been so few. But the audiences rose to respectable numbers the following year.

Geoffroy returned to the Egyptian fish skeletons in 1806 when the Commission of Sciences and Arts pressed him to complete his contributions to the monumental *Description of Egypt* (*Description de l'Égypte*), the series of books the Egyptian savants had been commissioned to write offering a comprehensive description of Egypt's ancient and modern history and its natural history. Laying out the preserved specimens of the Egyptian fish on the table in his rooms in the Museum of Natural History, Geoffroy must have remembered the excitement of the moment when, on a table in the Cairo palace, he had first seen their strangely human-shaped bronchioles and been so certain of his theory that there was a common body plan for all animals. With Cuvier's volumes nearby, he began looking for other homologies across the vertebrate-invertebrate border. To his amazement, he started to find structures common to vertebrates and invertebrates everywhere. The *furcula*, or wishbone, for instance, thought to be unique to birds, was, he discovered, also present in fish. All bodies did indeed seem to have a common abstract organizational plan.

Newly certain of his insights, Geoffroy set out a new methodology for comparative anatomical research centered on his idea of a common structural plan of organization. His papers met with approval from the Academy, from Cuvier, and from Napoleon himself. Cuvier continued to praise the facts of Geoffroy's work—the careful collection of weighty data—and to ignore the philosophical inferences that sometimes pulled dangerously toward transformism. Promotions followed. Geoffroy was appointed professor of zoology at the newly created Faculty of Sciences in the Imperial University in 1809 after Lamarck turned the position down for reasons of poor health.* Geoffroy's income increased; he gained an entourage of bright young students; his influence spread. Students set out to find common structures and patterns in the bodies of insects, spiders, and crustacea. They discovered how the mouthparts of caterpillars mirrored the nectar-sucking coiled tubules of butterflies. Even Cuvier's protégés were practicing philosophical anatomy. Then, in 1809, the year

*The Imperial University was not a physical place but rather a bureaucratic structure that organized all public instruction in France.

Geoffroy's twin daughters Stéphanie and Anaïs were born, his quest came to a standstill. Although he had found homologies for almost all bones across the vertebrate-invertebrate border, he had failed to find a homology for the bones of the gill cover in fish anywhere in the human body. He would keep on searching for a further eight years without success.

After 1802, through the years of Napoleon's rise and then spectacular fall at Waterloo in 1815 and through the coronation of a new French king placed on the throne of France by the Allies, as the priests and the police agents returned and the government became gradually more conservative under the Restoration, Cuvier became a medaled minister of state and eventually a baron in 1819, while Lamarck, working against time and ill health, doggedly continued to revise, clarify, and elaborate his theory, garnering as much proof as he could from the display cabinet drawers and vaults of the museum. In 1809 he published his most developed transformist work yet, *Zoological Philosophy.* In 1815, while Cuvier did his best to fend off the emissaries arriving in Paris to reclaim the natural history collections that Napoleon had stolen, Lamarck published the first of the seven volumes of *Natural History of Animals Without Backbones.* By the time the last volume appeared in 1822, he was completely blind.

Lamarck's eyesight, always poor, had deteriorated quickly since 1809, but in 1818, only a few months before his third wife died, his eyes failed completely, consigning him to ten years of utter dependency on three still unmarried and increasingly impoverished daughters, Rosalie, Cornélie, and Eugénie, then forty, twenty-six, and twenty-one years old respectively, and his partially deaf thirty-two-year-old artist son, Antoine. All four children still lived at home in the Maison de Buffon, on the grounds of the Jardin. His eldest son, André, a naval officer, had died in the Antilles in 1817; his third son, Charles René, had died before he reached adulthood; his youngest son, Aristide, mentally ill, was at the asylum at Charenton. Only his second son, Auguste, seemed to be forging an independent life for himself.

For those last ten years of Lamarck's life, while visitors lined up to see Baron Cuvier's spectacular new displays in the Gallery of Comparative Anatomy and celebrated the baron's achievements and genius, his daughter Rosalie struggled to run the Lamarck household and to find enough money to put food on the table and to pay the printers' and illustrators' bills. Cornélie sat upstairs with her blind father in an airless room battling to finish the final volumes of the *Natural History of Animals Without Backbones* as it was dictated to her, or received the visitors who came to view her father's treasured botanical collection that was now up for sale, or passed on instructions to the students who were now giving her father's lectures on his behalf. Then, in 1822, the year the Lamarck sisters finally sold their father's prized herbarium, Eugénie, the youngest, fell ill and died. She was only twenty-four. Auguste was now a successful engineer of bridges, but he was married with children of his own, and the salary of an engineer could not support two households. He often complained of his father's lack of familial responsibility. Lamarck had never been able to manage money or play the political games necessary to secure his power and influence in the hierarchy of the Jardin. Both his children and his science suffered as a consequence.

When Lamarck died in 1829, his children had to ask the Academy of Sciences to lend them money to pay for their father's funeral. Lamarck was buried in a temporary grave in the cemetery of Montparnasse; his bones were later dug up and, with thousands of other skeletons that had been disinterred from the overcrowded cemeteries of the city, scattered into the Parisian catacombs. There was very little money to spare in the museum coffers because the operating costs of the menageries, the expenses of natural history expeditions, and the upkeep of the buildings, as well as the extension of Cuvier's Gallery of Comparative Anatomy, had all been substantial.

Lamarck's reputation would rest on what Baron Cuvier now had to say in his obituary. When he went to visit the grieving family to gather details of Lamarck's early life, Cuvier must have been shocked to see how shabby the house had become, how threadbare and impoverished everything looked. When he suggested to Cornélie that she come for a walk with him, she was so overcome by the fresh air in the winter garden that he thought she would faint. But she gladly recounted her father's story, his struggles, his heroic years in the army, and the tale of his last stand at Fissinghausen as a seventeen-year-old soldier during the Pomeranian War. Cuvier was touched to see how devoted his children were.

But in the spring of 1830, when he might have been composing Lamarck's obituary, Cuvier had little time for anything but politics. The increasingly unpopular King Charles X had introduced the reactionary ultra-Catholic Polignac government in the summer of 1829 in order to pass more draconian and proaristocratic laws. Outraged republican groups formed in opposition. The Chamber of Deputies met in March 1830 to lodge a protest against the new government, but in May the king dissolved the Chamber and declared new elections and the suspension of freedom of the press. Riots broke out in the workers' sections of Paris. Heavily armed insurgents took control of the city over three days, attacking what remained of the Royal Guard, chopping down trees and pulling up paving stones to erect four thousand barricades across city streets, and raising the revolutionary tricolor flag over more and more buildings—the Louvre, the Archbishop's Palace, the Palais de Justice, and finally the Hôtel de Ville. Eventually, the revolutionaries declared Louis-Philippe, Duc d'Orléans, to be the new king.

In the Jardin, Cuvier faced his own upheavals. He had introduced a new classification system for dividing up the animal kingdom: every organism, he declared, belonged to one of four branches—vertebra, articulata, molluska, and radiata—and there were absolutely no connections between them. But now an alarming scientific paper appeared among the piles of Paris newspapers that his stepdaughter and assistant, Sophie Duvaucel, ensured were on his desk every day: a paper by two young naturalists from Lyons, Pierre-Stanislas Meyraux and Laurencet.* The two men claimed to have discovered homologies between two of Cuvier's divisions of the natural world, the mollusks and the vertebrates. If you bent the vertebrate backward so that the nape of the neck was attached to the buttocks, they wrote, then it was evident that the internal organs were arranged just like those of the mollusks.

When Geoffroy delivered a glowing report of the paper at the next meeting of the Academy, in which he declared this discovery to be proof of his own theory, mocked Cuvier, and declared the end of the era of mere facts and the beginning of a new philosophical zoology, Cuvier rose to his feet and demanded a retraction. There was a great deal at stake. Lamarck's death had changed the dynamics of power in the Jardin. For ten years Cuvier had refused to allow Geoffroy and his disciples to provoke

* Laurencet's given name has been lost from the historical record.

him into public debate about transformism, but now there were too many people watching, too many disciples, too many journalists with pens poised, for Cuvier to continue maintaining a pointed silence. If he was to keep his power base in the Jardin, he knew he had to win this fight.

Over the following two months the two professors slugged it out in the debating chamber of the National Institute of the Sciences and the Arts. Ostensibly, this was a debate not about transformism, but about grand theory versus dry fact. The presence of journalists and the public fanned the flames of their disagreement into a public conflagration. Cuvier called Geoffroy a dreamer. Geoffroy insisted that all his ideas were grounded in fact. The crowds of visitors increased session by session. Journalists from all over Europe took notes in the auditorium. Cuvier attacked every part of Geoffroy's theory. Geoffroy defended and enlarged his evidence. They were both right in part; their argument turned on time scales. Finally Cuvier shifted the debate away from bones to heresy, accusing Geoffroy of contradicting religious truth, of being a virtual transformist and a supporter of the German school of philosophical natural history, the *Naturphilosophie.* The audiences cheered and clapped at each twist and turn of the increasingly theatrical debate. There was no resolution. Geoffroy, frustrated that theology had been brought into the contest alongside zoology, withdrew from the field, publishing the original report and the papers that made up the debate.

Journalists everywhere declared that the future of zoology and science itself hung in the balance in Paris. Students, writers, and intellectuals, including the novelists Honoré de Balzac, Gustave Flaubert, and George Sand, read the newspaper descriptions of the debates and declared Geoffroy to be fighting single-handedly against the dark forces of conservatism. In Germany, the esteemed Johann Wolfgang von Goethe was more astonished at the news of the debate than at the news of the fall of Charles X. "The volcano has come to an eruption," he told a visitor; "everything is in flames. Now the debates are finally out from behind closed doors."

Through the months of 1830, through the abdication of the king and the fall of the government and the crowning of a new king, through the street riots and the building of barricades, Geoffroy and Cuvier continued to attack each other in their lectures, in their published papers, and in their jibes at each other at the Institute. Geoffroy, conscious that the eyes of the world were on him and that he had the support of Goethe and was corresponding with eminent intellectuals including George Sand and Honoré de Balzac, became increasingly and openly transformist in his ideas, claiming, for instance, in October that the famous fossil crocodiles that had recently been found at Caen were the ancient ancestors of modern crocodiles. Cuvier took the floor to protest, but he was careful not to be drawn back into direct combat. He was, however, preparing a subtle counterattack. In his long overdue obituary for Lamarck and his final series of lectures he determined to see off these philosophical speculators for good.

In the spring of 1832, Cuvier took to the lecturing platform again for the first time in fifteen years to give a series of lectures on the history of science from the Egyptians to the present day. Only facts, he declared, stood the test of time. At the same time he drafted and redrafted his memorial to Lamarck. Some esteemed men of science, he wrote—meaning, of course, Lamarck—had unfortunately neglected to collect evidence. "They have mingled many fanciful conceptions [with real discoveries]; and, believing themselves able to outstrip both experience and calculation, they have laboriously constructed vast edifices on imaginary foundations, resembling the enchanted palaces of our old romances, which vanished into air on the destruction of the talisman to which they owed their birth." Using a cruel caricature of birds' legs elongating over time simply because the birds could not bear to get their bodies wet, he mocked Lamarck's transformist ideas as the fantasies of a dreamer; such notions might "amuse the imagination of a poet," he wrote; "a metaphysician may derive from it an entirely new series of systems; but it cannot for a moment bear the examination of anyone who has dissected a hand, a viscus, or even a feather." If Lamarck was a dreamer and a fantasist, he implied, he was also obsessive, blinkered, unused to contradiction, reclusive, a spendthrift, and a gambler who had failed to support his family and had died in poverty. All this, of course, was expressed in Cuvier's most deferential and elegant prose.

But Cuvier did not live long enough to deliver his obituary. In May 1832, he took to his bed and died a few days later. His savage obituary of Lamarck was read at the Institute in his absence six months later.

—

Through the first decades of the nineteenth century, radical young men carried transformist ideas all over Europe. Some had attended Lamarck's lectures, others had read his books in the museum library or seen them summarized in the pages of scientific journals. One such young man was the brilliant young soldier, botanist, and algae expert Jean-Baptiste Bory de Saint-Vincent, who first visited Lamarck in the Maison de Buffon in the Jardin des Plantes in 1803 to discuss botany and followed his work closely thereafter. Having served Napoleon and then been forced into exile after the emperor's fall following Waterloo, he had wandered through Europe for several years to evade the king's secret police. For two years he lived in the caves of St. Peter, at Maastricht, in Holland, where he wrote a book of natural history, *Voyage Underground* (*Voyage souterrain, ou Description du plateau de Saint-Pierre de Maestricht et de ses vastes cryptes*), which described the caves and their fossils, strata, and animals from a Lamarckian point of view. All through his years of exile he published books and papers in which he deplored Lamarck's treatment at the hands of Cuvier, and in 1822 he launched an encyclopedia, published eventually in seventeen volumes, to promote a materialist system of natural philosophy inspired by both Lamarck and Geoffroy.

In 1827, as Bory continued to write his encyclopedia entries from a Parisian prison, where he had been incarcerated not for heresy but for debt, a giraffe arrived in the Jardin menagerie sent as a gift to France from the pasha of Egypt and accompanied on its journey across France by the fifty-five-year-old Geoffroy Saint-Hilaire. Bory, like most Parisians desperate to see the strange animal with his own eyes, called in a few favors from his friends at the Jardin and persuaded the warden to allow him to climb onto the roof of the prison. At a prearranged time, the giraffe's Egyptian keeper walked the animal from the menagerie in the Jardin des Plantes to the top of the domed hill called the Labyrinth. For half an hour, across the rooftops of Paris, Bory watched the movements of the giraffe through his telescope as it reached for overhanging leaves. Just think, he might have exclaimed to the guard who sat with him on the prison roof, how many thousands upon thousands of years of striving and stretching it must have taken for that neck to come into being!

In London, throughout the 1820s and 1830s, young medical students discussed Lamarck's and Geoffroy's theories, transposing their transformist ideas into a radical politics of reform. In the taverns and back rooms of student lodgings around Smithfield and in the pages of the new radical medical journals, they called for an appropriation of Church property, for working class suffrage, universal education, the abolition of the House of Lords, and the end of privilege. Transformism, as it had been called in France, transmutation or the theory of descent as it was beginning to be called in Britain, had always been political. In London in the 1830s it had become urgently so.

In 1836, in the state of Kentucky in America, boys threw stones at the window of a retired and increasingly eccentric elderly professor whom they called the French madman. Inside, in impoverished squalor, among thousands of books and cases of fossils and bones, Constantine Rafinesque was struggling to finish his life's work, a poem called *The World, or Instability,* in which he proposed that the world had evolved through millions of years, species evolving one after another from simple early forms. If you had asked him, he would have told you, because he liked to talk, that he had been born in Constantinople to a German mother and French father, that his parents had been forced out of France by the French Revolution, that he had been raised and educated in Tuscany by his grandmother and an Italian natural philosopher tutor she employed to teach him, and that he had sailed to America in 1803 when he was in his early twenties to make his fortune, reading precious books and papers by Jean-Baptiste Lamarck, the botanist Antoine de Jussieu, and the Comte de Buffon as he traveled across state border after state border, thousands of miles on foot into mountains and swamps and creeks in search of new plants and animals, always aware that his ideas and manuscripts were incomprehensible to most non-Europeanized Americans and almost always heretical. He would tell you about the shipwreck he had survived when he returned to America from Sicily in 1815 in which he lost every one of his fifty boxes of belongings—his library and all his botanical and natural history collections, including six hundred thousand shells—and he would tell you that he had worked as a university professor, part-time tutor, inventor, curiosity dealer, and journalist in Philadelphia for the better part of his later years.

He would add that he had always been misunderstood, that in America, the country he called home, his botanizing was admired but his theories ridiculed. It was the great age of the fact, he would complain. The age of the visionary naturalist-philosophers like Buffon and Lamarck was over. He had tried arguing his transformist ideas in his book *The Flora of North America;* he had tried arguing them in specialist articles and in papers given at scientific societies, but no one would engage with him or take his ideas seriously. Poetry was the way to reach people, he had decided; in his poem he would challenge scientific orthodoxies, bring light to the darkness, forge the ideas of Swedenborg, Geoffroy Saint-Hilaire, Erasmus Darwin, and Lamarck into a unique synthesis, a vision of the world progressing from its smoky beginnings through great tracts of time and metamorphosing species.

In Kentucky, transformism, imported from Paris, gave Constantine Rafinesque a new way of understanding the American Declaration of Independence, a model for human striving, diversification, and self-improvement and a way of understanding his own sequence of metamorphoses:

> Versatility of talents and of professions, is not uncommon in America; but those which I have exhibited in these few pages, may appear to exceed belief: and yet it is a positive fact that in knowledge I have been a Botanist, Naturalist, Geologist, Geographer, Historian, Poet, Philosopher, Philologist, Economist, Philanthropist . . . by profession a Traveller, Merchant, Manufacturer, Collector, Improver, Professor, Teacher, Surveyor, Draftsman, Architect, Engineer, Pulmist [lung specialist], Author, Editor, Bookseller, Librarian, Secretary . . . and I hardly know myself what I may not become as yet.

As it had for many of the radicalized medical men of London in the 1830s, or for Bory on the run across Europe assuming ever new identities and disguises, transmutation had become not only a scientific theory about animal-human kinship and descent, but also a way by which these men might explain change—both how they had come to be and what they might yet become.

10

The Sponge Philosopher

EDINBURGH, 1826

October 1826—a beach at the harbor port of Leith, near Edinburgh, in Scotland. Now that most of the summer visitors had returned to the cities, the bathing machines had been lined up in rows at the far end of the beach and covered in tarpaulins to protect them from winter winds and rain. Still work continued. Barefooted fisherwomen and their children levered mussels off the black rocks to bait their lines; in the fishermen's warehouses, men and women packed fish or salt or stacked boxes onto the backs of wagons; in the six cone-shaped brick kilns of the Leith Glasswork Company, glassworkers poured molten glass into bottle molds. Whale blubber that was being boiled to make soap and candles in the boiling houses fouled and thickened the air. Beyond the Martello Tower, built to defend the town against French invasion during the Napoleonic Wars, the masts of moored whaling and fishing boats looked from a distance like a forest against the great expanse of the Firth of Forth, which stretched out to the horizon, gray and steely in the winter light.

Charles Darwin, seventeen years old, his boots stained with salt, his coat pulled tight against the wind, stood among the black rocks on the sands. The red leather-bound diary he wrote in was full of notes on sea mice, cuttlefish, sea anemones, starfish, and the migration patterns of birds. Ever since he had enrolled as a medical student at the university a year earlier, he had skipped lectures to walk down the long Leith Road to the beach here every few days, first with his brother and now alone, to poke about in the rock pools along the shore, carrying sea creatures in glass jars back up the long road to Edinburgh to his rooms to put them under his microscope.

There were two other men on the beach at Leith carrying glass jars and dissecting equipment on this particular day. One of them was Robert Grant, a local doctor in his midthirties; the other was Grant's protégé, John Coldstream, a nineteen-year-old medical student at the university. They were collecting sea sponges in buckets. Something—mutual recognition and interest, curiosity, good manners—brought these three men together. They introduced themselves, shaking hands, voices raised against the biting wind. It took Grant only a few minutes to discover that this well-dressed young man was the grandson of the great Erasmus Darwin, author of *Zoonomia*. He was intrigued and impressed, and he explained to Darwin that he was in the middle of experiments, trying to determine once and for all whether sea sponges were plants or animals. His work was indebted, he said, to both Erasmus Darwin and Aristotle. Since the Greek philosopher had first examined sea sponges in the lagoons of the Greek islands, he added, no one had come to understand them any better. He showed young Darwin his dissecting equipment and microscope, explaining how important it was to be able to dissect and observe outside, right on the edge of the water, while the sponge was still alive. There were so many important questions to

Robert Edmond Grant aged fifty-nine.
Thomas Herbert Maguire (1852)

be asked, so many answers to be found, he declared. The men arranged to meet again.

So began Charles Darwin's friendship with one of the most remarkable men in Edinburgh; it was a relationship that would transform the way Darwin thought about the natural world. He was lucky. There was no one else like Grant in Edinburgh at the time, no one who had read so much, investigated so much, or thought so much, or so freely, about species, origins, and time. Cynical, clever, and reclusive, Robert Grant was a radical Lamarckian; he was absolutely certain that species had evolved from primitive aquatic organisms. The timing of their encounter could not have been better. Grant was coming to the end of an intense period of investigation, an attempt to discover the origins of life, a journey that had taken him from Edinburgh to Paris, across Europe and back to a boarded-up house a mile along the Edinburgh shore at the village of Prestonpans, where he had established a secret laboratory. By the time he met Darwin, he was turning his notes into articles and publishing them. Darwin claimed in his *Autobiography* that he was astonished by Grant's Lamarckian conclusions. But if he was astonished, he was also wildly curious about them.

Within weeks, Grant had enrolled Darwin in the student-run Plinian Natural History Society, a group of students who met to discuss questions related to the new biological sciences, and recruited him as an assistant for his own research. Two years later they would fall out and lose touch with each other, and Grant would disappear into eventual obscurity in London.

As they walked along the shores of the Firth of Forth over the next few weeks and months and as Grant taught him dissection techniques, Darwin asked endless questions about his new friend's work and travels. His great European journey, Grant told Darwin, had begun when he had read Erasmus Darwin's *Zoonomia*. From a large local family, he had enrolled as a medical student at Edinburgh. He had attended the lectures of Robert Jameson, Regius professor of natural history, who, fascinated by the history of the earth, taught some of his classes in rooms in the Royal Museum of the university, rooms lined with the enormous natural history collections he had inherited from his predecessors and since expanded. Jameson took his students to the hills around Edinburgh to speculate upon their formation and introduced them to new Continental ideas about comparative anatomy and earth history. He urged them to think, to debate, to experiment. Men of science all around the world, he assured them, were on the verge of great new discoveries.

Then, as Grant prepared for his dissertation in 1813 on the circulation of blood in the human fetus, he had fallen upon *Zoonomia, or The Laws of Organic Life* in the university library. He described to Darwin his excitement at discovering in those pages the chapter titled "Generation," in which Erasmus Darwin had argued that the shared patterns of structure that Grant had seen in animal bodies were the result of common parentage and that all life had started out as aquatic filaments in a universal sea. It had revolutionized the way Grant understood the natural world. But when he had enthused about the brilliance of *Zoonomia* to fellow students and teachers, he discovered that most of them had not read it, and those who had dismissed it as mere speculation. The author was a poet and an inventor, they said, not a trained man of science. Grant took detailed notes from *Zoonomia* nonetheless and kept returning to those notes years later.

When Grant graduated as a doctor in 1814 and came into a small inheritance, Jameson encouraged his promising pupil to go abroad—Germany and France were the places to go with the questions that preoccupied him about anatomical patterns. Paris was the center of comparative anatomy, and in 1815, the year after Grant graduated, the Napoleonic Wars had come to an end when the Allies had defeated and captured Napoleon at Waterloo, making the great French capital accessible again. By way of preparation for his journey, Grant spent months improving his French, German, and Greek. He wanted to be able to read Aristotle's *The History of Animals* in its original language.

Grant was twenty-two when he arrived in Paris at the beginning of the winter of 1815–16. The war was over; the Allies occupied the city, and Napoleon had been banished to the island of St. Helena. The city, with its wide boulevards and green parks, was full of men in uniform—Prussian soldiers, boastful and ebullient, filled the bars and dance halls; British soldiers, more sober and disciplined, had established camps down the Champs-Élysées; even Scottish soldiers in kilts were to be seen on street corners buying lemonade from the street sellers. Grant was one of only a few British medical students in the city. They shared lodgings and met in cafés, comparing notes, gossiping about the charismatic French professors or the surgeons who ran the wards in which they worked, improving their French, absorbing and questioning new ideas.

Reforms to medical practice, research, and hospital management in the wake of the French Revolution had made Paris the heart of the new medicine, characterized by systematic close observation of human anatomy. The vast Parisian hospitals, no longer in the hands of the Church but under the control and management of eminent anatomists, physicians, and surgeons, had been turned into centers of research and teaching; a new generation of young doctors was being trained to look more closely inside the human body to determine the structure and patterns of disease. Thirty thousand people a year were treated in the hospitals of Paris, and of those who died, four-fifths were dissected. In 1815, soldiers from Waterloo were still pouring into the city hospitals, many of them with limbs blackened by gangrene. That winter, all medical students had to learn how to amputate limbs without anesthetic. It was grueling work. The new generation of medical students was being trained to dissect, describe, and record, and to trust the evidence of their eyes above the authority of old medical textbooks.

Grant spent his first winter studying comparative anatomy in the Museum of Natural History at the Jardin des Plantes, attending lectures, reading everything he could in the enormous library, including Aristotle in Greek, and studying the great collections of bones and dried and bottled specimens. Letters of recommendation from Professor Jameson gained him access to the salon of the robed and magisterial Professor Cuvier. By contrast, Lamarck, professor of insects, worms, and microscopic animals, in his seventies in 1815 and completing his seven-volume *Natural History of Animals Without Backbones,* seemed a frail and shadowy figure. The old man's controversial transformist ideas, mocked by the great Cuvier, were, Grant realized, remarkably similar to those of Erasmus Darwin. But the number of students attending his lectures was declining fast. Grant did not attend. Instead he read everything he could by the three men, in French and in translation.

Lamarck's work, including the first volume of his *Natural History of Animals Without Backbones,* which had been published that year, made Grant think again about marine invertebrates. He returned to the extraordinary sea-sponge specimens and fossils arranged behind glass in the Museum of Natural History, admiring their strange, fanned, branched, and sometimes mysteriously bowl-like forms covered in tiny holes. He read Aristotle on sponges, then work by eighteenth- and nineteenth-century naturalists such as Zeller, Lamouroux, Gmelin, Peyssonel, Ellis, Montagu, Pallas, Guettard, Jussieu, Blumenbach, Lichtenstein, Schweigger, and Marsigli, translating and taking copious notes. "All facts known about the sponge," he concluded with some amazement, "have remained where Aristotle left them, or rather, in this branch of study, mankind has gone backward ever since his time." As he also remarked, "It is pleasing to observe that our forefathers, at such a remote period, were occupied, like ourselves, among the rocks of the sea shore, experimenting on this humble and apparently insignificant being."

If Lamarck and Erasmus Darwin had persuaded Grant that marine invertebrates were the key to understanding the origins of life, Cuvier and Geoffroy convinced him of the importance of dissection, of mapping and analyzing internal as well as external body structures in order to understand common patterns between apparently different organisms and life-forms. Trained in rigorous scientific experiment and observation, inspired both by Aristotle's work and by Parisian speculation about the nature and origins of life, and in possession of one of the finest microscopes he could buy, Grant determined to solve the problem of the sea sponge and use it to build a theory of the origin of species.

To a disciple of Lamarck and Geoffroy like Grant, sponges were particularly intriguing because the surviving fossil forms were so similar to living species; they had changed little since they had first evolved as multicellular organisms. If they were ancestral forms, as Lamarck's work implied, they might yield clues to the mutation of species in one of its earliest stages; yet, as Grant had discovered, they were so little understood. This was partly because, as Aristotle had found, they were difficult to observe and almost impossible to dissect, given that once they were taken out of the water they quickly lost their vibrant colors and died. No one in Grant's time seemed able to agree on whether they were animals or plants. If animal life was to be defined by sensitivity, the ability to move and the existence of digestion, then the sea sponge appeared to have few discernible signs of being an animal. It was an enigma as unreadable as the Egyptian hieroglyphs.

Grant reread Aristotle's passages about sponges, copying out the original Greek words the philosopher used, trying to work out more exact and accurate translations so as to understand more fully what Aristotle was doing and thinking. He determined to bring Aristotle, the great collector and observer of marine invertebrates, together with Lamarck and Erasmus Darwin, the great speculators on origins. He would do this by conducting his own extensive experiments until he knew everything there was to know about sponges. Once he had done that, he might be able to use the sponges as a way of opening up the history of species mutability.

With a clear set of philosophical problems in mind, Grant set off for southern Europe to collect new species and to visit as many European marine-invertebrate specialists, libraries, and natural history collections as he could before his inheritance ran out. These questions took him to Rome, Florence, Pisa, Padua, and Pavia, not for paintings or ruins but for scientific books and articles in French, German, and Italian and for beaches and rock pools. He collected and dissected Mediterranean sea sponges on the shores of Leghorn, Genoa, and Venice. He spent eighteen months in Germany, then traveled through Prague, Munich, Switzerland, back down to the south of France, and to the University of Montpellier. In 1820 he returned to the natural history and anatomy collections and libraries in Paris and then London, making last checks and rereading manuscripts. He arrived back in Edinburgh at the end of the year with crates of notebooks, copies of papers, and dried and bottled specimens. He kept his work to himself, however, anxious to avoid being associated with any form of materialist science. Like Erasmus Darwin before him, he knew he would have to establish and maintain a practice as a doctor in a small town, where reputation was everything.

By the time Grant returned to Edinburgh, he knew his time and his inheritance were running out. Preparing himself for earning a living as a doctor and as a part-time lecturer in comparative anatomy, he enrolled in Dr. Barclay's private anatomy school in 1821, training in anatomy during the day and conducting his sea sponge dissections and experiments in the anatomy school dissection rooms at night. Eventually an acquaintance offered him the use of a house in Prestonpans through the winter, a village on the Firth of Forth a few miles to the east of Edinburgh. There Grant set up his own secret laboratory.

The men and women of Prestonpans still worked mainly in the salt industry, boiling water dredged from the Firth in enormous pans. But by the early nineteenth century the village had also become a picturesque tourist attraction, offering newly fashionable bathing machines to the summer tourists who rented the large seafront houses. Grant had found Prestonpans Bay to be especially rich in sea sponges. Walford House, his new winter residence, had a high-walled garden with direct access through a gate to the beach. He could collect sponges on the shore early in the morning or after a night's storm and transfer them to shallow seawater basins in his study within a few minutes.

He began by trying to discover the purpose of the holes that covered the sea sponge's body. Were these, he speculated, the elementary forms of a digestive system? Some experts claimed that the holes were made by parasitic creatures burrowing into the soft flesh, others that sponges were vegetables and that the holes were like the pores on the surface of leaves, through which the vegetable absorbed water. Grant knew that the only way to settle this matter once and for all was to keep watch—to observe the holes of the sponge night and day under a microscope, just as Charles Bonnet had done with the aphid.

Discovering that the intense light of a single candle enabled him to view his specimens more effectively than daylight, Grant soon developed a habit of entirely nocturnal research at Prestonpans. He made his first major breakthrough some weeks after he began his watch, alone in the middle of the night, his eye pressed to the microscope, with only the sound of the waves and an occasional lone gull to break the silence. "On moving the watch-glass, so as to bring one of the apertures on the side of the sponge fully into view, I beheld, for the first time," he wrote,

> the splendid spectacle of this living fountain vomiting forth, from a circular cavity, an impetuous torrent of liquid matter, and hurl-

> ing along, in rapid succession, opaque masses, which it strewed everywhere around. The beauty and novelty of such a scene in the animal kingdom, long arrested my attention, but, after twenty-five minutes of constant observation, I was obliged to withdraw my eye from fatigue, without having seen the torrent for one instance change its direction, or diminish, in the slightest degree, the rapidity of its course. I continued to watch the same orifice, at short intervals, for five hours . . . but still the stream rolled on with constant and equal velocity.

Grant concluded that the hole could be described as a "fecal orifice." If this was a primitive anus, there had to be a digestive system of some kind. Next he had to find a mouth. For several nights he experimented with blocking the anal orifices of the sponges with a variety of domestic objects to determine the strength of the "fecal current"; he used pieces of chalk, cork, dry paper, soft bread, unburned black coal, almost anything at hand. He recorded all the experiments in detail in his notebooks. Only a drop of mercury, he discovered, was heavy enough to block the stream. So Grant concluded that while some of the holes on the body of the sea sponge were fecal orifices, the others were "pores" used for ingesting food, like mouths. The holes that led into the labyrinthine passages of the sponge were, he observed, a constant conduit of liquids: food passing in and opaque fecal liquids passing out. And Aristotle had known this, too, he realized when he consulted the manuscript. The Greek words Aristotle used for these holes meant both "pore" and "orifice."

The sponge had met the first of the three criteria for animalhood: it seemed to have a form of digestion. Next Grant had to determine if it had independent movement—animals move independently, plants do not. In his long nocturnal vigils in late autumn, Grant watched a sea sponge excrete eggs or ova through its holes, and under his microscope he saw to his amazement that the eggs were covered in small hairs, which they used to propel themselves along, away from the parent sponge. The parent sponges were inert, but the young had independent movement. Grant was sure there was a good reason for this: the propagation of the species depended upon the spontaneous motion of the young, their ability to reach new breeding grounds. The enigmatic and peculiar sponge was perfectly adapted, then, to survival in the deep sea—so perfectly that it had not needed to evolve any further, given that its aquatic conditions had remained relatively stable, as had been the case for Lamarck's mummified ibises. "This animal," Grant wrote, " . . . seems eminently calculated for an extensive distribution, from the remarkable simplicity of its structure, and the few elements required for its subsistence."

Next Grant had to prove that the sponge was sensitive, another important area of dispute among zoologists since Aristotle's time. Over the nights of three winters he failed to find any evidence of sensitivity. "I have plunged portions of the branched and sessile sponges alive into acids, alcohol and ammonia," he wrote in frustration, "in order to excite their bodies to some kind of visible contractile motions, but have not produced, by these powerful agents, any more effect upon the living specimens, than upon those which had long been dead." The sponge had failed the final criteria of animalhood, but for Grant that was a good result. The lack of sensitivity, he believed, proved it to be a transitional organism, just as he had hoped it might be. It shared both plant and animal characteristics.

During these experiments Grant recruited several local boys to help him collect sea sponges. One of them was John Coldstream, a young naturalist and medical student. Like Grant, Coldstream had been born and raised in Leith. But unlike Grant, Coldstream was a devout Christian, a member of the Leith Juvenile Bible Society. He had been collecting sea creatures on these shores since he was child. Grant, obsessive, clever, and utterly convinced of the transmutation of species, turned Coldstream's head. Beginning in 1823, when he came into Grant's orbit, Coldstream's diaries record a sudden conflict of beliefs. He was seventeen when he met the charismatic Leith doctor, and his journals from 1823 onward record a painful story of self-blame, torment, and physical self-loathing. On his eighteenth birthday, March 19, 1824, he wrote: "My praise is altogether an unclean thing; my glorifying of the Lord is filthiness before Him. From the dust do I cry unto thee, O God! Hear me, hear me. I earnestly beseech Thee to purify my heart. . . . I am at a time of life when the amusements of this little world lead me away into temptation; now, heavenly Father, point out to me in what measure I should best enjoy these, that all my conduct may be to Thy glory. Oh, were I prepared, how I would fly from the attractions of the flesh."

Yet Coldstream did not leave the Plinian Society or stop assisting Grant in his work. He continued to struggle with his religious doubts and with a certain mysterious sense of disgust about his body during these years. He became one of the Society's presidents in 1824–25 and one of the doctor's closest companions, collecting sea sponges for him and for the Museum of Natural History at the university and was always on the lookout for other young naturalists who might be recruited to the Society or to help Grant. In return, Grant trained his assistant in dissection techniques and encouraged him to publish natural history papers in the *Edinburgh Philosophical Journal,* articles on the springs of Ben Nevis, on the saltiness and transparency of the water of the ocean, on hoarfrost, on the aurora borealis, which he saw out in the Firth of Forth on his nineteenth birthday in 1825, and on the sea sponges and the zoophytes of the Firth. Though Coldstream struggled to maintain his faith, he was beginning to ask questions he did not want to ask.

In the spring of 1825, Grant was ready to unveil the results of his five-year investigation and moved back to Edinburgh. First he decided to test some of the basic premises of his philosophy on the members of the Wernerian Natural History Society, run by Professor Jameson. He turned up to give his very first paper to the Society on April 2, 1825, carrying buckets of dead cuttlefish recently dredged from the Firth of Forth. Cuvier had argued that no invertebrate had a pancreas; Grant opened up the stomach of one of his Firth cuttlefish and revealed to his audience that what had been thought to be the creature's ovarium was in fact a pancreas. He repeated the dissection several times until everyone agreed with his conclusions. There could be no doubt, he claimed, that the pancreas was to be found much lower in the scale of animals than had previously been believed. A cuttlefish had a pancreas just like a human; there were common structures shared by both vertebrates and invertebrates.

A few weeks later he was back with buckets of gastropods and sea slugs. These, he showed his audience, had a pancreas too. Grant was using cuttlefish and sea slugs to bring Cuvier's influential map of nature, fixed and divided into four absolute branches, into question. Cuvier had pronounced there to be no common structures shared between organisms belonging to the four branches; Grant, paper by paper, stage by stage, was proving Cuvier wrong.

By 1826, in the last of the carefully sequenced and stage-managed papers, Grant was claiming that his evidence proved that the sea sponge was so close to the boundary between the animal and vegetable kingdoms as to be virtually on it. By comparing sponge ova and those of other simple living organisms with the ova of the algae, he argued that there was a common monadic base for plants and animals. Somewhere in their ancestry there had been a meeting point, and the sponge represented that point.

And it was at this moment, just as Grant was in the middle of publishing his most controversial of ideas, that the grandson of Erasmus Darwin turned up on the beach at Leith.

Over the winter of 1826–27, Grant, Darwin, and Coldstream walked the shores of the Firth of Forth carrying microscopes and collection jars, poking about under rocks, lowering nets carefully into rock pools, collecting sponges, sea anemones, and sea squirts, and debating the philosophical implications of Grant's recent findings. In the fishing villages, they stopped to talk with fishermen and women selling fish or sitting mending nets on the steps of the red-tiled two-story houses. The fishermen knew Grant well; they kept the discarded contents of their nets for him to look through. Every few weeks, the three naturalists joined the fishermen on the dredging boats at dawn, sailing out from the harbor at Newhaven, the center of the oyster trade in the Firth and only a mile to the west of Leith, sometimes traveling as far as the coast of Fifeshire or out to the islands of May and Inchkeith looking for new marine organisms.

On Tuesday evenings, Grant accompanied Darwin and Coldstream to the Plinian Natural History Society meetings held in an underground room at the university. Many of the members were young medical students, some, like Darwin, still very young.* Grant was the only older man in the group. Members and invited guests gave short papers every week describing natural history expeditions or presented reports on new methods of obtaining bromine from soap boilers' waste, or on the capture of whales on the coast of the Shetlands, or on the oceanic and atmospheric currents. Occasionally students presented philosophically controversial ideas. When the seventeen-year-old William Greg gave a paper that would prove that "the lower animals possess every faculty & propensity of the human mind," the details of his paper were later struck from the record. Grant, accustomed to Continental freedom of

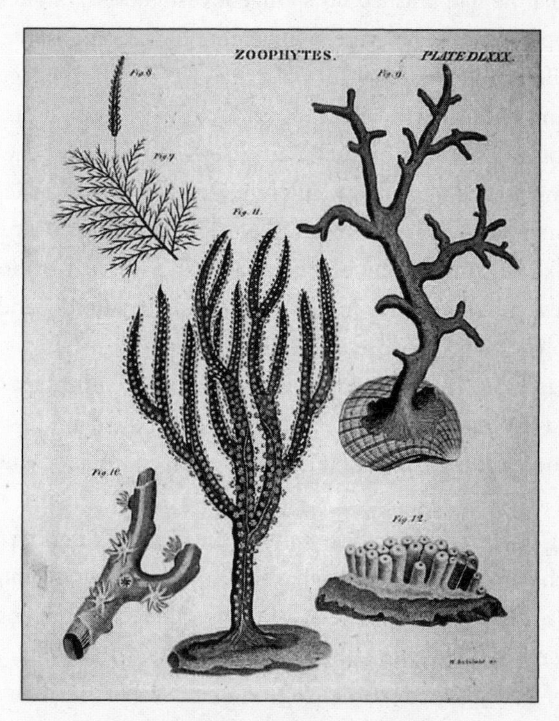

Zoophytes illustrated in an engraved plate from the *Encyclopaedia Britannica,* ca. 1813.

* Darwin was seventeen; William Kay from Liverpool and George Fife from Newcastle were nineteen; John Coldstream from Leith and William Ainsworth from Lancashire were twenty; William Browne from Stirling was twenty-two.

debate and speculation, encouraged the younger members to read widely and to challenge orthodoxies. Enlightenment, he told Darwin and Coldstream, depended on it.

In 1826, Grant announced his discovery that the eggs of the sea sponge were free-swimming, that they propelled themselves along by the vibration of small hairs, or cilia, that covered their bodies. This was a crucial discovery, an important part of the process of solving the riddle of the sea sponge and determining its place in nature. The sponge may not have movement of any kind, he wrote, but its eggs could move. He ended his essay with the words: "How far this law is general with zoophytes must be determined by future observations."

Now that he had finished his sponge investigations, Grant turned to other zoophytes, other marine animals that looked like plants, to see if he could determine whether they also had free-swimming eggs. Grant, Coldstream, and Darwin began to work closely together, dissecting in Grant's rooms and even on the beach itself, searching for the presence of swimming eggs in as many different apparently immobile sea creatures as they could. Darwin began to work on the *Flustra,* a pale brown seaweedlike organism known colloquially as a sea mat or hornwrack, which grew in colonies made up of hundreds of interdependent, connected polyps on rocks close to the shoreline.

On March 19, 1827, Darwin made a small but important discovery. He rushed to Grant's house in Prestonpans with the news as soon as the dredger moored in the harbor. He later described the moment:

> Having procured some specimens of the Flustra Carbocea (Lam.) from the dredge boats at Newhaven; I soon perceived without the aid of a microscope small yellow bodies studded in different directions on it.—They were of an oval shape & of the colour of the yolk of an egg, each occupying one cell. Whilst in their cells I could perceive no motion; but when left at rest in a watch glass, or shaken they glided to & fro with so rapid a motion, as at some distance to be distinctly visible to the naked eye. . . . That such ova had organs of motion does not appear to have been hitherto observed either by Lamarck Cuvier Lamouroux or any other author:—This fact although at first it may appear of little impor-

> tance yet by adducing one more to the already numerous examples will tend to generalize the law that the ova of all Zoophytes enjoy spontaneous motion.

Now that he knew how to look, what to look for, and how to wait and watch, Darwin quickly made a second discovery also described in his notebook:

> One frequently finds sticking to oyster and other old shells, small black globular bodies which the fishermen call great Pepper-corns. These have hitherto been always mistaken for the young Fucus Lorius . . . to which it bears a great resemblance. . . . But on examining some others I found that this fluid, acquiring by degrees a vermicular shape, when matured was the young Pontobdella Muricata (Lam.) which were in every respect perfect & in motion.

Darwin found swimming eggs in almost all the zoophytes he dissected, but the more evidence he collected, the more Grant felt the originality of his own work to be compromised. Grant announced the discovery of the free-swimming ova of the *Flustra* to the prestigious Wernerian Society on March 24, 1827. He also announced that he had found cilia on the young of other zoophytes and that he had discovered the mode of reproduction of the sea leech, *Pontobdella muricata.* Darwin gave a paper on the ova of the *Flustra* to the Plinian Society three days later, his very first scientific paper. He was eighteen.

The relations between Darwin and Grant cooled. This was Darwin's first experience of scientific territorialism, and it upset him. A note written by one of Darwin's daughters and allegedly found in a bundle of papers in 1947 (and since lost again) confirms this. Henrietta wrote:

> I then made him repeat what he had told me before, namely his first introduction to the jealousy of scientific men. When he was at Edinburgh he found out that the spermatozoa (?) / ova (?) of (things that grow on sea weed) / Flustra move. He rushed instantly to Prof. Grant who was working on the subject to tell him, thinking he wd be delighted by so curious a fact. But was con-

founded on being told that it was very unfair of him to work at Prof. G's subject and in fact that he shd take it ill if my Father published it. This made a very deep impression on my Father and he has always expressed the strongest contempt for all such little feelings—unworthy of searchers after truth.

These were also, perhaps thankfully, Darwin's last weeks in Edinburgh. Grant, too, was destined to leave Edinburgh that summer. He had been recommended for a chair at the new London University, founded to open up a university education to some of those who were denied it by the traditional universities (Dissenters, Catholics, and Jews were all excluded from Oxford and Cambridge) and to bring about reform in the nation's professions and institutions. Darwin, on the other hand, was destined for Cambridge. His father had decided that if he was not going to qualify as a doctor, he should try for the Church. His uncle Josiah Wedgwood (in one of the many Wedgwood-Darwin alliances, in 1796 Erasmus Darwin's son, Robert Waring Darwin, had married the first Josiah Wedgwood's daughter, Susannah—this Josiah was Susannah's brother) reminded his rather dejected nephew that a clergyman might also be a naturalist.

John Coldstream had decided to go to Paris to continue his medical studies and to follow Grant's example; in fact, several graduating members of the Plinian Society set off for Paris in the summer of 1827, and Grant joined them there later in the summer on his annual visit to the Paris museums and libraries. Darwin also traveled to Paris in June and July of that summer. His uncle Josiah was journeying to Geneva to pick up his daughters, Fanny and Emma Wedgwood;* he offered to take Charles as far as Paris. While his uncle traveled on to Geneva, Charles was left in Paris to occupy himself for several weeks. Now that Grant had cut him adrift, he wandered on his own around the displays of the Museum of Natural History in the Jardin des Plantes thinking through a fascinating set of questions.

Darwin found Coldstream unhappy, even in that first summer of his Paris studies. Parisian science, Coldstream admitted, unsettled him even

*In another Wedgwood-Darwin alliance, Emma Wedgwood would later become Darwin's wife.

more than Grant's ideas had. William Mackenzie of Mission House in Passy, in whom Coldstream confided, recorded that "he was troubled with doubts arising from certain Materialist views, which are alas! too common among medical students. He spoke to me of his doubts, and manifested anxiety on the subject of religion." Soon Coldstream was in a Swiss sanatorium recovering from an emotional collapse. Convalescing, he made decisions. There would be no further Continental study, no further materialist temptations. He would return to Leith in 1828 and work simply as a local doctor, healing the sick. He would shun the company of heretical naturalists such as Grant and other members of the Plinian Society. "In our day the majority of naturalists, I fear are infidels," he wrote in 1829.

A broken man, Coldstream moved back into his parents' house in Leith, still haunted by festering bodies and moral corruption. "A fair exterior covers a perfect sink of iniquity," he wrote of himself on January 1, 1830, blaming his mental weakness on lack of discipline and the temptations of the flesh. "My present condition of mind," he wrote, "is much inferior in strength and solidity to what it might have been had I not given loose reins to my lustful appetites. I have been ruined and enervated by a life of effeminacy and slothful indulgence."

From Cambridge, Darwin wrote to Coldstream expressing sympathy for his friend's illness and grief that he had given up marine zoology, arguing that "no pursuit is more becoming for a physician than Nat. Hist." Coldstream agreed, but added cheerfully that he had decided to dedicate himself to "useful knowledge." He had also resolved, he wrote, to give up dissecting live creatures—it seemed to be against nature. Nonetheless he asked Darwin to pass on his regards to Dr. Grant, and he ended: "Be so good as to write me again soon, and tell me something of the present state of Natural History in Cambridge. Have you had any opportunity of studying marine Zoology since you left this?"

Darwin had done no marine zoology since leaving Edinburgh. In Cambridge, eighty miles from the sea, his attention had turned to beetles. The Reverend John Henslow, his new botany teacher and walking companion, was already remarking, "What a fellow that D. is for asking questions." Passing his exams in 1831, Darwin was now ready to take on a parish and settle down like Henslow. Instead, his imagination full of the travel tales of Alexander von Humbolt in South America, his thoughts turned to tropical expeditions. He wrote to his cousin William Darwin Fox in 1831: "It strikes me, that all our knowledge about the structure of our Earth is very much like what an old hen wd know of the hundred-acre field in a corner of which she is scratching."

The answer was to leave the corner of the field and scratch elsewhere; but what would his father have to say to this? In September 1831, the opportunity presented itself in the form of an invitation to accept the position of ship's naturalist on a mapmaking voyage of coastal South America on board HMS *Beagle.* His father, exasperated by his son's apparent flightiness, refused to agree to the voyage, but when his uncle intervened, arguing that "Natural History . . . is very suitable to a Clergyman," eventually Darwin's father agreed. Darwin wrote one last time to Coldstream, asking for advice and information about deep sea dredging and techniques for meteorological observations. In particular, he asked his old friend to draw him a diagram of an oyster dredger in such a way that he could have one designed for the *Beagle* voyage. Coldstream drew careful instructions and diagrams and urged Darwin to contact Robert Grant in London for further advice.

Darwin and Grant met in London in late 1831 in the weeks before the *Beagle* sailed. What must Grant have felt as he drank tea with his old protégé? Grant had once been the great traveler, crossing European mountain ranges and walking Mediterranean coastlines in the footsteps of Aristotle; now, without the luxury of a private income like Darwin's, and living a long way from the sea, he struggled to make ends meet, his days entirely taken up with writing lectures, marking essays, and giving tutorials. Darwin came to him, wealthy, twenty-two years old, asking for precise instructions on collecting marine creatures in exotic tropical seas.

Grant's London University position had not turned out to be what he had expected. Though the university had given him freedom to teach what and how he liked, that freedom came at a cost. The university was badly organized and in poor financial shape. On his arrival, Grant had to put together a museum collection for his students single-handedly, prepare three lecture courses, give two hundred lectures a year, and even buy his own dissection materials. Comparative anatomy and zoology were not compulsory subjects in the curriculum, so he never had enough fee-paying students to make up a steady salary, nor could he practice medicine to supplement his small teaching income because the medical corporations in London stipulated that as a Scotsman, he would have to take a new set of exams in order to practice, and Grant absolutely refused to do that. He was an angry, tired, and disappointed man.

There were, however, compensations. Grant had found allies among the London medical reformers when he arrived in the city, radicals who were campaigning against the nepotism and the monopoly of the Anglican landed classes over the medical profession. They were a famously outspoken group, making their demands for reform in Thomas Wakley's journal *The Lancet,* founded in 1823. Wakley particularly admired Grant's support for Continental ideas and his lack of deference toward the Church; he defiantly promoted his friend as the greatest professor in Europe.

Grant's course of London lectures promoting French materialist science, weaving together the ideas of Lamarck and Geoffroy, quickly became a thorn in the side of the Anglican scientific establishment in London. "While myriads of individuals appear and disappear, like passing shadows in rapid succession," he declaimed from the lectern podium,

> the species . . . are still prolonged on the earth. The species, however, like the individuals which compose them, have also their limits of duration. The life of animals exhibits a continued series of changes, which occupy so short a period, that we can generally trace their entire order of succession, and perceive the whole chain of their metamorphoses. But the metamorphoses of species proceed so slowly with regard to us, that we can neither perceive their origin, their maturity nor their decay, and we ascribe to them a kind of perpetuity on the earth. A slight inspection of the organic relicts deposited in the crust of the globe, shows that the forms of species, and the whole zoology of our planet, have been constantly changing and that the organic kingdoms, like the surface they inhabit, have been gradually developed from a simpler state to their present condition.

London was politically volatile in 1831; Britain was in a period of economic depression, high prices, and high unemployment. The uprising in the streets of Paris in 1830 had terrified British conservatives and excited the radicals and dissenters. In 1820, a group of British revolutionaries had planned to start a general uprising by assassinating the entire cabinet; in 1819, a large crowd at a reform meeting in St. Peter's Field in Manchester were killed by cavalrymen in what became known as the Peterloo Massacre.

Natural science in Britain in the 1830s, and particularly in Cambridge and Oxford, was fundamentally different from that in France or Germany. Most of the teaching of scientific subjects in the two major English universities was undertaken by ordained ministers; it had come to be dominated by the forces of conservative natural theology. British naturalists tended to write about the wonderful adaptations of living creatures, each designed by God to adorn and enrich the earth and man's husbandry of it. This tradition of natural theology—nature study as both an act of worship and a means of proving the existence and bounty of God—was at its peak in the 1830s; William Paley's elegantly written and widely read *Natural Theology, or Evidences of the Existence and Attributes of the Deity Collected from the Appearances of Nature* could be heard quoted in lecture theaters and churches and found on the bookshelves of most undergraduates, including those of the young Charles Darwin.

Darwin knew that Grant was taking risks in this atmosphere. Infidel men of science before him had been mauled by the Anglican establishment. The radical young surgeon William Lawrence, a brash and sarcastic friend of Wakley's and a well-connected companion of radicals including the Shelleys, had given a course of lectures in 1816 at the Royal College of Surgeons in which he had argued that life depended on physiological structures, not souls. When in 1819 he published those lectures as *Lectures on Physiology, Zoology and the Natural History of Man,* the Lord Chancellor declared the book blasphemous. When his job, practice, and hospital positions were threatened, Lawrence resigned from the Royal College of Surgeons and withdrew his book from sale, but when he failed to prevent radical presses from publishing pirated versions in 1822, he was suspended from his surgeon's post and obliged to write a retraction. A combination of career threats and professional inducements over the subsequent decade deradicalized him. The German physiologist Carl Gustav Carus, visiting England in the 1840s, complained: "He appears to have allowed himself to be frightened and is now merely a practising surgeon, who keeps his Sunday in the old English fashion, and has let physiology and psychology alone for the present."

Darwin may have wondered how Grant could hold his small dissenting platform against the Anglican-dominated medical and scientific establishment of London, but he was on board the *Beagle* in 1836 when Grant came under direct attack. The backlash came, perhaps inevitably, from close quarters, from the hands of the brilliant young comparative anatomist Richard Owen, eleven years younger than Grant and a devout Anglican. Owen had been impressed by Grant's passionate Lamarckianism when the two men first met; they had lodged at the same hotel in Paris in the summer of 1831. Grant had introduced Owen to all the important French philosophers at the Jardin des Plantes; the two men had talked late night after night about homologies and evolution. At the same time, Owen listened to what Cuvier had to say about Lamarckianism and about the limitations of Geoffroy's theories. The dangers of anticlericalism would also have been firmly impressed on him by the sight of the burned-out Archbishop's Palace, next to Notre Dame, set on fire in the street fighting of the previous year. The first revolutions of France, some whispered, had much to do with the writings of infidel philosophers that should have been more aggressively suppressed.

Another young scientist learned the lessons of Paris during these years: Charles Lyell. After witnessing fighting in the Paris streets during the violent July Revolution of 1830, he wrote home about the immense ages that would be required for our "Ourang-Outangs to become men on Lamarckian principles." Returning to Britain during the Reform Bill crises of 1831–32, and "distracted by the disturbed state of politics" in London, he began to write the pages that would become the two volumes of the groundbreaking *Principles of Geology*, the second volume of which, published in 1832, set out to demolish Lamarckian transmutation over forty eloquently argued pages.

The success of Lyell's second volume assured a blanket opposition to Lamarckianism in Britain; but, ironically, by refuting Lamarck's theories in such minute detail and across so many pages, Lyell also disseminated knowledge of those ideas much more widely. In *Principles,* Lyell distanced himself from French ideas in an attempt to depoliticize himself and his claims. Studying fossils and strata did not have to be blasphemous or support materialism, he implied; landscapes might have shifted infinitely slowly through long periods of time, but there was absolutely no reason to believe that species had evolved. Lamarck, he insisted politely and respectfully, had been quite wrong to "[degrade] Man from his high Estate." But *Principles* was not strong enough, Lyell's colleagues complained, to silence the transmutationists. Lamarckian ideas were proliferating wildly among the medical men of London while Robert Grant remained unchecked.

Richard Owen knew he had to put some distance between himself and his former friend Grant if he was to make a career for himself in London. Grant's lecture audiences were growing in the early 1830s, and despite his ideas, he was also gradually making an inroad into institutional positions outside the university; he had been appointed to the Linnaean, Geological, and Zoological Societies. At the same time, the *Medical Gazette* in 1833 accused Grant and Wakley "of ribald jestings on holy things and blasphemous derision of the sacred truths of Christianity." Owen bided his time. Any direct attack on Grant or his ideas, he knew, would only give the older man more publicity.

Things came to a head when Robert Grant and Richard Owen were both elected to the Council of the Zoological Society in 1832, just as Lyell's second, anti-Lamarckian volume of *Principles of Geology* appeared in the bookshops. For two years the two men jostled for power within the council committees, attempting to undermine each other beneath a veneer of politeness. Grant, keen to build a wider platform for himself within the corridors of power in London, managed to talk his way into giving a course of lectures to the Fellows in 1835. Owen, exasperated by Grant's audacity and by the weakness of the Fellows, launched a campaign to get him balloted out of the Society in the April elections. It worked. Grant was ousted. At one stroke he was robbed of credibility, of a platform for his ideas, and of the ready supply of specimens for dissection that he needed in order to continue his work. It was a very public humiliation, reported even on the front page of *The Times.* Grant in turn declared he would have nothing more to do with the Zoological Society.

—

In the 1840s, after Darwin returned from the *Beagle* voyage, he chose not to call on his former mentor. Grant was now in dire financial circumstances. He still refused to take the London medical exams that would enable him to practice in England. He had finally begun to publish his lectures as a book called *Outlines of Comparative Anatomy,* issuing it in parts, but the series stopped abruptly in 1841 with no conclusion and he never published again. When the number of his students declined dramatically in 1850, the university granted him a small stipend of £100 per annum. Wakley launched a campaign and awarded him a further annuity of £50. Grant continued giving lectures, but his students found him bitter and melancholic. In 1849, when one of his university colleagues, John Beddoe, visited Grant in London, he found him living in a slum in Camden Town. When he urged him to find new lodgings, Grant declared bitterly, "I have found the world to be chiefly composed of knaves and harlots, and I would as lief live among the one as the other."

Through the 1840s and 1850s, Charles Darwin, who had locked away his species theory in his desk drawer and had turned instead to describing uncontroversial barnacle taxonomies, must have listened for further news of Grant's decline and humiliation with considerable fascination and alarm. His transformist mentor had become, like Lamarck, virtually a recluse. No one listened to him. People described him flitting from his house near Euston station to his lecture rooms, dressed still in his old, worn swallowtail coat, both vexed and aggressive when spoken to. He was yet another ghost to be laid to rest before Darwin could publish. But there was much to be admired, too, in the man. Without a private income, Grant had forged a professional living for himself on the basis of his own merits. That was impressive. He was right to denounce privilege and nepotism, right to denounce the Anglican Oxbridge clique who controlled all the scientific establishments of London. But his failure to compromise, his aggressive attitude toward the establishment, had been disastrous for him personally, for the reform cause, and for the reputation of Continental scientific ideas. Robert Grant had become a laughingstock. Reform would not come by these means. Other ways, Darwin knew, must be found.

11
The Encyclopedist

EDINBURGH, 1844

It was November 1844. In the city of Edinburgh, a man in a frock coat and top hat walked slowly along a wide thoroughfare in the new part of the city. Around him carriages clattered past; street sellers pushed handcarts piled high with fruit or flowers or fish past elegantly dressed men and women walking their dogs; students clutching books jostled one another, making their way to the anatomy theaters, lecture halls, or infirmaries in adjoining streets; high school boys darted through clusters of lecturers and professors in black gowns passing through the classical façade of the Medical Faculty of Edinburgh University. Opposite the entrance to the Faculty, the man stopped outside the shop front carrying the sign MacLachlan, Stewart and Co., booksellers to the university.

Inside the warm shop filled with customers, the bookseller nodded in the man's direction, recognizing him as Robert Chambers, the journalist and wealthy proprietor of the publishing house he ran with his brother William, famous now for the bestselling and influential *Chambers's Edinburgh Journal*. Chambers did not want to be recognized; he

Robert Chambers, aged around fifty.
Steel engraving by T. Brown from an oil portrait by Sir J. Wilson Gordon. Frontispiece to the twelfth edition of Vestiges.

wanted to observe. Inside the shop the tables were piled with new publications, books in calf leather or bound with cloth, books on physiognomy or anatomy, translations of Continental books on chemistry or comparative anatomy, books on diseases, as well as novels, travel books, and histories. The volumes of Lyell's *Principles of Geology* were stacked here, too, he noted, still selling well alongside Barclay's course of lectures on anatomy and George Combe's books on phrenology. Then Chambers's eye lit on the pile of red books on the central table, copies of a new work called *Vestiges of the Natural History of Creation,* which had been anonymously published. The pile was diminishing rapidly. Only four people in the country know that he—Robert Chambers—was the book's secret author. If he had his way, no one else would ever know.

Men and women from the upper and middle classes, medical students, liveried servants buying for wealthy men waiting in carriages outside, carried the red book to the sales desk. Chambers noticed that a few them were also carrying copies of the *Examiner,* a weekly reform newspaper, which had just published an enthusiastic and lengthy review of *Vestiges.* "In this small and unpretending volume," the reviewer had written,

> we have found so many great results of knowledge and reflection, that we cannot too earnestly recommend it to the attention of thoughtful men. It is the first attempt that has been made to connect the natural sciences into a history of creation. An attempt which presupposed learning, extensive and various; but not the

> large and liberal wisdom, the profound philosophical suggestion, the lofty spirit of beneficence, and the exquisite grace of manner which make up the charm of this extraordinary book.

Vestiges was certainly "small and unpretending," but in its arguments and propositions it was also deeply heretical. The author had kept his identity a secret for good reason.

Making every effort to look uninterested, to avoid drawing attention to himself, Chambers wondered what the men and women buying his book would make of it. He had written *Vestiges* for just these people. The reviewer had recognized that, noting that the anonymous author had written with style and color and passion; he was not writing for specialists—medical men or philosophers—but for the middle classes who took an interest in the world around them, in the great mysteries of nature, ordinary people who were curious enough to ask big questions. How and when did the earth begin? How has it changed? How did life start out? Now Chambers had given them answers, told them a story

South Bridge, Edinburgh, ca. 1829.
W. H. Lizzars (ca. 1829)

of the earth from its spectacular birth in a nebula fire-mist through the origins of life to the evolution and metamorphosis of new and ever more diverse species as aquatic creatures became reptiles as reptiles became birds and birds humans. To engage them, he had written it almost like a novel, almost like a tale by Walter Scott.

Chambers had worked hard to keep his authorship secret; even the book's publisher did not know who had written it. What Chambers did not yet know was the scale and range of the scandal his book would ignite: that in London the bishops and priests would soon be calling it heresy and its author an infidel, that some men of science would declare it nonsense, dismiss it as full of mistakes of fact and interpretation, that everywhere at the dinner tables of the aristocracy and the public houses of market towns people would be talking about it for years. They would talk about what man had been and what he would become, about the age of the earth and the origins of time. Chambers did not know that the poet Alfred, Lord Tennyson, soon to be the poet laureate, had already ordered his copy of *Vestiges* from his bookseller and that he would soon be publishing his complex and eclectic ideas about deep time and the origins of life and the future extinction of man, ideas gleaned from both Lyell's *Principles* and Chambers's *Vestiges,* in a long poem that Queen Victoria would keep by her bedside for a decade, the bestselling *In Memoriam.*

Chambers would have been delighted but surprised to hear that in the late months of winter, in Balmoral, Prince Albert would begin reading it aloud to the young Queen Victoria and that they would be discussing it with their royal guests among the glitter of chandeliers and silver. Nor did he know that in 1848, walking through the streets of Paris during another uprising in this most incendiary of cities, a second poet, Arthur Hugh Clough, would discuss the book and praise the revolution with the great American philosopher Ralph Waldo Emerson, and that both would thread ideas about evolution derived from his book through their subsequent essays and poetry.

Though *Vestiges* was one of the greatest publishing sensations of the decade and outsold Darwin's *Origin,* Chambers always publicly denied that he wrote it. It was just too dangerous to take responsibility for its authorship.

—

The hush and smell of bookshops had always excited Robert Chambers. In Peebles, the small Scottish market town in which he grew up, life was often brutal and aggressive; "violence held rule almost everywhere," his brother remembered.

> [Boys] were flogged and buffeted unmercifully, both at home and at school; and they in turn beat and domineered over each other according to their capacity, harried birds' nests, pelted cats, and exercised every other species of cruelty within their power. A coarse bustling carter in Peebles, known by the facetious nickname of "Puddle Michty," used to leave his old worn-out and much abused horses to die on the public green, and there, without incurring reprobation, the boys amused themselves by, day after day, battering the poor prostrate animals with showers of stones till life was extinct.

As a bookish child, Chambers had taken refuge in the local bookshop, where the bookseller Alexander Elder kept the family cow behind bookshelves stacked with expensively bound classics, school slates, notebooks, and reams of paper. Upstairs above the shop, Elder ran a circulating library where local residents could rent and return books. Here, books piled in front of him, the smell of warm manure and cowhide rising from below, Chambers found all the history, geography, and science that was so desperately missing from his school lessons. Sandy's library looked out on the High Street, and Chambers later remembered seeing soldiers on the street outside, parading recruits raised for Wellington's Peninsular Campaign. He gorged himself on words and ideas, happy to go for long hours without food rather than lose his place in his book.

Robert's father shared his love of books, though the hours he spent in the local public house meant he did not always remember things as clearly as he once had. The Chambers family enthusiastically borrowed *Gulliver's Travels, Don Quixote, Peregrine Pickle,* and Pope's translation of the *Iliad* and read them aloud in the evenings after Robert's father had played his flute. Robert Chambers could never read fast enough. He and his brother would borrow a book and read it simultaneously, lying stretched out on the floor and alternating the page turning. Robert's father had a telescope, and he and his father read books on astronomy, trying to make out what they saw in the skies.

Chambers was an intellectual roamer, a dilettante, a natural polymath. The bookshop was a place where he wandered freely, memorized, reflected, made connections, and framed new questions. Years later, when he wrote his memoirs and reflected on who he was, he could not write enough about books. He wanted to remind his children that he had always had to struggle to get them, and that reading meant sacrifice. He wanted them to know that books gave him more pleasure than anything else, that he lived inside their covers, that they sustained him through all his frustrations and deprivations and disappointments.

He remembered with a thrill of pleasure coming across a chest of books among stacks of cotton wefts and meal arks (chests for storing dried food for the winter) in his attic as a young boy and finding inside a full set of the *Encyclopaedia Britannica*. His father had bought it from Alexander Elder on a whim when it had first arrived, realized that the volumes were too big for the shelves, stored them in the attic chest, and then forgotten about them. "From that time for weeks," Chambers wrote,

> all my spare time was spent beside the chest. It was a new world to me. I felt a religious thankfulness that such a convenient collection of human knowledge existed, and that here it was spread out like a well-plenished table before me. What the gift of a whole toy-shop would have been to most children, this book was to me. I plunged into it. I roamed through it like a bee. I hardly could be patient enough to read any one article, while so many others remained to be looked into. In the one on Astronomy, the constitution of the material universe was all at once revealed to me. Henceforth, I knew—what no other boy in the town dreamed of—that there were infinite numbers of worlds besides our own,

> which was by comparison a very insignificant one. . . . I pitied my companions who remained ignorant of what became to me familiar knowledge.

Familiar knowledge became secret knowledge: Chambers's developing ideas about the birth of planets, nurtured in the dark attic by lamplight, would have been scandalous in conservative, superstitious, God-fearing Peebles.

But Robert's secret attic treasure was suddenly gone a year later, the contents of the chest sold to pay off his father's debts. When James Chambers's weaving business failed as new machinery replaced the old hand looms, he was forced to sell the old house and its belongings in order to set himself up in a drapery business that would itself soon fail. Robert, his head full of wild geological and astronomical ideas, was glad to be distracted by the arrival of a troop of French prisoners of war, captured in Spain, who had been sent to the town on parole. The soldiers spent their days in a derelict once-grand house playing billiards and performing plays in a makeshift theater they built in an old ballroom. The brothers began to breed rabbits, selling them for eightpence a pair to the French prisoner-chef who cooked for the soldiers; they used the money to buy books. The soldiers taught the boys French and cookery; they performed Molière with them; back in the village on their way to church on Sundays, the boys listened to the scandal the soldiers raised in the village when they continued to perform their plays and music on the Sabbath. For the two boys and their intellectual and feckless father, the French prisoners of war were exotic and colorful. Both Robert and William Chambers, listening to the droning of the preacher in the pulpit, knew where they would rather be.

When the entire community of prisoners was suddenly moved out to Dumfriesshire, leaving an empty house full of discarded tins, books, and a broken stage set, the boys were devastated. But the soldiers that Robert's father had befriended and lent money to had also failed to pay their substantial debts, forcing the family to move to Edinburgh, where Robert attended school and persuaded the local booksellers to allow him to sit quietly reading in back rooms in return for an errand or two. He read voraciously in the Agency Office book auction room opposite the university, talking to the booksellers about books, authors, and the book trade. It was warm and well lit and smelled of new furniture; it was as good as a reading room. His father was now running a saltwork yard in Joppa Pans, a salt panning village adjoining Portobello, and the rest of the family moved out to live there, leaving William and Robert in lodgings in Edinburgh. William was learning the book trade, living, like thousands of others around them, in genteel poverty while scratching out an existence.

The Peebles bookseller, the *Encyclopaedia* in the dark attic chest, the Sabbath-breaking French prisoners of war—they were all powerful influences on Robert, but there were others. An aged porter named James Alexander lived in an obscure corner of Carlton Street in Edinburgh, making a small living from mending china. "He was great in electricity," Robert remembered excitedly, "and had once fabricated a machine for producing that fluid. His house, which resembled the den of an alchemist or magician—so full was it of all kinds of odd and unaccountable implements—was also resorted to by two young men of the name of King, one of whom was shopkeeper to a seedsman, while the other was a draper, and who were really most ingenious young men." Here, among the machinery, electrical and chemical experiments, and engineering schemes, Robert developed his questions about astronomy, the movement of the planets, and the origins of life.

By 1818 the Chambers brothers, now eighteen and seventeen, had established their own bookstalls and then bookshops with circulating libraries on the same street in Edinburgh, Leith Walk, where they passed the time with many disaffected radicalized young men angry about privilege and the power of church and king. The Peterloo Massacre of 1819 had further enraged an already aggressively alienated underclass. The possibility of a revolutionary uprising in the streets of Edinburgh or London alarmed Robert. He was passionate about reform, but he believed it had to come about without violence. It made him think again about the power of books, journals, and magazines to inform, argue, and persuade.

By 1821 the brothers had taken their first foray into the publishing business, producing a literary journal called *Kaleidoscope,* hand-printed by William on an old repaired printing press and almost entirely written by Robert. Later Robert told his readers that "it was my design from the first to be the essayist of the middle class—that in which I was born, and to which I continued to belong. I therefore do not treat their manners and habits as one looking *de haut en bas,* which is the usual style of essayists, but as one looking round among the firesides of my friends. . . . Everywhere I have sought less to attain elegance or observe refinement, than to avoid that last of literary sins—dullness."

In 1830 the air was full of talk of reform and the political implications of the proposed extension of the franchise to the middle classes. Robert, remembering the radical young men on Leith Walk and the customers who bought the penny magazines in his bookshop, knew that those lower and middle classes wanted to be educated, and that they were seeking out the kind of easily digested knowledge that he had found so seductive in the attic and in the bookshop but that was, he knew, always a struggle to afford for someone living on a tradesman's salary. There were a few cheaply priced magazines like the *Kaleidoscope* on sale in Edinburgh, but most were poorly written or inconsistent in quality. The libraries and the museums of Edinburgh University were closed to people like him and the clever, intellectually curious King brothers. Those well-heeled professors were suspicious, convinced that a working man in shabby clothes, however genteel, however seemingly well educated, would cause trouble, even steal or deface precious objects and books.

But for Robert Chambers, fiery radicalism was as dangerous and backward-looking as evangelicalism. The street demonstrations in Paris and other European cities and in London in the run-up to the Reform Bill that he read about in the papers terrified him. He had written to Walter Scott in 1830 about his fears and convictions. "This fervour is as fatal to literature as the irruption of the Goths," he lamented. "Nor do I think it near an end: it is rather at a beginning. People formerly had a maxim, which history in all ages showed to be good, that the great object of informed and civilised society was to keep the mob in check; but now the maxim is, that the government must reside in the mob. . . . The fiend, say I, take all the fools who are now hurrying us on to revolution and vandalism. But this way madness lies; I must to business." Chambers had begun to believe that it was time to leave antiquarianism and historical nostalgia behind and begin to respond instead to the present by finding new ways of bringing knowledge to the people.

Robert's publishing ambitions had taken on an increasingly political edge. He had seen how, as his family fell further and further into poverty, doors had closed against them. It made him angry. While libraries and universities and museums refused entry to the workingman, he told his fiancée, Anne Kirkpatrick, while books were too expensive to buy, while ignorance prevailed, there could be no social change or progress. The workingman would remain no more than an animal, a slave to the factory and the shop and a slave to the men who had the monopoly on knowledge, the men of the Church.

So when William suggested to his newly married brother that they produce a serial magazine affordable to shopkeepers, drapery assistants, and insurance clerks, Robert poured all of his educational zeal into the manifesto. The aim of *Chambers's Edinburgh Journal* was, he wrote, to "take advantage of the universal appetite for instruction which at present exists; to supply to that appetite food of the best kind." It was one of several such ventures, such as the *Penny Magazine* published by the Society for the Diffusion of Useful Knowledge, which had been founded in London in 1826 to supply clearly explained scientific ideas for the rapidly expanding reading public; but none was as successful or as long-lived as *Chambers's Edinburgh Journal*. The sixteen-page magazine was designed to be collected. It carried short, colorfully written articles, many by Robert Chambers, on a range of historical, philosophical, scientific, and literary subjects, and cost just a penny. The first number sold thirty thousand copies in Scotland; the third number sold eighty thousand in bookshops around the country. It brought the Chambers brothers wealth almost overnight. "All previous hardships and experiences," William wrote, "seemed to be but a training in strict adaptation for the course of life opened up to us in 1832." Within a decade the circulation of the *Journal* made W. & R. Chambers one of the biggest publishing houses in the world.

For the brothers, it was imperative that the knowledge they were disseminating to the lower and middle classes be secular and nonpartisan. It was a point of principle with them. "It has been a matter of congratulation," William wrote later, "that Chambers' Journal owed nothing, in its inception or at any part of its career, to the special patronage or approval of any sect, party or individual." But it was not easy to maintain such neutrality. As soon as the journal had this kind of circulation, the evangelicals, deeply concerned about growing radicalism and growing secularism among the urban poor, began to take an interest in its contents. Soon the interest had turned into a kind of persecution. "On all hands," William wrote,

> we were beset with requests to give it the character of a "religious publication." It was in vain for us to state that that was not our role; that our work was addressed to persons of all shades of thinking, religious and secular, and that we could not, without violation of our original profession, take a side with any one in particular. We only got abused, and were called names. The era of this species of persecution, for such it was, grotesque and ridiculous, extended for nearly twenty years after the commencement of the work.

The evangelicals were enormously powerful in Scotland during the reform crisis, and the pulpit denunciations worked. Robert was enraged by the campaign being waged against the *Journal,* but it made him more defiant. When his own vicar, the Reverend Dickson, denounced *Chambers's Edinburgh Journal* from his pulpit, waving it in the air, declaiming that knowledge without God was useless, even dangerous, Robert and his wife, Anne, walked out of the church with their infant children. They never returned.

The more that evangelicals criticized the godlessness of the *Journal,* the more successful it became. In 1833 the brothers began to print and distribute the *Journal* in Ireland. The poet Allan Cunningham described the shepherds of Galloway passing copies of the magazine between them: "The shepherds, who are scattered there at the rate of one to every four miles square, read it constantly, and they circulate it this way: the first shepherd who gets it reads it, and at an understood hour places it under a stone on a certain hill-top; then shepherd the second in his own time finds it, reads it, and carries it to another hill, where it is found like Ossian's chief under its own gray stone by shepherd the third, and so it passes on its way, scattering information over the land."

In January 1835, Robert told his readers proudly that the weekly magazine "still penetrates into every remote nook of the country; still travels from hand to hand over pastoral wastes—*the fiery cross of knowledge*—conveying pictures of life, and snatches of science, and lessons of morality, where scarcely any such things were ever received before." They reported that in one Glasgow mill, no fewer than eighty-four copies were regularly purchased by workers, that it "reaches the drawing-rooms of the most exalted persons in the country, and the libraries of the most learned; that, in the large towns, a vast proportion of the mercantile and professional persons of every rank and order are its regular purchasers . . . it pervades the whole of society."

In using the phrase "the fiery cross of knowledge," Robert Chambers was being deliberately provocative. The fiery cross, or Crann Tara, had been used by the Highland clans for centuries as a declaration of war. A small wooden cross was first dipped in the blood of a goat, then set on fire and carried from town to town. At sight of the Fiery Cross, every man capable of bearing arms, from sixteen years old to sixty, had to arm himself immediately and travel to the gathering place. Anyone who failed to appear would suffer death by sword or fire. It had been used during the Jacobite uprising in 1745; Walter Scott described its effects and importance many times in his poetry and novels. Its meaning would have been perfectly understood by all of Chambers's Scottish readers: Robert Chambers was announcing a secular crusade and gathering the men and women of the Scottish working and middle classes to resist the persecution of the Church, to take a stand against ignorance, prejudice, and privilege. The *Journal* became more outspoken by the year. Robert wrote to his friend Alexander Ireland around this time: "I believe this liberal view is advancing, but we are still far from being able to fight those dogs of clergy."

Since he had walked out of the church, Robert Chambers had made new friends among the phrenologists of Edinburgh, a fiery, articulate, reformist group of men and women who promoted and practiced this new science. Phrenologists believed that certain areas of the brain governed particular kinds of behavior; thus, by analyzing the precise shape of the head, it was possible to determine particular personality types.* The group gathered around George Combe, fourteen years older than Robert, whose book *The Constitution of Man,* published in 1828, had been denounced as materialist and atheist for arguing that mental qualities were determined by the size and shape of the brain and not by the soul.

George Combe and the members of the Phrenological Society of Edinburgh, which contained almost as many clever and educated women as it did men—women such as the novelist Catherine Crowe and Combe's wife, Cecilia, daughter of the celebrated actress Sarah Siddons—met to discuss ideas. They lent Chambers recommended books, explained new ideas and theories to him, and shared his conviction that progress and reform would begin only when the people were given new knowledge. Phrenology would liberate people, Combe argued; it was a way of life. If you could persuade people to understand and live by natural physiological laws, to understand that mental functions and characteristics were controlled by different parts of the brain, the world would be a fairer, more just place; self-help could proceed on a rational basis.

Excited by the power of such "a new gospel," Chambers became for a while a proselytizer. He published *Introduction to the Sciences* in 1832, a book that promoted core phrenological principles, although he was careful to avoid using the word "phrenology" itself because of its heretical associations. The way to overturn long-entrenched beliefs, he told Combe, was not to preach but to persuade. It would take time. "When we reflect," he wrote to Combe, "that some of the forms of heathenism survived in Scotland till the close of the last century, perhaps eight hundred years after Christianity had been acknowledged to all intents and purposes, we must not fret at the slow progress that phrenology makes."

Chambers was at first alarmed to discover that transmutation, which he had seen so powerfully rebutted by Charles Lyell in *Principles of Geology,* was discussed widely and enthusiastically in this new circle. Trans-

* Although it had an influence on the development of later sciences such as neuroscience and psychology, phrenology is now regarded as a pseudoscience.

mutation took many forms by the 1830s—Lamarckian, Geoffroyian, and hybrids of the two. Ideas circulated among medical men in London and Edinburgh, some of whom had sat in on Lamarck's lectures in Paris. George Combe was, however, set against all such ideas. He was more uncompromisingly materialist than Chambers, but he was afraid of transmutation. It was too French, and the French were too violent. He did not want to be tarred with that particular materialist brush. He denounced the "revolutionary ruffians" in his *Constitution of Man* and their program of "fraud, robbery, blasphemy, and murder."

Until the late 1830s, *Chambers's Edinburgh Journal* had consistently denounced transmutation as fantastic and absurd. Ignorance of the facts, Chambers wrote in an article published in the *Journal* in 1832, had "tempted philosophers to hazard the absurd opinion that man had his beginning as a minute animalcule, and has attained his present perfect condition from progressive improvement by reproduction." He lamented the fact that "even so late as the year 1803, one of the greatest scholars, and ablest medical men of his day, Dr Darwin, espoused these false doctrines." But though Chambers believed the doctrines false and absurd, they were interesting enough for him to quote from Erasmus Darwin's poem *The Temple of Nature* at length. He was clearly struggling to resist Erasmus's arguments: "The acute and anatomical knowledge of the Doctor, and his deeply sophisticated arguments, have a strong tendency to seduce the less philosophical reader into his baseless doctrines. . . . But views like these can never be entertained by healthy minds and it requires but little reflection to dispel such absurd theories."

As late as 1835, when Robert was discussing transmutation in the drawing rooms of the phrenologists, he was still confident enough to declare in the *Journal* that Charles Lyell's rebuttal of Lamarck in *Principles of Geology* was "so satisfactory as to require us to say nothing in addition" and to express renewed amazement that "some very eminent philosophers" had claimed that "man himself, Socrates, Shakespeare and Newton, were merely zoophytes in a state of high improvement and cultivation!"

By the mid-1830s, Robert Chambers had begun to change his mind about transmutation. In 1835 he started writing a treatise on phrenology, and this shaped new questions and sets of connections in his mind.

He became more reclusive, ordering book after book from MacLachlan, Stewart and Co., or rummaging through their shelves, newly impressed by the kind of books that he had previously been suspicious of, large-scale theories of the earth written for sizable, nonspecialist audiences, books like John Pringle Nichol's *Views of the Architecture of the Heavens,* published in 1837. Chambers knew Nichol well. The two of them toured poverty-stricken Ireland together that summer, talking about the forces of conservatism, the enforced ignorance of the poor, science, progress and reform, and undertaking geological experiments. Nichol's book, which described in vivid and colorful detail the evolution of the universe from the formation of galaxies and stars and was intended to be the first of six volumes in which Nichol would demystify "the mechanism of Nature," had been an instant bestseller.

The number of science articles in *Chambers's Edinburgh Journal* increased dramatically that summer as Robert read voraciously and turned in one article after another on zoology, botany, spontaneous generation, origins of races, nations, languages, and civilizations, always searching for the laws that held all these different sciences together. Inspired by Nichol's *Architecture of the Heavens,* Chambers now expanded his book on phrenology to take in the origins of time, the earth, and species, and he even attempted to predict the future progress and mutation of the human race. Now "preoccupied with speculative theories" and with the search for new explanations for the history of the earth and of species, Chambers had himself elected to the Royal Society of Edinburgh, which considerably expanded the circle of scientists around him.

By the late 1830s, Chambers was buying or borrowing books on earth history, zoology, botany, and geology at an extraordinary pace, reading them in his office among the thundering sounds of the printing presses in the publishing house. Then, sometime in 1842, the Chambers family suddenly moved from Edinburgh to a house in Abbey Park, on the outskirts of the coastal university town of St. Andrews. Robert seems to have suffered a kind of breakdown, brought on by overwork and the noise of printing presses. Convalescing, he returned to his manuscript with a new sense of clarity as he gathered his books around him in a study looking out to open space and sky. The move also protected him from prying eyes. At Abbey Park, his daughter recalled, he could "work at his secret with all the security of a criminal unrecognized in the midst of the police."

Chambers destroyed all the papers and notes and letters he wrote during these years in St. Andrews, so the only evidence of what he read as he assembled *Vestiges* is in the book itself. He quoted more than eighty authorities to support the different parts of his case. He was certainly rereading Lyell's *Principles of Geology* as he expanded his geological knowledge, which took him back to Lyell's detailed rebuttal of Lamarck. The more medical and physiological books Chambers read for his phrenological project, the more he encountered transmutationist ideas at every turn, transmutationism either roundly rebutted as it had been by Lyell or modified to accommodate religious explanations of creation. Now he was ready to take Nichol's brilliant first volume of *Architecture of the Heavens* forward in time, to tell a story that stretched from the birth of the planets to the birth and metamorphosis of species.

In 1844, his manuscript complete, Chambers and his family moved back to Edinburgh. He summoned his friend the journalist Alexander Ireland and told him he was about to publish a dangerous book and needed his help in approaching the London publisher he had chosen: John Churchill, publisher of *The Lancet*. "I do not think Churchill is likely to boggle," he wrote in explaining his choice, "for publishers of that class are a little used to such things." Ireland subsequently wrote to Churchill on June 27, 1844, offering a book on behalf of an anonymous friend. Robert's wife, Anne, began the laborious task of copying out the manuscript so as not to give away Robert's identity, sending the pages to Churchill in batches through July and August 1844. Churchill told Ireland he proposed publishing a thousand copies; later, fearing financial risk, they renegotiated a print run of 750 copies. The book was originally to be called *The Natural History of Creation,* but Chambers insisted on softening the title to *Vestiges of the Natural History of Creation* to protect himself from charges of blasphemy. By adding the word "vestiges," meaning fragments or traces, the book might begin, he reasoned, to look a little like an antiquarian work and the author like an archaeologist or a classicist solving a puzzle.

By the time *Vestiges* reached bookshops in October 1844, a total of 150 advance copies had already been delivered to leading men of science in London, Oxford, and Cambridge, to major libraries of universities, to mechanics' institutes, to literary and philosophical institutions, and to writers and politicians. Churchill placed advertisements for the book in newspapers and weeklies and periodicals across the religious spectrum. Only four people knew the name of the book's author: the author's wife, Chambers's brother William, Alexander Ireland, and his closest friend, Robert Cox, Combe's nephew and the editor of the *Phrenological Journal*. Three others were allowed into the secret later so that they could help with specific questions about science after the book was attacked: David Page, editorial assistant on *Chambers's Journal*, who knew a good deal about geology; Neil Arnott, a member of the Royal Society and physician extraordinary to the queen, who could advise on general scientific issues; and the Glasgow professor of astronomy John Pringle Nichol, who would help on matters of astronomy. None of the Chambers children was told, and Robert kept all matters relating to the book in a locked drawer in his study. Years later, when his son-in-law asked him why the secret had had to be maintained for so long, Chambers "pointed to his house in which he had eleven children and then slowly added, 'I have eleven reasons.' "

Chambers enjoyed the excitement and frisson of anonymity; he took pleasure in sitting at dinner tables joining in with speculation about the identity of the author. It was only two months before his name became linked to the book, but in February 1845 he was still only one of several proposed authors, the list including Richard Vyvyan, Ada Lovelace, and even the scientifically minded Prince Albert. What Robert had not anticipated was that his friends might be implicated. George Combe, Catherine Crowe, and Neil Arnott were all at one time or another in the firing line for authorship. By April 1845, Chambers was alarmed at the small but growing number of letters he received from readers who addressed him directly as the author; but, he wrote to Ireland, "they can but suspect and surmise."

Meanwhile, *Vestiges* was being read across the country and was making an impact. The first edition sold out in weeks. Another thousand copies were printed and sold out within a month. The third edition of fifteen hundred sold out on the day of publication in mid-February 1845. In April, Churchill printed a further two thousand copies. Few people would have read the book without a frisson of anticipation and a thrill of transgression, for it was denounced by reviewers everywhere. The anonymous author, speaking in a voice that was warm and respectful and never sermonizing, unfolded the story of the earth's history, supporting his claims with facts garnered from scores of authorities, offering to tie together all the new discoveries in zoology, anatomy, and geology to explain the great mysteries of time and creation. The author and his readers were embarking on a voyage of discovery together, he promised. And he told his readers again and again that they had a right to be curious about creation. Uncovering the answers to a series of perfectly innocent qustions about how life had come to be was as natural as a child's asking questions at his mother's knee.

Chapter by chapter the reader's eyes were opened to the "facts" of the birth of the earth in a great fire-mist, to the "facts" of the early spinning earth, gradually emerging, bubbling and molten from a primal ocean, right up to the controversial material on the emergence of species, a chapter humbly entitled "General considerations respecting the origin of the animated tribes." Chambers wrote in the margins of the manuscript he sent to the publisher: "The great plot comes out here." Everything had led to this great question. How did life come to be? "Would it be too bold to imagine?" Erasmus Darwin had asked, 450 pages into *Zoonomia,* as he began to broach the issue of the origin of species. Chambers tackled the question only 152 pages into *Vestiges.* There was, he reassured the reader, no danger in asking the question. It was entirely legitimate and there was no blasphemy in asking it:

> A candid consideration of all these circumstances can scarcely fail to introduce into our minds a somewhat different idea of organic creation from what has hitherto been generally entertained. That God created animated beings, as well as the terraqueous theatre of their being, is a fact so powerfully evinced, and so universally received, that I at once take it for granted. But in the particularity

> of this so highly supported idea, we surely here see cause for some re-consideration. It may now be inquired—In what way was the creation of the animated beings effected?

Throughout *Vestiges* the author appealed repeatedly to the reader's knowledge, to his or her common sense, tying new discoveries from many different sciences together to make a plausible, verifiable, chronological history of the earth and species. He made concessions everywhere to religious feeling, trying again and again in rereading the manuscript in its final stages to imagine how different people might balk at the ideas, trying to find ways to smooth potentially ruffled feathers, modifying the effect of speculative claims with familiar and everyday observations. He tried to disarm at every point. "I believe my doctrines to be in the main true," he wrote; "I believe all truth to be valuable, and its dissemination a blessing." He even added passages to reassure a religious reader. "Every effort is made that reason and common sense would at all admit of to keep smooth with the sticklers—though I daresay I shall not succeed with the extreme ones," Chambers wrote to Ireland. "I am happy to say that I have been able at the end to introduce some views about religion which will help greatly, I think, to keep the book on tolerable terms with the public, without compromising any important doctrine."

Most important of all, the author of *Vestiges* helped his readers at every turn to reconcile these new ideas with their belief in a Christian God. *Vestiges* was not an atheist book. Rather, the narrator insisted that all this extraordinary change and transformation was God's work. God had created nature to do his work for him, to work through the laws that he had made. If Chambers had not presented transmutation as driven in some vague way by divine law, *Vestiges* would never have found its way into Buckingham Palace or into as many libraries and living rooms as it did. It enabled his readers to come to terms with the new science, but though it retained the place of God in the universe, it also needed them to accept that though God had set the earth and its laws spinning, he did not intervene.

Many embraced this new, more optimistic portrayal of nature's processes with enthusiasm. Alfred Tennyson, whose closest friend, the talented poet and intellectual Arthur Hallam, had died from a stroke at the age of twenty-two leaving him, after reading Lyell's *Principles,* doubly despairing about the seeming indifference of nature. He had described nature in his famous elegy to Hallam, *In Memoriam,* as no longer a benevolent and tender mother, but one who abandoned her offspring, left them to fend for themselves, allowed whole species to die out; not only was nature "red in tooth and claw," he concluded, but "Time a maniac flinging dust" and "Life, a Fury slinging flame." Man, too, would go the way of the dinosaurs, "blown about the desert dust / Or seal'd within the iron hills."* Reading *Vestiges* provided Tennyson with a new confidence and optimism about nature's ways and the future of mankind; he began a new romantic-comic narrative poem full of ideas of reform, "The Princess: A Medley" (1847), and added a new and optimistic ending to the still-evolving *In Memoriam.*

The very first reviews of *Vestiges* showed no indication of the storm to come; the book was praised for its prose style and energy, and no mention was made of its more controversial ideas. The first of the religious monthlies and quarterlies predictably stressed the heretical nature of the idea of progressive development and warned its readers against it. But few of the prestigious quarterlies featured any reviews at all. There was an odd silence for several months. Potential reviewers and churchmen read the pages, shook their heads, and looked to the scientists and professors at Oxford and Cambridge to make a judgment, waiting for someone else to take the lead. By April 1845, *Vestiges* was in its fourth edition.

The silence was broken first at a meeting of the British Association for the Advancement of Science in May in Cambridge. Anne and Robert Chambers traveled there by train, nervous with anticipation, determined to watch and listen to what Chambers hoped would be a rallying of men of science around his controversial book. Instead, the couple watched the gathering of conservative men of science close ranks against it. The scientific community, enraged by an anonymous impos-

*The first use of the word "dinosaur" was by Richard Owen in 1842; he coined the term *Dinosauria* to refer to the "distinct tribe or sub-order of Saurian Reptiles" that were being dug up from mine, canal, and railway workings and quarries around the world and assembled in museum galleries.

tor using scientific papers to endorse an unproved and dangerous set of speculations, denounced the book not as blasphemy but as just bad science. Chambers and his wife heard the great astronomer John Herschel deliver a devastating attack on the book as they sat in the front row of a large crowd at Senate House.

Until May 1845, the Oxford and Cambridge Anglican men of science could not decide whether this anonymously authored, metropolitan, unphilosophical book deserved the honor of being taken seriously. In November 1845, the liberal Tory diplomat and geologist George W. Featherstonhaugh wrote to the Reverend Adam Sedgwick, Woodwardian professor of geology, vice-master of Trinity College, and canon of Norwich Cathedral, warning him about the impact *Vestiges* was having in London. "I think you could smash him and I wish you would," he wrote. "Already some consider this book as the signal of the Revolt against the Church." But Sedgwick replied that though he hated the idea of such a book, he had no time, had not read it, and did not write for the quarterlies.

It was March before Sedgwick read *Vestiges* and April before he made his outrage public, at a breakfast for clerics in the cathedral town of Ely, when he denounced the book as a work of "rank materialism" almost certainly written by a woman. Though the book undoubtedly had a "charm of manner & good dressing," he wrote to Charles Lyell two days later, he was now determined to "strip off the outer covering and show its inner deformity and foulness." *Vestiges* was, he wrote to Macvey Napier, a "rank pill of asafoetida and arsenic, covered with gold leaf." He feared for its impact on young women in particular: "You have no conception what mischief the book has done & is doing among our London blues [bluestockings], & God willing I will strive to abate the evil." He would, he promised Napier, stamp with an "iron heel upon the head of the filthy abortion, and put an end to its crawlings."

For the canon of Norwich Cathedral, the professor of geology, the vice-master of Trinity College, *Vestiges* was the whore of Babylon, a feathered and bejeweled crone, a snake, and a not-quite-dead abortion. He set to work on the book, taking his pencil to page after page in fury. His ferocious annotations survive him. Although his published review was more measured, his sense of moral apocalypse was barely modified.

He warned his readers that *Vestiges* "comes before them with a bright, polished, and many-coloured surface, and the serpent coils of a false philosophy, and asks them to stretch out their hands and pluck forbidden fruit." The review he began in May grew to be eighty-five pages before he had finally vented his spleen and his misogyny.

Sedgwick was fighting a series of battles, not only against the spread of materialism and the increasing articulacy of atheists, but against the popularization of science, the rise of the nonspecialist man of science, Catholicism, the power of women, and the corrupting effect of urbanization on the intelligentsia. The result was venomous.

Members of the editorial boards of the Christian journals sighed with relief. The Anglo-Catholic *Christian Remembrancer* praised Sedgwick's "masterly essay" for placing the dangers of the development hypothesis "beyond all further controversy." In Scotland, where the Scottish evangelical party had broken away from the Established Church of Scotland to form the Free Church, *Vestiges* was perceived to be an even greater threat to the authority of the Church. The split was seen as having weakened the Churches against the forces of secular liberalism. When Chambers published *Vestiges* in 1844, the Combes and their friends, reformist freethinkers in Edinburgh, felt themselves an increasingly beleaguered minority surrounded by evangelicals. "We seem to be standing on the verge of a vast volcano," wrote one Scottish journalist in 1847, "ready to explode and overwhelm us with terrible destruction." The Scottish Church called on fine writers such as the evangelical geologist Hugh Miller to refute *Vestiges,* to challenge the anonymous author on his own ground. In September 1845, Miller referred to *Vestiges* in the *Witness* as "one of the most insidious pieces of practical atheism that has appeared in Britain during the present century." Not content with refuting the book in a review, he published a deeply hostile book, *Foot-prints of the Creator,* in 1849.

Miller's rhetoric was quite different from Sedgwick's. He saw no whores of Babylon, no foul crones, no abortions to be stamped upon. Instead he saw swamps, infection, and sinking sands. He declared that "the lower levels of society had sunk into a miasmatic marsh out of which poor law assessment, fierce revolutionary outbreaks, plagues, and pestilence, threaten to arise and envelope in indiscriminate ruin the classes above." The Church must make its most valiant attempts at "draining and purifying the bog."

Infidelity was the common enemy of all the churches of Scotland, high and low; they used a language of warfare to articulate their sense of being under siege. *Vestiges,* churchmen complained, had become a banner that the infidels rallied beneath; it had become their fiery cross. The Reverend Graham Mitchell dedicated a whole chapter to denouncing the book in his *Young Man's Guide Against Infidelity* (1848); other pulpit refutations followed. The Reverend Dr. John Brown, in a keynote speech for the formation of the Scottish Association for Opposing Prevalent Errors, declared to the alarm of his audience: "Infidelity is of the spirit of the present and of the spirit of the future. It is coming in stronger and stronger." And the Reverend Andrew Thompson declaimed, "Infidelity . . . like the spider, was hanging out its webs on many a beautiful tree of knowledge, and but all too often, alas! it succeeded in ensnaring the unwary passer by."

Vestiges raised hackles for decades, drawing the most aggressive rhetoric from otherwise restrained men. As late as 1854, when the tenth edition of the book was published, a decade after its first publication, Thomas Henry Huxley, who, though he would later become Darwin's most outspoken and ardent supporter, was then a sworn enemy of transmutationism, wrote a review. He began with the lines from Shakespeare's *Macbeth,* words uttered by Macbeth as Banquo's ghost appears before him yet again: "Time was, that when the brains were out, the man would die." Comparing the book to Banquo's unsilenceable ghost was not quite the same as comparing it to a not-quite-dead abortion, but Huxley meant much the same thing: Why does the hand of transmutationism still rise from the lake we have tried to drown it in? Why will it not die?

But *Vestiges* also inspired people, challenged them, enlarged their sense of the dimensions of the past and of the future, persuaded them that transmutation was possible, even probable, and that it need be neither godless nor dangerous to think so. In 1846 Florence Nightingale, then aged twenty-six and trying to educate herself in anatomy in preparation for a vocation in nursing, visited the museum of the Royal College of Surgeons with a friend. The two women stopped to admire the series of skeletons of flightless birds from New Zealand and took the time to trace out anatomical connections between the bone structures of the recently found fossil skeleton of the giant moa, an extinct bird bigger than the ostrich, and smaller living species. "The thing that was most curious of all," she wrote to her cousin at school, "was to see how the species ran into one another, as *Vestiges* would have it." In London, few people huffed and puffed like Sedgwick. Instead they were fascinated by the new understanding of time *Vestiges* had given them.

With good manners and careful social strategies, Chambers remained unidentified. Through the 1840s he continued tending his still growing family and his printing presses, taking geological trips, and trying to establish himself as a serious writer on geology. It was not until November 1848, four years after *Vestiges* was published, that his reputation suffered a serious blow. And he himself might have admitted that it was due to a lapse of judgment on his part. He had allowed his name to go forward for the position of lord provost of Edinburgh, the city's highest civic honor.

Within a few days of the announcement, a long letter in the *Witness* in November 1848 expressed "grave objections" to his candidacy. "Every one knows," the anonymous letter writer declared,

> that Mr Chambers studiously excludes all religious subjects and references from his periodicals; and that notwithstanding the vast multitude of papers of all kinds which he has written or published, it would be difficult to gather from any one of them that a God exists, or that a way of salvation from sin has been revealed. He is, indeed, the great representative of our non-religious periodical press. This is of itself seriously criminal. But he is charged with even worse than this. He is charged with writing and sending forth to the world a work which, not to speak of its false and superficial science, "expels the Almighty from the Universe" and renders "the revelation of His will an incredible superstition" (*North British Review*); which "tells us that our Bible is a fable, when it teaches us that man was made in the image of God" (*Edinburgh Review*). I do not say that Mr Chambers *is* the author of this revolting production; but he is *alleged* to be the author.

Chambers, worried for the reputation of the publishing house and horrified at his continuing persecution by the Church, withdrew his candidacy in disgust.

Charles Darwin, in his thirties with his theory of natural selection locked away in his study drawer, watched from the wings and bided his time. His own ideas about species would have to wait, he realized, if he was to avoid the opprobrium being meted out to the anonymous author of *Vestiges.* The book was lively enough, he noted to himself, but it was full of flaws. Only when there was a weight of evidence, only when there was a theory, a mechanism for how species have adapted and evolved, would anyone treat the theory with any respect. For now, Darwin determined to keep his drawer locked.

His closest friend, the botanist Joseph Dalton Hooker, wrote to Darwin on December 30, 1844, declaring that he had "been delighted with *Vestiges,* from the multiplicity of facts he brings together, though I do [not] agree with his conclusions at all, he must be a funny fellow: somehow the book looks more like a 9 days wonder than a lasting work: it certainly is 'filling at the price.'—I mean the price its reading costs, for it is dear enough otherwise; he has lots of errors." Darwin replied that he had been "somewhat less amused at it . . . the writing & arrangement are certainly admirable, but his geology strikes me as bad, & his zoology far worse." In October 1845, Darwin wrote to Charles Lyell that Sedgwick's review was a "grand piece of argument against mutability of species" that he had read with "fear & trembling," but had been "well pleased to find" that he had anticipated Sedgwick's objections and "had not overlooked any of the arguments."

Vestiges became both a thorn in Charles Darwin's side and a warning. If *Vestiges* was a fiery cross, a call to arms, then Darwin failed to respond to it. Instead he resolved to wait to see how the book fared, to let others rally to its cry. He read Chambers's supplementary essay "Explanations" early in 1846 and thought "the spirit of [it], though not the facts, ought to shame Sedgwick." In April 1847, having met Chambers and received a gift of *Vestiges* anonymously a few days later, he was now convinced that Chambers was the author.

For ten years Darwin shifted positions uncomfortably, grappling with his own conscience, unable to decide whether he belonged with the jeering authorities of science and the Church who were throwing stones or with the man who belonged to the clan he knew he also owed allegiance to, the group he had called "us transmutationists" in a letter to Joseph Hooker in 1847. When Huxley sneered at the author of *Vestiges* in his review of the tenth edition in 1854, Darwin, who was cultivating Huxley as a valuable and influential friend, told him that he thought the review had been "rather hard on the poor author." At the very least "such a book if it does no other good," he persisted, "spreads the taste for natural science." And he worried for the reception of his own book, which, he confessed to Huxley, would be "almost as unorthodox about species as *Vestiges* itself, though I hope not *quite* so unphilosophical." In the introduction to *Origin,* Darwin mocked rather than defended the book: "The author of the 'Vestiges of Creation' would, I presume, say that, after a certain unknown number of generations, some bird had given birth to a woodpecker, and some plant to the mistletoe, and that these had been produced perfect as we now see them; but this assumption seems to me to be no explanation, for it leaves the case of the coadaptations of organic beings to each other and to their physical conditions of life, untouched and unexplained."

When the still anonymous author cried unfair in the next edition of *Vestiges,* declaring that Darwin's *Origin* "expresses substantially the same general ideas [as *Vestiges*]. . . . The difference seems to be in words, not in facts or effects," Darwin, his conscience spiked, removed the offending passage from the third edition of *Origin* (1861) and all subsequent editions. In his "Historical Sketch," he tried to make amends, writing in more measured tones:

> The author . . . argues with much force on general grounds that species are not immutable productions. But *I cannot see* how the two supposed "impulses" account in a scientific sense for the numerous and beautiful co-adaptations which we see throughout nature; *I cannot see* that we thus gain any insight how, for instance, a woodpecker has become adapted to its peculiar habits of life. The work, from its powerful and brilliant style, though displaying

> in the earlier editions little accurate knowledge and a great want of scientific caution, immediately had a very wide circulation. . . . In my opinion it has done excellent service in this country in calling attention to the subject, in removing prejudice, and in thus preparing the ground for the reception of analogous views.

Darwin was not just being polite. He was right. *Vestiges* had brought transmutation to the people; its author had shown his readers how fascinating it all was, persuaded them how incomprehensibly old the earth must be, how the edges of bodies, even the apparently solid structures of skeletons, were not fixed in stone, but were adaptable, that animals and humans were all joined up, part of an extraordinary moving, spinning, webbed, and interconnected world. He showed them that to hold these thoughts, to imagine those transformations, was neither godless nor dangerous. He had gotten some of his facts badly wrong; in the interest of making a gripping and continuous narrative, he had overstated some things and understated others; and he had failed to find a mechanism by which species might have evolved and diversified through time. But after *Vestiges* there was no suppressing any of those questions about origins and time. It is difficult to see how Darwin's *Origin* might have fared in the world without the ruffling of theological feathers, the raising of dreams, the firing of imaginations and conversations that *Vestiges* brought into being.

12

Alfred Wallace's Fevered Dreams

MALAY ARCHIPELAGO, 1858

In a thatched hut on a volcanic star-shaped island rising leafy green from the sea off the coast of northeastern Indonesia, a young English specimen collector lay wrapped in blankets struggling with the intermittent fevers of a malaria attack. For ten years he had depended on the sale of the birds and insects he collected out on these remote islands; from dawn to dusk he had hunted, killed, set and opened traps, then stuffed and skinned, pinned and wired scores of rare specimens every day—butterflies, birds, beetles, lizards, small mammals—labeling them carefully and crating them up to be sent home to his natural history dealer in London. Now he could do almost nothing. Every day fever descended on him around the same time, like a visitation; for hours every afternoon he shivered, convulsed with cold, then burned up, dripping with sweat. Strange waking dreams tormented him. Cold, hot, wet. He waited it out, marking off the stages as they passed. He had the worst kind of malaria, his eighteen-year-old Malay assistant, Ali, told him, wiping his head, the kind people died from. He had to rest.

Photographic portrait of Ali, Wallace's Malay assistant, aged around eighteen, and a portrait of Wallace himself, aged around twenty-four.
Alfred Russel Wallace, My Life *(1905)*

But Alfred Russel Wallace did not sleep. As the afternoon sun dipped, and as Ali continued to label and record a few insects and then slipped away for evening prayer, Wallace shook and sweated, pulling blankets off and on as his temperature rose and fell, his mind wandering over the set of questions that had been pressing upon him for years now, bringing him first to Brazil and now to the Malay Archipelago. Nothing else excited him as much as this great mystery: How did all these wondrously diverse species *come to be*?

Sometimes in these fevered hallucinations Wallace talked aloud to himself, or to Ali, or to absent friends, acquaintances, and correspondents—to his fellow collector and friend Henry Bates, who was somewhere on the Amazon, or to the naturalists Charles Darwin and Charles Lyell as he imagined them back in England in their clubs or walking in their rose-heavy English gardens, or to Lamarck or the anonymous author of *Vestiges*. Imagined and familiar conversations branched off in all directions. *All species have evolved, adapted, and mutated. All living organisms have descended from earlier forms.* All of his work as a species collector and as a geologist had confirmed transmutation to be true, a self-evident fact. He had published a paper about the species question; it had drawn the attention of some important people. But the *mechanism,* the *means by which* evolution happened, still evaded him. It had, he believed, evaded everyone.

Years later Wallace remembered the feverish bolt-from-the-blue moment of realization with absolute clarity. In the midst of a daydream, the air of his hut thick with the smell of dead insects and exotic birds, "somehow my thoughts turned," he wrote, to a book he had read in a drafty public library in Leicester fourteen years earlier: Thomas Malthus's *An Essay on the Principle of Population*. For some reason he remembered Malthus's clear description of the "checks" that stopped populations from growing—disease, accidents, war, famine—"which keep down the population of savage races to so much lower an average than that of more civilised peoples."

A violent history of earthquakes, floods, volcanic eruptions, malaria, and famine had left its scars on all the islands of the Malay Archipelago and had periodically decimated human and animal populations. "Vaguely thinking over the enormous and constant destruction which this implied," he continued,

> it occurred to me to ask the question, *Why do some die and some live?* And the answer was clearly, that on the whole the best fitted survive. From the effects of disease the most healthy escaped; from enemies, the strongest, the swiftest, or the most cunning; from famine, the best hunters or those with the best digestion; and so on. Then it suddenly flashed upon me that this self-acting process would necessarily *improve the race,* because in every generation the inferior would inevitably be killed off and the superior would remain—that is, *the fittest would survive.* Then at once I seemed to see the whole effect of this. . . . The more I thought over it the more I became convinced that I had at length found the long-sought-for law of nature that solved the problem of the origin of species.

Wallace had to wait until the fever began to pass before he could drag himself from under the mosquito net to his desk, put on his glasses, and try to get the idea down on paper in sentences that did not tangle. It is easy to imagine the young naturalist's stupefaction, his hands shaking, inky words blurring on the paper in front of him, asking himself the driving question one last time: How had so many species come into being? The answer was breathtakingly simple: Only the fittest survived. The others fell by the wayside, destroyed in vast numbers in famines, plagues, and earthquakes. In this way, generation by generation, animal bodies modified, adapted to changing conditions, to hotter, wetter, or colder climates. New species came into being as the furriest, or the scaliest, or the tallest, or the individuals with the largest eyes survived and passed the thicker skin or fur or largest eyes on to the next generation.

Did Wallace try to explain to Ali what he could now see? *Could* he explain it? Even in England there were only a handful of people who had read enough, thought and puzzled enough, to be able to understand the immense significance of his new idea. Wallace's friend the entomologist Henry Bates would understand; so would the naturalist and explorer Charles Darwin, who was writing a book about the origin of species. Then there was the geologist Charles Lyell, knighted in 1848, who, despite his passionate opposition to transmutation, understood the species question. But out here in this hut in the Malay Archipelago, there was only Ali. Ali was smart, a fine collector, a hard worker. What did he make of his master's ideas about species, this man who was also his friend and teacher? Ali was a Muslim, and Wallace's ideas about the mutability of species would have contradicted the creation story in the Qur'an, that God created the earth in one breath and breathed life into man at the very beginning, marking him out from the rest of creation as special and divinely ordained. Man, Islam had taught Ali, was different from the animals. Did Ali challenge his master's assumptions, or, if he could see the truth of Wallace's arguments, did he find a way to square evolution with his belief in the Qur'an?

Wallace knew it was Sir Charles Lyell he must persuade above all others. By vigorously rebutting Lamarck's ideas, Lyell had persuaded British men of science that there was no more to be said on transmutation. He had become the gatekeeper of the species question. No one would listen or be won over until Lyell listened and was won over. Over the course of two nights, as Wallace waited for the fever to lift and his hands to stop shaking, he wrote his idea into a short essay that engaged in imagined conversation with Lyell in every line, an essay tailored for publication in a British journal of science. Then he and Ali sailed back across the sea to Ternate.

Conversations with people in England were always behindhand. Although Wallace was thrilled to recognize Darwin's handwriting on a letter in the pile that had arrived in Ternate in his absence, he knew the letter had been written three months earlier. Darwin's letter was full of encouragement for Wallace's work. If there were spots of occasional reticence and a slightly patronizing tone here and there, Wallace put that down to reserve rather than territorialism. Darwin reassured him that the right people—including Sir Charles Lyell himself—were reading and admiring Wallace's work. He told him that his own book on species was going very slowly and that he had no plans to discuss the origins of man. Nothing could have been better news for Wallace. He wrote out the words "Ternate, February 1858" on the essay, addressed it to Charles Darwin at Down House, Kent, and asked him to send the essay on to Lyell. "I said that I hoped," he remembered later, that the idea "would be as new to him as it was to me, and that it would supply the missing factor to explain the origin of species."

Over the next fifty years, Wallace told that story so often that it became a kind of personal myth: the fever, the chills, the remote location, the sudden, searing understanding of how the fittest survived, how he had to wait for the shivering to pass before he could put pen to paper. There is no reason to doubt any of his story—Wallace was honest, careful with detail, and an accurate record and diary keeper; he had a sharp memory. It is certain that in February 1858 in the Malay Archipelago, the final piece of a fourteen-year-long piece of detective work had fallen into place for him. Once the letter containing his paper had been safely delivered onto the mail boat to England, Wallace knew he would have to wait months for a reaction.

The consequences are well known. Receiving Wallace's letter, Darwin was devastated; he was certain that though he had discovered natural selection twenty years earlier, Wallace would publish before him and claim the prize. He called on Charles Lyell and his friend the eminent botanist Joseph Hooker and explained the situation. Lyell and Hooker presented the two papers at the Linnaean Society, asking the assembled members to make a judgment as to which man had discovered natural selection first. On July 1, 1858, the Linnaean Society members considered the evidence and gave their verdict: though both men were to be congratulated, they declared, Darwin had first recorded the idea and had it witnessed by others in 1844 in an unpublished essay. A judgment had been made about priority; nothing more. No bishop lost sleep or denounced Darwin or Wallace from the pulpit. All of that noise and outrage would not begin until Darwin's now-rushed-to-print book, *On the Origin of Species by Natural Selection,* reached the bookshops on November 24, 1859.

Wallace always described his idea as a revelation, a road-to-Damascus moment that "flashed upon" him. But though the idea surfaced from his fevered brain all in one piece like a wreck from the sea, he had been diving for it for years. Remembering Malthus's theories of natural population control had given him the missing piece of the jigsaw, the constellation point, but there had been other similar moments of startlingly new understanding along the way that had contributed incrementally to Wallace's remarkable discovery.

For as long as he could remember, Wallace had been interested in the differences between human races and in the geographical borders that marked the edges of those gradations. He grew up on the border between Wales and England. The hills ten miles off near Abergavenny, which he could see from his bedroom window, marked the beginning of "the unknown land of Wales," a place where people talked differently and where they looked different, too, though it was always difficult to say exactly where one country began and the other ended because of migrations and intermarriages. Because he was tall and had strikingly flaxen hair, the people of Usk called Alfred, the youngest of the Wallace children, the "little Saxon." When he asked about that word and what it meant, no doubt someone explained about the different races that lived in Britain—the Celts, the Saxons, and the Anglo-Saxons—and how they had arrived and settled in different places and how Scotland, Wales, and Ireland were mostly occupied by Celts and the rest of the country by Anglo-Saxons. The inseparability of geography and race impressed itself on him, just as it had struck Jahiz a thousand years earlier as he watched the Bedouins coming into the marketplace or the pilgrims on the road to Mecca or the black slaves working out in the salt marshes and wondered how these physical differences had come to be so closely bound up with landscape.

Like many lower-middle-class families in the 1830s and 1840s who were forced to survive by following work, the Wallace family migrated from town to town, educating their children as well as they could along the way. With the best of intentions, Wallace's father, a small-time lawyer, speculated on railways and other financial ventures, but he was neither lucky nor a good judge of investments, and the family income was always precarious. Wallace's education was patchy. But as literacy grew, book clubs and provincial libraries flourished. Wallace's father belonged to a book club for a few years through which, Wallace later recorded, "we had a constant stream of interesting books, many of which [my father] used to read aloud in the evening": travel books, novels, poetry, Swift, Scott, Defoe.

Of all the books his father read aloud by candlelight in the family living room in Hertford, it was Daniel Defoe's *History of the Great Plague* that Wallace remembered most vividly later. With the shocking detail of an eyewitness, Defoe described the three weeks in 1665 when plague wiped out a third of the population of London. He described the anguish with which the survivors searched for a meaning for the devastation they had seen. "Nothing but the immediate finger of God, nothing but omnipotent power, could have done it," Defoe concluded. "The contagion despised all medicine; death raged in every corner; and had it gone on as it did then, a few weeks more would have cleared the town of all, and everything that had a soul. Men everywhere began to despair; every heart failed them for fear; people were made desperate through the anguish of their souls, and the terrors of death sat in the very faces and countenances of the people." Wallace would later come to understand this tragic account in Malthusian terms.

The Wallace family attended church, but only once did Alfred find himself moved to anything he would have called religious fervor; churchgoing was nothing more than social ritual to him. Thus Defoe's descriptions of plague as a divine punishment seemed both superstitious and inadequate as explanations for such violent decimations of life. When his father took a job in a local lending library, Alfred spent every spare moment of his days there, "reading, squatting down on the floor in a corner." He read adventure stories, novels by Marryat, Cooper, Bulwer, Smollett, Godwin, and Fielding, poetry by Byron, Scott, Pope, and Milton. He was, like so many clever young men of his generation, self-educated and following his changing curiosities at his own pace.

London gave the young Wallace a political education. Sent to the capital to live for a few months with his elder brother John in Hampstead when he was only fourteen, Wallace sat among the wood shavings in the workshop where John was apprenticed as a joiner making staircases and doors and windows, listening to the young carpenters talk and argue about change, reform, and class. It was 1837. The working classes of London would rarely be so actively radical and united again. The first Reform Bill of 1832 had granted the vote to the middle classes but not the working classes; now angry young men were demanding a further extension of the franchise. In meetings in the Hall of Science, off Tottenham Court Road, which the teenage Wallace boys attended most nights, young men passed around books by the radical social reformer and socialist Robert Owen; lecturers explained how Owen's reforms would bring freedom for the enslaved masses and release them from the bigotry of religion. All men are products of their environment, they preached, responsible for their own destinies. Wallace was intoxicated.

Such an itinerant early life, led among self-educated, politicized British working men, made Wallace a freethinker, highly moral, socialist, and secular. When he left London a few months later to be apprenticed as a land surveyor in Bedfordshire to the eldest of his brothers, William, Wallace found that there was "a pervading spirit of scepticism, or free-thought as it was then called," in his elder brother's circle too, "which strengthened and confirmed my doubts as to the truth or value of all ordinary religious teaching." William borrowed controversial books from friends or from the local mechanics' institute that Alfred read surreptitiously, including a book of lectures on David Friedrich

Strauss's scandalous book *The Life of Jesus, Critically Examined,* which argued that the miracles in the New Testament were no more than the myths that inevitably grow up around the lives of great men. The Earl of Shaftesbury would later call *The Life of Jesus* "the most pestilential book ever vomited out of the jaws of hell." This was dangerous territory: although Wallace's family were moderate in their religious observance, "the word 'atheist' had always been," he wrote, "used with bated breath as pertaining to a being too debased almost for human society." By adulthood, all of the Wallace boys were religious skeptics or atheists passionately committed to some form of socialism and to the education of the masses.

In Wales, where the Wallace boys picked up more land-surveying work in the late 1830s, they witnessed desperate struggles for land and for survival. Through the newly passed Enclosure Act and the Tithe Act, the government was systematically carving up and appropriating common land, land owned by the people, so as to tax its use, and they were using land surveyors like the Wallace boys to implement it. In Wales these new acts were putting impossible pressures on desperate people at a time of widespread famine after failed harvests, making them pay taxes for their land and tolls for access to roads and robbing them of their common grazing land. With the benefit of hindsight Wallace would later call the Enclosure Acts land theft, but at the time he and William were simply doing a job. As they worked, mobs of young farmers and agricultural workers across Wales smashed up tollgates at night, attacked workhouses, and set fire to the homes of wealthy landowners. Wallace, still not yet twenty, zealous and naive, drew up plans and procedures for bringing a mechanics' institute and science library to Neath, convinced that these farmers needed education, not violence. *Education is power,* he would repeat to anyone who would listen.

The two brothers read voraciously in the evenings, either in local libraries or workingmen's clubs or in their lodgings, borrowing books from circulating libraries or buying cheap editions, when they could afford to, from local booksellers. A bookseller in Neath, Charles Hayward, charmed by Alfred's boyish enthusiasm and curiosity, introduced him to books, journals, and magazines on all aspects of science. Alfred bought a book on botany and started identifying local plants in the streets and fields and on the banks of rivers, keeping notebooks, looking for patterns of distribution. Within a few months he had become a competent amateur botanist with a good broad knowledge of British flora. Soon he was reading William Swainson's *Treatise on the Geography and Classification of Animals,* which made him think more precisely about the relationship of species to landscape.

In Leicester, where Wallace moved in 1844 to take up a position as a teacher at the Collegiate School when his brother's business began to decline, he spent all his spare time in the public library, which had the finest collection of books he had come across in his travels thus far. Here in the cold and drafty library, he reread *The Voyage of the Beagle,* astonished by Darwin's descriptions of South American forests, birds, flowers, and strange peoples; then, inspired by Darwin's enthusiasm for Humboldt, he read Humboldt's *Personal Narrative of Travels in South America* and Prescott's *History of the Conquests of Mexico and Peru.* While teaching interminable rote-learned Latin lessons, in which he barely kept ahead of his students, Wallace dreamed of rain forests, of putting himself in Darwin's shoes, feeling the barrel of a gun in his hand, taking aim at imaginary birds, riding on horseback up into remote mountain ranges in search of new species.

Sometime in 1844, Wallace took down Malthus's *Essay on the Principle of Population* from the shelves of the Leicester library, a book that made a bleak contrast with the color and exoticism of Darwin's South American landscapes. "It was," Wallace wrote later, "the first work I had yet read treating of any of the problems of philosophical biology and its main principles remained with me as a permanent possession, and twenty years later gave me the long-sought clue to the effective agent in the evolution of organic species."

Wallace was lucky. At the Leicester public library he not only found Malthus but also met another reader, Henry Walter Bates, the clever, bookish nineteen-year-old son of a local hosiery manufacturer. Bates was more interested in natural philosophy in general and beetles in particular than he was in running his father's factory. Eager to impress his new friend, he challenged Wallace to guess how many different beetles might be found in the small district near a town. Wallace, who now knew a good deal about local plant diversity, knew very little about beetles. He guessed fifty. Bates took Wallace to his father's elegant redbrick house in Queen's Street, where he showed him the hundreds of different beetles he had found and identified in the streets and parks of the city and had pinned and carefully arranged behind glass in his study. There were, he told Wallace breathlessly, perhaps a thousand different types within ten miles of this single town. He opened a copy of a thick volume on his shelves to show Wallace descriptions of more than three thousand British species.

Wallace was fascinated by the close attention Bates had given in his collection to the location and range of the Leicester beetles. Bates was already a biogeographer, though neither he nor Wallace would have used the word. In ninth-century Basra, Jahiz, making lists of species attracted to the light from a night campfire, had been one, too. Wallace was a convert to Bates's methods. He invested in a collecting bottle, pins, and a storage box and roamed the local bookshops for a copy of Stephens's *Manual of British Coleoptera* at a wholesale price. Every Wednesday and Saturday afternoon, as the summer blossomed in Leicestershire, he took his pupils away from their Latin lessons and walked them to the grounds of Bradgate Park, a "wild, neglected park with the ruins of a mansion," where they collected and identified beetles with Bates and his younger brother Frederick, setting sugared traps, turning over leaves, and kicking over rotted logs and rocks. Every scuttling new find was carefully numbered, labeled, pinned, and added to a map.

Wallace never settled anywhere for long. Compelled to return to Wales in 1846 to attend to William's business after his brother's sudden death from pneumonia, Wallace extended his search to Welsh insects, collecting them through the summer along the line of levels up the Vale of Neath to Merthyr Tydfil, where he took a job as a surveyor for the new railway. He carried beetle jars and boxes in his bag with his surveying equipment and wrote every few days to Bates, sending him rare specimens and details of his finds. Other family members migrated back to Wales that summer, but Wallace felt restless. His sister, recently returned from teaching at a college in Macon, Georgia, told him stories of cotton fields and cotton boats, the coming of the railways, the slaves and the native Americans and unimaginable beetles and butterflies for which she had no name. They all talked of emigration.

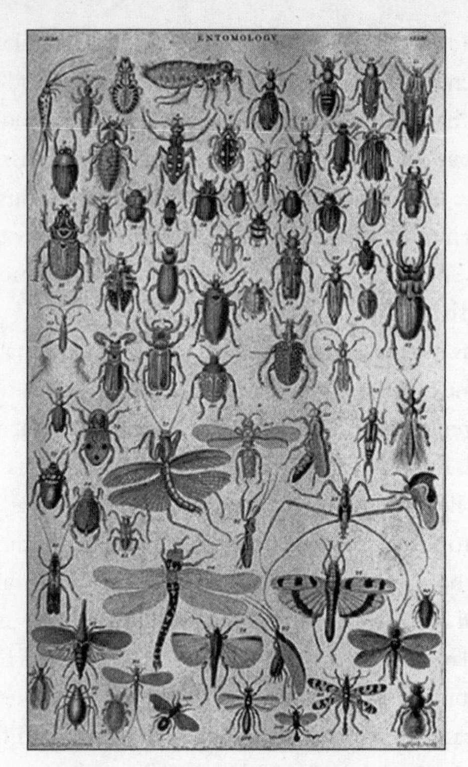

Insects arranged as they would be in a cabinet
(nineteenth-century engraving).
Engraving by R. Scott after T. Brown,
Wellcome Library, London

Wallace may have been restless in Wales, impatient for a clear sense of vocation and direction, but he was not bored. The scandalous and much discussed red-leathered book *Vestiges of the Natural History of Creation* arrived in Neath sometime in 1845. In Welsh scientific circles, as it was in London, *Vestiges* was the book of the season. Wallace was utterly beguiled by the book's speculations; the implications of species transformation were breathtaking. He wrote to Bates asking whether he had read it, only to discover to his disappointment that Bates thought the book generalized. Wallace replied in robust defense:

> I do not consider it as a hasty generalization, but rather as an ingenious speculation strongly supported by some striking facts

> and analogies but which remains to be proved by more facts & the additional light which future researches may throw up on the subject— It at all events furnishes a subject for every observer of nature to turn his attention to; every fact he observes must make either for or against it, and it thus furnishes both an incitement to the collection of facts & an object to which to apply them when collected.—I would observe that many eminent writers give great support to their theory of the progressive development of species in animals & plants.

He told Bates to read Lawrence's *Lectures on Man.* If *Vestiges* would not persuade him about the "progressive development of species," perhaps Lawrence's essay would.

Vestiges was indeed an incitement to action. It seemed to Wallace as though he had been waiting for this book all his life. It put together everything he had been baffled by, drew all his interests and questions together—geology, plant and animal distribution, astronomy, species diversification, fossils, variation—into a narrative of unstoppable species change that was utterly persuasive. It was self-evident. If there wasn't enough evidence to support the *Vestiges* hypothesis yet, someone had to find it. Here was his vocation, his reason for being. He would collect, not as a pastime, not in his spare time, not in the fields of Wales, but on islands and in jungles and river valleys where no one had been before. He would collect evidence to prove species mutation, and Bates would come with him.

Wallace plunged into preparations for a journey even before he and Bates had agreed on a destination. When Bates came to visit, an excited conversation about W. H. Edwards's recently published *A Voyage Up the Amazon* settled them both on Brazil; now they had to find ways to fund the journey. Neither of them had any money. Bates was working as a clerk for Allsopps, the brewers at Burton-on-Trent. His father thought his travel plans were preposterous and refused to fund the journey, so they began to wonder if they might make their living from collecting and selling specimens along the way. Fanny Wallace, a keen Francophile and traveler, took Wallace to London and then to Paris to show her brother the collections in the British Museum and the Jardin des Plantes and served as interpreter in his conversations with professional collectors, agents, and taxidermists there. "I begin to feel rather dissatisfied with a mere local collection," he wrote to Bates, "—little is to be learnt by it. I shd. like to take some one family, to study thoroughly—principally with a view to the theory of the origin of species. By that means I am strongly of opinion that some definite results might be arrived at."

The two men went to meet Samuel Stevens, a natural history agent and dealer based on Bloomsbury Street, and offered him their services. They talked to as many potential buyers and museum curators as would give them the time of day, but getting buyers to take them seriously was not always easy—they were, after all, the sons of tradesmen and barely out of boyhood. Finally, when they were confident that Samuel Stevens would buy specimens from them, they bought their tickets for the boat, packed up small trunks of belongings, and prepared to depart.

The two collectors found Brazil difficult at first, even disappointing. The rain forests seemed oddly empty. "On my first walk into the forest," Wallace wrote, "I looked about, expecting to see monkeys as plentiful as the Zoological Gardens, with humming-birds and parrots in profusion." They were just not visible. As the months went by and their eyes readjusted, they began to see—and catch—more specimens. Stevens was delighted with the first crates of insects that arrived in London and gave the two collectors good prices for them. The two friends separated for a few months, heading in different directions in order to cover more territory. "The more I see of this country, *the more I want to,*" Wallace told Stevens. It took Wallace most of his time in Brazil to adjust to the cultural differences, to learn his trade, to experiment with methods of preservation, storage, and trapping, and to find ways of getting around, sleeping, eating, communicating, and dealing with ants, fevers, and loneliness.

Everywhere he went he seemed to be treated like a rare species himself: "One of the most disagreeable features of travelling or residing in this country is the excessive terror I invariably excite," he wrote.

> Wherever I go dogs bark, children scream, women run & men stare with astonishment as though I were some strange & terrible cannibal monster. . . . One day when in the forest an old man

> stopped to look at me catching an insect. He stood very quiet till I had captured, pinned and put it in my collecting box when he could contain himself no longer, but bent himself almost double & enjoyed a hearty roar of laughter.

Wallace rarely complained. Even after he lost his entire cargo of exotic species, his notes, and everything he owned when the ship in which he was returning to England caught fire and sank, he tried to be optimistic. "And now everything was gone," he wrote, "and I had not one specimen to illustrate the unknown lands I had trod, or to call back the recollection of the wild scenes I had beheld! But such regrets I knew were vain, and I tried to think as little as possible about what might have been and to occupy myself with the state of things that actually existed." It was a temperament that would shield him from disappointment later.

Back in England, his notes now scattered somewhere on the ocean floor, among broken crates of beetles and birds, Wallace was forced to distill and summarize, to look for the general laws beyond the particular facts he had gathered. He published two books, his *Narrative of Travels on the Amazon and Rio Negro* and a shorter book, *Palm Trees of the Amazon and Their Uses.* As the details fell away he began to think on a larger scale about the geographical limits of species. Since he had seen Bates's beetle collection, he had always been scrupulous about geographical precision in his collecting. Species range was of utmost importance to him. All the different species of monkeys had a specific geographical area in which they roamed, he told his audiences, the edges of which were marked by rivers or mountain ranges. "I soon found," he told the Zoological Society back in London, "that the Amazon, the Rio Negro and the Madeira formed the limits beyond which certain species never passed."

By the time Wallace reached the Malay Archipelago in 1854, five years after he had determined to become the proof gatherer for the still unproved theories of progressive species development outlined in the scandalous *Vestiges,* he was in a unique situation among European naturalists: poor but financially independent, he was free from obligations to an institution, patron, or business, and he had no family to protect from embarrassment or personal wealth to risk. He had come to see and respect the diversity of religious belief and ritual practiced all around him in the jungles and villages of the Amazon, but he did not have to reconcile his discoveries with a personal creed or God. He had long accepted transmutation as an explanation for species diversity, but he was still looking for the mechanism by which species had diversified. Brazil had not given him the weight or density of species he needed to find that mechanism, and the sheer labor of collecting and surviving had kept him distracted by detail.

But things were about to change.

Islands were crucial to the discovery of natural selection for Wallace, as they were for Darwin. Wallace had come to the Malay Archipelago, a scattered string of islands off Java, primarily in search of birds of paradise and other rarities that would fetch high prices back in London, but these islands, each with its own unique ecosystem, would also provide him with a series of natural laboratories in which to test his developing theories. What Wallace was looking for—evidence of geographical species range that would help him understand how species had come to be—was much more intensely illustrated in the Malay Archipelago than it had been on the Amazon. He had found his way to the right place.

Wallace's mind and imagination moved constantly between the large-scale and the small-scale. During the day he worked to identify the minute differences between species, attending to plumage or butterfly wing or beetle thorax patterns and colors; at night he pored over maps, studying landmasses that covered thousands of miles, plotting out the edges and overlaps between different species ranges and thinking back deep into time. Like Jahiz, he was interested in the relationship between species and location, asking, for instance, why one moth lived in the desert while another lived only in the mountains. But as a biogeographer, Wallace was doing something quite different from Jahiz. Jahiz had had no thesis to prove. The diversity and range of moths that gathered in the light of the desert fire at night, the fact that each moth or insect seemed to occupy its own unique landscape, demonstrated for him the ubiquity of Allah's grace and the brilliance of his natural design. When Wallace mapped out the territories of primates, however, noting that there seemed to be a "boundary line" that the great apes "never pass," he was looking for clues that would help him understand how species had diversified over millions of years.

Wallace was getting closer to an answer to the species question. In 1855, while in Sarawak, he decided to nail his colors to the mast and publish a paper on transformism. The paper was published in the *Annals and Magazine of Natural History* in 1855. No one wrote to him about it. No one seemed even to have noticed it. Wallace felt the silence keenly.

In 1856, he made a major breakthrough. Sailing for the eastern end of the archipelago, he missed a connection and was stranded in Lombok and Bali for two months with Ali. The two islands, he pointed out to Ali as they recorded and labeled hundreds of specimens, though they were very close geographically and similar in terrain, seemed to be two "quite distinct zoological provinces." One set of birds belonged zoologically—and, by inference, by descent—to Australia, the other to Asia. Wallace drew a line on the map laid out in his hut, a line that formed the boundary between what we would now call ecozones, large areas of the earth's surface where animals and plants developed in relative isolation for thousands of years, the landmasses on which they lived separated from one another by oceans or deserts or mountain ranges that prevented animals and plants from migrating. The line he had drawn, the Wallace Line, would make his name famous. Lombok and its animals and plants lay on one side of it and Bali on the other. Unknown to Wallace, the line followed the course of a deep-sea channel marking the contour of the continental shelf miles beneath the surface of the sea.

Wallace's line divided the ranges of human races, too. In January 1857, when the monsoons turned the rice fields of the coastal plains around Macassar into swamps and rotted all his specimens, a half-Dutch, half-Malay sea captain offered to take Wallace out to the Aru Islands, a tiny and legendary cluster of islands a thousand miles east, and bring him back six months later. As Wallace's boat arrived at the island of Ke, off the south coast of New Guinea, his Malay crew were mobbed by a boatload of indigenous Papuans, "forty black, naked, mop-headed savages . . . intoxicated with joy and excitement." Comparing the groups "side by side" he realized in "less than five minutes" that they "belonged to two of the most distinct and strongly marked races" on earth. "Had I been blind, I could have been certain that these islanders were not Malays," he wrote. The two peoples, geographical neighbors, had descended from quite distinct ancestors who had lived and evolved on different landmasses. The beach at Ke marked the beginning of "a new world, inhabited by a strange people." For six months, when he was not paralyzed by the ulcerated wounds from the bites of insects that he was convinced were taking their revenge on him, he roamed across these scattered islands with a team of Papuan boy hunters led and trained by Ali, observing the characteristics and behaviors of the different races closely, amazed by the gulf of difference.

When he arrived back in Macassar, Wallace found two important letters among the pile that had accumulated in his seven-month absence. Finally someone had read his species paper. Bates wrote to congratulate him on his courage in going into print. "I was startled at first to see you already ripe for the enunciation of the theory," he confessed; "the idea is like truth itself, so simple and obvious that those who read and understand it will be struck by its simplicity; and yet it is perfectly original." A second letter, from Darwin, offered more muted praise. He told Wallace that he had been working on the species question for twenty years: "By your letter & even still more by your paper in Annals, a year or more ago, I can plainly see that we have thought much alike & to a certain extent have come to similar conclusions. In regard to the Paper in Annals, I agree to the truth of almost every word of your paper; & I daresay that you will agree with me that it is very rare to find oneself agreeing pretty closely with any theoretical paper; for it is lamentable how each man draws his own different conclusions from the very same fact."

Whether Darwin had meant to fire a shot across the young collector's bows, or whether he was being genuinely encouraging, is impossible to say for certain. But Wallace, always slow to pick up the signs of rudeness, was delighted to have received a letter from the naturalist he most admired. For him, Darwin was an ally, and Sir Charles Lyell, with his clever rebuttals of Lamarckian transmutation, a dragon to be slain. Wallace's entire mission depended now on being able to overturn Lyell's arguments.

Darwin's letter, the very fact of it, galvanized Wallace into writing. He composed four essays for publication in July 1857, all tackling different aspects of the species question, challenging Lyell, presenting evidence, and paving the way for a bolder set of claims. The Sarawak paper was only a statement of the theory, he wrote to Bates, not its development. That was still to come. He was, he added, now writing a book on the subject. It was the first time he had told anyone about the book. He wrote the four essays at white-hot speed while preparing and preserving specimens for hours during the day and keeping swarms of ants at bay. His specimens had to be stored on shelves suspended by ropes from the ceiling to keep the ants from attacking the skins and eye sockets of the dead animals as they awaited preservation. When Ali accidentally left a palm frond touching the shelves, the ants climbed it and swarmed over the treasures, ravaging everything.

In "On the Natural History of the Aru Islands" Wallace attacked Lyell's theory of special creation head-on. But though Wallace's ideas were thrilling, heretical, and revolutionary in themselves, they were buried in dense technical prose, and his essays bore dull, unpromising titles, perhaps because he was acutely aware of his outside status as a collector rather than a theorist. He was trying to sound right: professional, detached, objective. In "Note on the Theory of Permanent and Geographical Varieties" he trod very carefully. "As this subject is now attracting much attention among naturalists and particularly among entomologists," he wrote, "I venture to offer the following observations, which, without advocating either side of the question, are intended to point out a difficulty, or rather a dilemma, its advocates do not appear to have perceived."

Having set his house in order, sorted his files, and sent his letters and papers ahead, Wallace sailed for the Spice Islands on November 19, 1857, on a Dutch steamer, his book forming in his mind and still engaged in an imagined conversation with Charles Lyell, the question of the origin of species continuing to torment him. He was again struck by the marked contrast of vegetation and animal life on the archipelago's islands: those to the east of Bali were barren with a few low, scrubby plants, while the Spice Islands were covered with dense green forests. He had chosen as his base Ternate, a small island that had been the center of the Spanish and Portuguese clove trade since the sixteenth century. It was a landscape scattered with the overgrown and earthquake-shattered ruins of ancient forts, mosques, and sultans' palaces.

The "king" of Ternate, a wealthy Dutchman named Duivenboden, who owned half of the main town, also called Ternate, and ships and more than a hundred slaves, gave Wallace a house five minutes from the market and the beach. The house was in need of repair and furniture, but it had a deep well and a wilderness of fruit trees. Through January 1858, Wallace and Ali feasted on mangoes, milk, fresh bread, meat, and vegetables and tallied up their collections and specimens from the archipelago. Wallace sent the list to Bates proudly. It included 2,000 moths and 3,700 beetles and amounted to 8,540 species in total.

But though Ternate was comfortable, well stocked with food and resources, and had provided Wallace with a perfect base for his hunting, it was the neighboring Gilolo that captured his imagination, an island shaped like a starfish with four densely forested arms. No one had ever collected there. In early January, Wallace sailed seven miles across the narrow channel to spend a month collecting on the island, finally settling in Dodinga, a village located on a narrow isthmus separating the north and south peninsulas, directly opposite Ternate. It was guarded by a Dutch corporal and four Javanese soldiers who lived in the remains of a small ancient Portuguese fort that had been rent almost in two by earthquakes. One of the first things that struck Wallace, again, was the absolute difference between the two primary indigenous races—the Malay and the Papuan. "Here then," he wrote, "I had discovered the exact boundary line between the Malay and the Papuan races, and at a spot where no other writer had expected it."

But though the month of February may have promised the most fruitful collecting expedition Wallace had yet undertaken, he had not counted on malaria. Within days of his arrival, he experienced the first fevers of the recurrent attacks that would paralyze him physically for the best part of the month and force him to leave the physical work to Ali. His mind wandered every day while he waited for the fever to pass; memories crowded into his brain, alongside fragments of books he had once read in dusty, drafty libraries and bookshops in England and Wales. The debilitating malaria attacks worked like alchemy in his brain, stirring the depths of his memory, providing new arrangements and juxtapositions of ideas. The answer surfaced suddenly like a wreck from the ocean bed.

The revolutionary essay he produced over the following evenings, still weak from fever, was written in imagined and excited conversation with both Darwin and Lyell and was called "On the Tendency of Varieties to Depart Indefinitely from the Original Type." He could have sent it directly to one of the zoological journals that might have published it quickly, but he knew it needed an advocate, someone who would defend it. In those long days of illness, Darwin had become a fellow traveler to him, a correspondent and ally.

Wallace had learned to be patient. He knew he could not receive a reply from Darwin for at least three months. As soon as he had recovered his strength, he returned to work, making plans and gathering provisions for the next expedition—beeswax, spoons, string, penknives, wide-mouthed collecting jars, and food. In April, as the boat carrying his letter moved ever closer to Europe, Wallace sailed to an island called Dorey, off the mainland of northern New Guinea. There he suffered the usual cluster of minor disasters that made him feel increasingly as if the wildlife of the Malay Archipelago were taking its revenge on him. "I was the first European who had lived alone on this great island," he wrote proudly, "but partly owing to an accident which confined me to the house for a month, and partly because the locality was not a good one, I did not get the rare species of birds of paradise I had expected. . . . The weather had been unusually wet, and the place was unhealthy." A wound on his ankle ulcerated in the wet conditions, and one by one his assistants fell ill with malaria. Immobilized again, this time for a further month, Wallace read Dumas's novels and back copies of the *Family Herald.* When the schooner finally took the weakened men back to Ternate, there were no letters waiting for him from Kent.

Back in England, when Wallace's letter finally reached Darwin at Down House in June, Darwin was devastated. "Your words have come true with a vengeance—that I should be forestalled," he wrote to Lyell, his syntax contorted with anxiety. "I never saw a more striking coincidence; if Wallace had my MS sketch written out in 1842 [Darwin began his first "pencil sketch" of his species theory in 1842 and completed the full 230-page abstract in 1844] he could not have made a better short abstract! Even his terms now stand as heads of my chapters. . . . So all my originality will be smashed, though my book, if it will ever have value, will not be deteriorated; as all the labour consists in the application of the theory." When no letter came from Lyell in reply, he wrote again a week later asking for advice—should he publish quickly; could he do so honorably? "I would far rather burn my whole book," he wrote, "than that he or any other man should think that I had behaved in a paltry spirit. Do you not think his having sent me this sketch ties my hands? . . . If I could honourably publish I would state that I was induced now to publish a sketch . . . from Wallace having sent me an outline of his conclusions. We differ only [in] that I was led to my views from what artificial selection had done for domestic animals."

He added a postscript: "It seems hard on me that I should be thus compelled to lose my priority of many years' standing, but I cannot feel at all sure that this alters the justice of the case." He appealed to Lyell to determine where that justice lay. Lyell sent the paper to Hooker and assembled an emergency meeting of the Linnaean Society in which to present the material of both men for publication: Wallace's paper of 1858 and Darwin's unpublished 230-page essay of 1844.

Meanwhile, out in New Guinea in the middle of June, Wallace, who had recovered from his ulcerated ankle and returned to collecting beetles, fell sick with another fever that was followed by "such a soreness of the whole inside of my mouth, tongue and gums, that for many days I could put nothing solid between my lips." Two of his Malay hunters and assistants fell ill, one with fever, the other with dysentery. As he and Ali tended the sick boys, ants continued to swarm, building a nest on the roof, making papery tunnels down every post, carrying away insects from under his nose as he worked on them, and biting him day and night; blowflies settled in swarms on his bird skins, laying eggs that hatched maggots in hours. Jumaat, an eighteen-year-old Muslim boy hunter from Bouton, died from a fever on June 26. The sea journey back to Ternate, which should have taken five days, took seventeen because the winds would not fill the sails or blew from the west rather than the east.

In England, Darwin's baby son, Charles, was dangerously ill with scarlet fever; he died on June 28. His wife and daughter were ill with diphtheria. Darwin asked others to make the necessary decisions about Wallace's paper. He had no strength or heart to do so. When both sets of papers were presented to the Linnaean Society for a judgment, the assembled members agreed that Darwin had priority. That same day, an exhausted and seriously thin Wallace was walking the deck of a prau watching for a wind that would not come.

When Joseph Hooker remembered the occasion many years later, he recalled feeling that a battle had begun there in the Linnaean Society that day:

> The interest excited was intense, but the subject was too novel and too ominous for the old school to enter the lists, before armouring. After the meeting it was talked over with bated breath: Lyell's approval, and perhaps in a small way mine, as his lieutenant in the affair, rather overawed the Fellows, who would otherwise have flown out against the doctrine. We had, too, the vantage ground of being familiar with the authors and their theme.

When Hooker wrote to him to explain, Wallace's reaction to the verdict was famously measured and professional. If he felt badly treated, we have no record of it. He wrote to Hooker from Ternate on October 6, 1858:

> Allow me in the first place sincerely to thank yourself and Sir Charles Lyell for your kind offices on this occasion, and to assure you of the gratification afforded me both by the course you have pursued, and the favourable opinions of my essay which you so kindly expressed. I cannot but consider myself a favoured party in this matter, because it has hitherto been too much the practice in cases of this sort to impute all merit to the first discoverer of a new fact or a new theory, and little or none to any other party who may, quite independently, have arrived at the same result a

> few years or a few hours later. . . . It is evident that the time has now arrived when these and similar views will be promulgated and must be fairly discussed.

Wallace appears to have been delighted to have been part of the process by which Darwin had been forced to publish; he had no doubt that Darwin's book was more thorough than his own and that Darwin had the right to claim priority.

All of Wallace's early letters express satisfaction at being recognized by such eminent men. It seems certain that given the choice between living through the public opprobrium that Darwin suffered, the outrage of bishops, and the war that had now been openly declared between naturalists in England, and continuing to hunt for terra incognita or as yet undiscovered species, Wallace would have chosen the latter. Over the years, Wallace told the story of his discovery of natural selection and its fate at the hands of the gentlemen of the Linnaean Society so many times that he rarely questioned it or considered that there might have been anything unfair about what had happened. In his autobiography he called his discovery a "sudden intuition" that had been "hastily written" and that bore no comparison to the "prolonged labours of Darwin, who had reached the same point twenty years before me, and had worked continuously during that long period in order that he might be able to present the theory to the world with such a body of systematized facts and arguments as would compel conviction." It was the compelling of conviction that was yet to be accomplished, increment by increment, argument by argument. Both men were heretics and infidels after all, just different kinds who behaved in different ways.

And what of Ali? Once Wallace had left Sumatra by mail steamer in 1862, what did Ali remember of his former master? Did he talk of his godlessness, or his heresy, or his fine hunting skills? To Ali, Wallace was a man like himself, making a living by hunting and selling animal skins, but also a man of wisdom and knowledge who asked questions that made the world bigger and older. If Wallace had talked to him about how new beetles and lizards were coming into being all the time, through a process of constant destruction and adaptation too slow for the eye to see, if he had told Ali that animals and humans were all in

A tree fern from Wallace's *The Malay Archipelago.*
Alfred Russel Wallace, The Malay Archipelago *(1869)*

competition for land, that nature worked through competition for food and resources, and that was how new species came into being, infinitely slowly, limbs lengthening, feet webbing, beaks curving, could he, as a young Muslim boy, have accepted that way of seeing as truth? No doubt Ali held all of those competing explanations together in his head and continued to puzzle it out for himself; he might not have thought of Wallace's theory and the truth of the Qur'an as in opposition but might instead have considered them both possible truths, possible miracles. Birds of paradise were gifts of Allah, given to adorn the earth for man; but did Ali ever lie awake wondering and imagining how they had come to be over millions of years?

In 1907, when the twenty-three-year-old Harvard zoologist Thomas Barbour traveled to the Dutch East Indies with his wife, they had an unexpected encounter in Ternate:

> I was stopped in the street one day as my wife and I were preparing to climb up to the Crater Lake. With us were Ah Woo with his butterfly net, Indit and Bandoung, our well-trained Javanese collectors, with shotguns, cloth bags and a vasculum for carrying the birds. We were stopped by a wizened old Malayan. I can see him now, with a faded blue fez on his head. He said, "I am Ali Wallace." I knew at once that there stood before me Wallace's faithful companion of many years, who not only helped him collect but nursed him when he was sick. We took his photograph and sent it to Wallace when we got home. He wrote me a delightful letter acknowledging it and reminiscing over the time when Ali had saved his life, nursing him through a terrific attack of malaria. This letter I have managed to lose, to my eternal chagrin.

Epilogue

If Charles Darwin had been able to follow some of the men on his list into ancient Lesbos or eighteenth-century Paris or Cairo, if he had had time to linger there and ask them questions, he would have been struck by the astonishing kinship that existed between himself and his predecessors; perhaps he might even have been moved enough by their hard work, courage, and iconoclasm to become less anxious about his own claim to priority. As he collected information about them for his "Historical Sketch" between 1860 and 1863 and measured their ideas about species and mutability against his own, he did not know that within a hundred years almost all of them would have become virtually invisible to history, and that their invisibility would be directly related to his own rise to scientific sainthood.

Darwin would have recognized his own story in the struggles of his predecessors. Like Aristotle and Jahiz and Grant and Wallace, Darwin was a maverick; he was an intellectual gadfly whose curiosity was insatiable; he had a passion for collecting beetles or barnacles, and he pursued the answers to large-scale questions in the minutiae of species difference. In his search for answers he roamed across disciplines, for he knew that new knowledge often appeared in strange and unpredictable places. He valued his friendships with similarly obsessive men and women who would spar with him, challenge his conclusions, and not take him too seriously. Like Aristotle, Maillet, Geoffroy, Grant, and Wallace, everything he had previously understood about nature had been transformed by a long sea voyage. He knew about the mysteries of serendipity; he knew about the profound influence of chance encounters with passionate men such as Robert Grant; and like Abraham Trembley, he knew how powerful microscopes could make new truths visible.

Just as Aristotle, Jahiz, Leonardo, Maillet, Grant, and Wallace did, Darwin frequently valued the specialized knowledge of local people—beekeepers, pigeon breeders, orchid growers—above the theorized knowledge of scholars. He understood the value and beauty of a hunch; he knew how to pursue such hunches when they were no more than half glimpsed; and he knew that no matter how much he feared the consequences of publishing his heretical ideas, and as much as that fear often made him procrastinate, there would be no stopping him from pursuing the truth of his theory into eventual controversial publication.

Darwin might have noticed differences between himself and them, too. He would have acknowledged that many of his predecessors did not have substantial private incomes like his own that gave him time to work and intellectual independence. Some, like Wallace, Jahiz, Chambers, and Grant, struggled with poverty in their early years and had to pursue their scientific obsessions late at night or against the thundering of a printing press. He would have admired their achievements all the more for that. Most of his predecessors also lived in large cosmopolitan cities that gave them opportunities to converse with foreigners who had different ideas or ways of seeing the world. Darwin, often struggling with illness or simply shy of company, had to bring such people to him at Down House in Kent or maintain far-ranging conversations across vast distances by letter.

I like to think that Darwin might have recognized structural patterns in this long history of evolution and that it would have given him pleasure to see that the process of discovery did not travel in a straight line, a historical progression moving inexorably toward a final truth. Instead, like the history of species as he understood it, the story of the discovery of natural selection is a story of meanderings and false starts, of outgrowths, adaptations, and atrophies, of movements backward as well as forward, of sudden jumps and accelerations and convergences. The final stages of this story can perhaps intriguingly also be understood as mirroring what biologists now call "convergent evolution," a process by which unrelated species sometimes acquire similar body structures; just such a convergence took place in 1858 when Darwin received Wallace's essay and realized they had discovered natural selection simultaneously. The history of evolution ultimately testifies to the fertility of nature and its production, not only of a variety of forms and species, but also of a variety of ideas that can endlessly take new twists and turns.

"There is grandeur in this view of life," wrote Darwin in *Origin*, "with its several powers, having been originally breathed into a few forms or into one; and that, whilst this planet has gone cycling on according to the fixed law of gravity, from so simple a beginning endless forms most beautiful and most wonderful have been, and are being, evolved."

If Darwin and his predecessors had the power to move forward in time, to wander in the zoology departments or laboratories of Cambridge or California, to ask questions and perhaps conduct a few experiments, they would be astonished to learn of the computer simulations and data sets and powers of verifiability that have become possible. Think what they might make of genetics, the genome project, neuroscience and cloning, of the extraordinary things that we now know to be true, via the mapping of genomes, about our kinship with animals. But think, too, of how mystified they might be, given how well they understood the great strides that are often made as a result of crossing disciplinary boundaries, by the increasing narrowness of scientific specialties; and given that all of them, except Darwin and his grandfather Erasmus, Robert Chambers, and Alfred Russel Wallace, depended financially upon powerful and influential patrons who understood what they were doing and rarely interfered with their work or asked them to account for its usefulness or applicability, they might be baffled to hear about the hours modern scientists spend filling out funding application forms and negotiating complex institutional politics. They might have asked how mavericks or iconoclasts might flourish in such conditions, or whether serendipity might happen much here. And the answer might be a predictable one: Innovative thinkers do not disappear from the history of science as the conditions of scientific discovery change. They merely adapt; they mutate into new forms.

Appendix: An Historical Sketch of the Recent Progress of Opinion on the Origin of Species

BY CHARLES DARWIN

From the fourth edition of *On the Origin of Species by Natural Selection,* 1866, xiii–xxii

I will here give a brief, but imperfect, sketch of the progress of opinion on the Origin of Species. The great majority of naturalists believe that species are immutable productions, and have been separately created. This view has been ably maintained by many authors. Some few naturalists, on the other hand, believe that species undergo modification, and that the existing forms of life are the descendants by true generation of pre-existing forms. Passing over allusions to the subject in the classical writers,* the first author who in modern times has treated it in a scien-

*I have taken the date of the first publication of Lamarck from Isid. Geoffroy Saint Hilaire's ("Hist. Nat. Générale," tom. ii. p. 405, 1859) excellent history of opinion on this subject. In this work a full account is given of Buffon's conclusions on the same subject. It is curious how largely my grandfather, Dr. Erasmus Darwin, anticipated the views and erroneous grounds of opinion of Lamarck in his "Zoonomia" (vol. i, pp. 500–510), published in 1794. According to Isid. Geoffroy there is no doubt that Goethe was an extreme partisan of similar views, as shown in the Introduction to a work written in 1794 and 1795, but not published till long afterwards: he has pointedly remarked (Goethe als Naturforscher, von Dr. Karl Meding, s. 34) that the future question for naturalists will be how, for instance, cattle got their horns, and not for what they are used. It is rather a singular instance of the manner in which similar views arise at about the same time, that Goethe in Germany, Dr. Darwin in England, and Geoffroy Saint Hilaire (as we shall immediately see) in France, came to the same conclusion on the origin of species, in the years 1794–95.

tific spirit was Buffon. But as his opinions fluctuated greatly at different periods, and as he does not enter on the causes or means of the transformation of species, I need not here enter on details.

Lamarck was the first man whose conclusions on the subject excited much attention. This justly-celebrated naturalist first published his views in 1801; he much enlarged them in 1809 in his "Philosophie Zoologique," and subsequently, in 1815, in the Introduction to his "Hist. Nat. des Animaux sans Vertèbres." In these works he upholds the doctrine that all species, including man, are descended from other species. He first did the eminent service of arousing attention to the probability of all change in the organic, as well as in the inorganic world, being the result of law, and not of miraculous interposition. Lamarck seems to have been chiefly led to his conclusion on the gradual change of species, by the difficulty of distinguishing species and varieties, by the almost perfect gradation of forms in certain groups, and by the analogy of domestic productions. With respect to the means of modification, he attributed something to the direct action of the physical conditions of life, something to the crossing of already existing forms, and much to use and disuse, that is, to the effects of habit. To this latter agency he seems to attribute all the beautiful adaptations in nature;—such as the long neck of the giraffe for browsing on the branches of trees. But he likewise believed in a law of progressive development; and as all the forms of life thus tend to progress, in order to account for the existence at the present day of simple productions, he maintains that such forms are now spontaneously generated.* Geoffroy Saint Hilaire, as is stated in his "Life," written by his son, suspected, as early as 1795, that what we call species are various degenerations of the same type. It was not until 1828 that he published his conviction that the same forms have not been perpetuated since the origin of all things. Geoffroy seems to have relied chiefly on the conditions of life, or the "*monde ambiant*," as the cause of

* From references in Bronn's "Untersuchungen über die Entwickelungs-Gesetze" it appears that the celebrated botanist and palaeontologist Unger published, in 1852, his belief that species undergo development and modification. D'Alton, likewise, in Pander and d'Alton's work on Fossil Sloths, expressed, in 1821, a similar belief. Similar views have, as is well known, been maintained by Oken in his mystical "Natur-Philosophie [*sic*]." From other references in Godron's work "Sur l'Espèce," it seems that Bory St. Vincent, Burdach, Poiret, and Fries, have all admitted that new species are continually being produced.

change. He was cautious in drawing conclusions, and did not believe that existing species are now undergoing modification; and, as his son adds, "C'est donc un problème à réserver entièrement à l'avenir, supposé même que l'avenir doive avoir prise sur lui."

In 1813 Dr. W. C. Wells read before the Royal Society "An Account of a White Female, part of whose Skin resembles that of a Negro"; but his paper was not published until his famous "Two Essays upon Dew and Single Vision" appeared in 1818. In this paper he distinctly recognises the principle of natural selection, and this is the first recognition which has been indicated; but he applies it only to the races of man, and to certain characters alone. After remarking that negroes and mulattoes enjoy an immunity from certain tropical diseases, he observes, firstly, that all animals tend to vary in some degree, and, secondly, that agriculturists improve their domesticated animals by selection; and then, he adds, but what is done in this latter case "by art, seems to be done with equal efficacy, though more slowly, by nature, in the formation of varieties of mankind, fitted for the country which they inhabit. Of the accidental varieties of man, which would occur among the first few and scattered inhabitants of the middle regions of Africa, some one would be better fitted than the others to bear the diseases of the country. This race would consequently multiply, while the others would decrease; not only from their inability to sustain the attacks of disease, but from their incapacity of contending with their more vigorous neighbours. The colour of this vigorous race I take for granted, from what has been already said, would be dark. But the same disposition to form varieties still existing, a darker and a darker race would in the course of time occur; and as the darkest would be the best fitted for the climate, this would at length become the most prevalent, if not the only race, in the particular country in which it had originated." He then extends these same views to the white inhabitants of colder climates. I am indebted to the Rev. Mr. Brace, of the United States, for having called my attention to the above passage in Dr. Wells' work.

The Hon. and Rev. W. Herbert, afterwards Dean of Manchester, in the fourth volume of the "Horticultural Transactions," 1822, and in his work on the "Amaryllidaceae" (1837, p. 19, 339), declares that "horticultural experiments have established, beyond the possibility of refutation, that botanical species are only a higher and more permanent class of varieties." He extends the same view to animals. The Dean believes that single species of each genus were created in an originally highly plastic condition, and that these have produced, chiefly by intercrossing, but likewise by variation, all our existing species.

In 1826 Professor Grant, in the concluding paragraph in his well-known paper ("Edinburgh Philosophical Journal," vol. xiv, p. 283) on the Spongilla, clearly declares his belief that species are descended from other species, and that they become improved in the course of modification. This same view was given in his 55th Lecture, published in the "Lancet" in 1834.

In 1831 Mr. Patrick Matthew published his work on "Naval Timber and Arboriculture," in which he gives precisely the same view on the origin of species as that (presently to be alluded to) propounded by Mr. Wallace and myself in the "Linnaean Journal," and as that enlarged in the present volume. Unfortunately the view was given by Mr. Matthew very briefly in scattered passages in an Appendix to a work on a different subject, so that it remained unnoticed until Mr. Matthew himself drew attention to it in the "Gardener's Chronicle," on April 7th, 1860. The differences of Mr. Matthew's view from mine are not of much importance: he seems to consider that the world was nearly depopulated at successive periods, and then re-stocked; and he gives, as an alternative, that new forms may be generated "without the presence of any mould or germ of former aggregates." I am not sure that I understand some passages; but it seems that he attributes much influence to the direct action of the conditions of life. He clearly saw, however, the full force of the principle of natural selection.

The celebrated geologist and naturalist, Von Buch, in his excellent "Description Physique des Iles Canaries" (1836, p. 147), clearly expresses his belief that varieties slowly become changed into permanent species, which are no longer capable of intercrossing.

Rafinesque, in his "New Flora of North America," published in 1836, wrote (p. 6) as follows: —"All species might have been varieties once, and many varieties are gradually becoming species by assuming constant and peculiar characters": but farther on (p. 18) he adds, "except the original types of ancestors of the genus."

In 1843–44 Professor Haldeman (Boston Journal of Nat. Hist., U. States, vol. iv, p. 468) has ably given the arguments for and against the hypothesis of the development and modification of species: he seems to lean towards the side of change.

The "Vestiges of Creation" appeared in 1844. In the tenth and much improved edition (1853) the anonymous author says (p. 155): —"The proposition determined on after much consideration is, that the several series of animated beings, from the simplest and oldest up to the highest and most recent, are, under the providence of God, the results, *first,* of an impulse which has been imparted to the forms of life, advancing them, in definite times, by generation, through grades of organisation terminating in the highest dicotyledons and vertebrata, these grades being few in number, and generally marked by intervals of organic character, which we find to be a practical difficulty in ascertaining affinities; *second,* of another impulse connected with the vital forces, tending, in the course of generations, to modify organic structures in accordance with external circumstances, as food, the nature of the habitat, and the meteoric agencies, these being the 'adaptations' of the natural theologian." The author apparently believes that organisation progresses by sudden leaps, but that the effects produced by the conditions of life are gradual. He argues with much force on general grounds that species are not immutable productions. But I cannot see how the two supposed "impulses" account in a scientific sense for the numerous and beautiful co-adaptations which we see throughout nature; I cannot see that we thus gain any insight how, for instance, a woodpecker has become adapted to its peculiar habits of life. The work, from its powerful and brilliant style, though displaying in the earlier editions little accurate knowledge and a great want of scientific caution, immediately had a very wide circulation. In my opinion it has done excellent service in this country in calling attention to the subject, in removing prejudice, and in thus preparing the ground for the reception of analogous views. In 1846 the veteran geologist M. J. d'Omalius d'Halloy published in an excellent, though short paper ("Bulletins de l'Acad. Roy. Bruxelles," tom. xiii, p. 581), his opinion that it is more probable that new species have been produced by descent with modification, than that they have been separately created: the author first promulgated this opinion in 1831.

Professor Owen, in 1849 ("Nature of Limbs," p. 86), wrote as follows: —"The archetypal idea was manifested in the flesh under diverse such modifications, upon this planet, long prior to the existence of those animal species that actually exemplify it. To what natural laws or secondary causes the orderly succession and progression of such organic phenomena may have been committed, we, as yet, are ignorant." In his Address to the British Association, in 1858, he speaks (p. li.) of "the axiom of the continuous operation of creative power, or of the ordained becoming of living things." Farther on (p. xc.), after referring to geographical distribution, he adds, "These phenomena shake our confidence in the conclusion that the Apteryx of New Zealand and the Red Grouse of England were distinct creations in and for those islands respectively. Always, also, it may be well to bear in mind that by the word 'creation' the zoologist means 'a process he knows not what.' " He amplifies this idea by adding, that when such cases as that of the Red Grouse are "enumerated by the zoologist as evidence of distinct creation of the bird in and for such islands, he chiefly expresses that he knows not how the Red Grouse came to be there, and there exclusively; signifying also by this mode of expressing such ignorance his belief, that both the bird and the islands owed their origin to a great first Creative Cause." If we interpret these sentences given in the same Address, one by the other, it appears that this eminent philosopher felt in 1858 his confidence shaken that the Apteryx and the Red Grouse first appeared in their respective homes, "he knew not how," or by some process "he knew not what." Since the publication in 1859 of my work on the "Origin of Species," but whether in consequence of it is doubtful, Professor Owen has clearly expressed his belief that species have not been separately created, and are not immutable productions; but he still ("Anatomy of the Vertebrates," 1866) denies that we know the natural laws or secondary causes of the successive appearance of species; yet he at the same time admits that natural selection may have done something towards this end. It is surprising that this admission should not have been made earlier, as Professor Owen now believes that he promulgated the theory of natural selection in a passage read before the Zoological Society in February, 1850 ("Transact." vol. iv, p. 15); for in a letter to the "London Review" (May 5, 1866, p. 516), commenting on some of the reviewer's criticisms, he says, "No naturalist can dissent from the truth of your perception of the essential identity of the passage cited with the basis of that [the so-called Darwinian] theory, the power, viz., of species to accommodate themselves, or bow to the influences of surrounding circumstances." Further on in the same letter he speaks of himself as "the author of the same theory at the earlier date of 1850." This belief in Professor Owen that he then gave to the world the theory of natural selection will surprise all those who are acquainted with the several passages in his works, reviews, and lectures, published since the "Origin," in which he strenuously opposes the theory; and it will please all those who are interested on this side of the question, as it may be presumed that his opposition will now cease. It should, however, be stated that the passage above referred to in the "Zoological Transactions," as I find on consulting it, applies exclusively to the extermination and preservation of animals, and in no way to their gradual modification, origination, or natural selection. So far is this from being the case that Professor Owen actually begins the first of the two paragraphs (vol. iv, p. 15) with the following words: —"We have not a particle of evidence that any species of bird or beast that lived during the pliocene period has had its characters modified in any respect by the influence of time or of change of external circumstances."

M. Isidore Geoffroy Saint Hilaire, in his Lectures delivered in 1850 (of which a Résumé appeared in the "Revue et Mag. de Zoolog.," Jan. 1851), briefly gives his reason for believing that specific characters "sont fixes, pour chaque espèce, tant qu'elle se perpétue au milieu des mêmes circonstances: ils se modifient, si les circonstances ambiantes viennent à changer." "En résumé *l'observation* des animaux sauvages démontre déjà la variabilité *limitée* des espèces. Les *expériences* sur les animaux sauvages devenus domestiques, et sur les animaux domestiques redevenus sauvages, la démontrent plus clairement encore. Ces mêmes expériences prouvent, de plus, que les différences produites peuvent être de *valeur générique*." In his "Hist. Nat. Générale" (tom. ii, p. 430, 1859) he amplifies analogous conclusions.

From a circular lately issued it appears that Dr. Freke, in 1851 ("Dublin Medical Press," p. 322), propounded the doctrine that all organic beings have descended from one primordial form. His grounds of belief and treatment of the subject are wholly different from mine; but as Dr.

Freke has now (1861) published his Essay on "the Origin of Species by means of Organic Affinity," the difficult attempt to give any idea of his views would be superfluous on my part.

Mr. Herbert Spencer, in an Essay (originally published in the "Leader," March 1852, and republished in his "Essays" in 1858), has contrasted the theories of the Creation and the Development of organic beings with remarkable skill and force. He argues from the analogy of domestic productions, from the changes which the embryos of many species undergo, from the difficulty of distinguishing species and varieties, and from the principle of general gradation, that species have been modified; and he attributes the modification to the change of circumstances. The author (1855) has also treated Psychology on the principle of the necessary acquirement of each mental power and capacity by gradation.

In 1852 M. Naudin, a distinguished botanist, expressly stated, in an admirable paper on the Origin of Species ("Revue Horticole," p. 102; since partly republished in the "Nouvelles Archives du Muséum," tom. i, p. 171), his belief that species are formed in an analogous manner as varieties are under cultivation; and the latter process he attributes to man's power of selection. But he does not show how selection acts under nature. He believes, like Dean Herbert, that species, when nascent, were more plastic than at present. He lays weight on what he calls the principle of finality, "puissance mystérieuse, indéterminée; fatalité pour les uns; pour les autres, volonté providentielle, dont l'action incessante sur les êtres vivants détermine, à toutes les époques de l'existence du monde, la forme, le volume, et la durée de chacun d'eux, en raison de sa destinée dans l'ordre de choses dont il fait partie. C'est cette puissance qui harmonise chaque membre à l'ensemble en l'appropriant à la fonction qu'il doit remplier dans l'organisme général de la nature, fonction qui est pour lui sa raison d'être."

In 1853 a celebrated geologist, Count Keyserling ("Bulletin de la Soc. Géolog.," 2nd Ser., tom. x, p. 357), suggested that as new diseases, supposed to have been caused by some miasma, have arisen and spread over the world, so at certain periods the germs of existing species may have been chemically affected by circumambient molecules of a particular nature, and thus have given rise to new forms.

In this same year, 1853, Dr. Schaaffhausen published an excellent pamphlet ("Verhand. des Naturhist. Vereins der Preuss. Rheinlands," &c.), in which he maintains the progressive development of organic forms on the earth. He infers that many species have kept true for long periods, whereas a few have become modified. The distinction of species he explains by the destruction of intermediate graduated forms. "Thus living plants and animals are not separated from the extinct by new creations, but are to be regarded as their descendants through continued reproduction."

A well-known French botanist, M. Lecoq, writes in 1854 ("Etudes sur Géograph. Bot.," tom. i, p. 250), "On voit que nos recherches sur la fixité ou la variation de l'espèce, nous conduisent directement aux idées émises par deux hommes justement célèbres, Geoffroy Saint-Hilaire et Goethe." Some other passages scattered through M. Lecoq's large work, make it a little doubtful how far he extends his views on the modification of species.

The "Philosophy of Creation" has been treated in a masterly manner by the Rev. Baden Powell, in his "Essays on the Unity of Worlds," 1855. Nothing can be more striking than the manner in which he shows that the introduction of new species is "a regular, not a casual phenomenon," or, as Sir John Herschel expresses it, "a natural in contradistinction to a miraculous process."

The third volume of the "Journal of the Linnaean Society" contains papers, read July 1st, 1858, by Mr. Wallace and myself, in which, as stated in the introductory remarks to this volume, the theory of Natural Selection is promulgated by Mr. Wallace with admirable force and clearness.

Von Baer, towards whom all zoologists feel so profound a respect, expressed about the year 1859 (see Prof. Rudolph Wagner, "Zoologisch-Anthropologische Untersuchungen," 1861, s. 51) his conviction, chiefly grounded on the laws of geographical distribution, that forms now perfectly distinct have descended from a single parent-form.

I may add, that of the thirty-four authors named in this Historical Sketch, who believe in the modification of species, or at least disbelieve in separate acts of creation, twenty-seven have written on special branches of natural history or geology.

In June, 1859, Professor Huxley gave a lecture before the Royal Institution on the "Persistent Types of Animal Life." Referring to such cases, he remarks, "It is difficult to comprehend the meaning of such facts as these, if we suppose that each species of animal and plant, or each great type of organisation, was formed and placed upon the surface of the globe at long intervals by a distinct act of creative power; and it is well to recollect that such an assumption is as unsupported by tradition or revelation as it is opposed to the general analogy of nature. If, on the other hand, we view 'Persistent Types' in relation to that hypothesis which supposes the species living at any time to be the result of the gradual modification of pre-existing species—a hypothesis which, though unproven, and sadly damaged by some of its supporters, is yet the only one to which physiology lends any countenance; their existence would seem to show that the amount of modification which living beings have undergone during geological time is but very small in relation to the whole series of changes which they have suffered."

In December, 1859, Dr. Hooker published his "Introduction to the Australian Flora." In the first part of this great work he admits the truth of the descent and modification of species, and supports this doctrine by many original observations.

Acknowledgments

In writing this book I have depended on the generosity of leading scholars in their fields who have shared their expertise, explained issues and controversies, steered me through my historical investigations, and read and corrected my chapters. I could not have written this book without them or without their trust in the integrity of my project. Curtis N. Johnson, professor of government at Lewis and Clark College in Portland, Oregon, who wrote the definitive essay on Darwin's struggle to write a list of his predecessors, read and corrected my attempt to understand Darwin's frustrations and anxieties about them. Professor Sir Geoffrey Lloyd, emeritus professor at the Needham Institute in Cambridge, historian, philosopher, polymath, and author of a clutch of marvelous books on Aristotle and Greek science including *Early Greek Science, Greek Science After Aristotle,* and *Aristotelian Explorations,* gave me the confidence to believe that it was possible—with caution—to reconstruct something of Aristotle's life, zoological investigations, and cosmological frameworks. Professor James Montgomery, professor of classical Arabic at Cambridge University and the world's leading authority on Jahiz, explained the complex theological and literary contexts of the ninth-century Abbasid Empire, enabling me to understand a world and time that until then had been entirely beyond my reach; he also generously sought out and translated or retranslated key passages of Jahiz's *Treatise of Living Beings* and gave me access to the manuscript of his remarkable book-in-progress on Jahiz. Professor Michael Jeannert, now distinguished visiting professor at Johns Hopkins University, previously professor of French literature at the University of Geneva and author of a dazzling book on Renaissance mutability titled *Perpetual Motion,* read and corrected my chapter on Leonardo. Dr. Marc Ratcliff of the University of Geneva, author of several groundbreaking books on Enlightenment science and microscopes, read and corrected my work on Abraham Trembley. Dr. James Fowler, Diderot scholar at the School of European Culture and Languages at the University of Kent, generously assisted with my research on Diderot and Holbach; Dr. Patricia Fara and Professor Jim Secord of the peerless Department of the History and Philosophy of Science at Cambridge University read and corrected my work on Erasmus Darwin and Robert Chambers, respectively. Dorinda Outram, professor of history at the University of Rochester, New York, and author of the definitive book on George Cuvier as well as several groundbreaking books on Enlightenment science, and Professor Richard W. Burkhardt, professor emeritus of history at the University of Illinois and author of an important book on Jean-Baptiste Lamarck as well as a book in progress on the daily life of the Jardin, read several versions of my chapter on the Jardin des Plantes. Dr. Peter Raby of Homerton College, Cambridge University, author of the fine biography *Alfred Russel Wallace: A Life,* read my chapter on Alfred Russel Wallace. I am also grateful for the help of Adrian Desmond, who assisted with my work on Robert Grant several years ago when I first began to investigate his life and work for my book on Darwin, *Darwin and the Barnacle.* While I have been blessed with the finest scholarly readers I could have asked for, any mistakes or misinterpretations that remain are my responsibility alone.

I have great pleasure in thanking the Faculty of Humanities at the University of East Anglia, who provided me with a year's leave from teaching to complete the book, and to my colleagues who invariably had to shoulder some of the burdens of that leave. The entire book was written on a desk in the West Room of Cambridge University Library. Without this beautiful copyright library, its extraordinary and rare resources, helpful staff, and guarantee of quiet, this book would not exist. It is also my great pleasure to thank all the readers and editors who shaped the book as it neared completion—the geneticist Kate Downes, who helped me see my material through the eyes of a contemporary biologist; Anna Whitelock, historian and mentor; and the outstanding editorial team at Bloomsbury, particularly Michael Fishwick and Anna Simpson, who steered the book so masterfully and patiently into its final shape. And thanks to Bloomsbury's copy editor, Peter James, and my American editor, Cindy Spiegel, and her team, particularly Hana Landes, for dauntless and skillful final edits, and copy editor Emily DeHuff. I particularly thank Michael Fishwick of Bloomsbury and Cindy Spiegel of Spiegel & Grau, who backed the ambition and audacity of the project from the start, as did my fine literary agent, Faith Evans, who always sees the connections between my books so astutely, and my U.S. agent, Emma Sweeney.

Finally, I would like to thank my children, family, friends, and rowing crew for tolerating a degree of glassy-eyed distractedness at times when my mind has been absolutely elsewhere.

Notes

PREFACE

xiii **Many of Darwin's predecessors were called infidels:** See Edward Royle, *Victorian Infidels: The Origin of the British Secularist Movement, 1791–1866* (Manchester: Manchester University Press, 1974), and Andrew Wheatcroft, *Infidels: A History of the Conflict Between Christendom and Islam* (London: Penguin, 2004).

xiv **"us transmutationists":** Charles Darwin to Joseph Hooker, April 18, 1847, Letter 1082, DCP. I will be citing from the online Darwin Correspondence Project (hereinafter referred to as DCP) throughout the book.

1. DARWIN'S LIST

3 **The letters, he lamented to his wife:** Darwin used the word "swarms" to describe the disapproving letters he received immediately after the publication of *Origin* in a letter he wrote to Alfred Russel Wallace on May 18, 1860: Letter 2807, DCP.

4 **"We shall soon be a good body of working men":** Charles Darwin to Joseph Hooker, December 14, 1859, Letter 2583, DCP.

4 **the Reverend Baden Powell:** In 1855, Baden Powell had published a book called *Essays on the Unity of Worlds* in which he defended and tried to extend the ideas about species change argued by an anonymously published but highly controversial and bestselling book, *Vestiges of the Natural History of Creation,* published in 1844. For a study of Baden Powell, see Pietro Corsi, *Science and Religion: Baden Powell and the Anglican Debate, 1800–1860* (Cambridge and New York: Cambridge University Press, 1988).

4 **Of all the letters in that day's pile:** Powell's letter to Darwin has not survived, but the editors of the DCP have established its contents through Darwin's detailed reply to it written on January 8, 1860. I am indebted throughout this chapter to a long, detailed, and thoughtful article published by Curtis N. Johnson, "The Preface to Darwin's *Origin of Species:* The Curious History of the 'Historical Sketch,' " *Journal of the History of Biology* 40 (2007): 529–56.

5 **He should have included a short preface:** Charles Lyell had included a "Historical Sketch" in his *Principles of Geology* (1830 and 1832).

6 **aware that he was a poor scholar of history:** Darwin confessed to having poor historical skills when he wrote to Baden Powell in 1860: "The task [of writing a historical preface] would have been not a little difficult, and belongs rather to the Historian of Science than to me" (Charles Darwin to Baden Powell, January 18, [1860], Letter 2654, DCP); he repeated the idea in the second letter to Powell (Charles Darwin to Baden Powell, January 18, [1860], Letter 2655, DCP.

6 **He had even written to Hooker:** Alfred Wallace to Joseph Hooker, October 6, 1858, Letter 2337, DCP.

7 **"utterly knocked up & cannot rally":** Charles Darwin to Joseph Hooker, [December] 21, [1859], Letter 2591, DCP.

7 **a heavy volume of a French scientific journal:** Charles Victor Naudin, "Considérations philosophiques sur l'espèce et la variété," *Revue Horticole,* 4th series, 1 (1852): 102–9.

8 **He read and reread Naudin's paper:** "I am a very poor French scholar, though I read it with fluency," Charles Darwin to Edward Crecy, January 20, [1860], Letter 2657, DCP.

8 **"I cannot find one word like the Struggle for existence":** Charles Darwin to Joseph Hooker, [December] 23, [1859], Letter 2595, DCP.

8 **"I shall not write to Decaisne":** Charles Darwin to Joseph Hooker, [December] 25, [1859], Letter 2602, DCP.

9 **terrible storms lashed the country:** *Annual Register: A Record of World Events* 102 (1860).

9 **"Lenny has got the Measles":** Charles Darwin to Joseph Hooker, January 3, [1860], Letter 2635, DCP.

10 **"my health was so poor, whilst I wrote the Book":** Charles Darwin to Baden Powell, January 18, [1860], Letter 2654, DCP.

11 **"The manner in which [the reviewer] drags in immortality":** Charles Darwin to Joseph Hooker, [November 22, 1859], Letter 2542, DCP.

12 **"It is like confessing a murder":** Charles Darwin to Hugh Falconer, November 11, [1859], Letter 2524, DCP.

13 **"The stones are beginning to fly":** Charles Darwin to Joseph Hooker, February 20, [1860], Letter 2705, DCP.

13 **"all these attacks will make me only more determinately fight":** Charles Darwin to Alfred Russel Wallace, May 18, [1860], Letter 2807, DCP.

13 **"I will buckle on my armour & fight my best":** Charles Darwin to Asa Gray, May 18, [1860], Letter 2808, DCP.

13 **"I freely acknowledge that Mr. Matthew has anticipated":** Darwin letter in *Gardeners' Chronicle,* April 21, 1860.

14 **"ill-written unintelligible rubbish":** Charles Darwin to Joseph Hooker, January 15, [1861], Letter 3047, DCP.

16 **Darwin had read Aristotle at school:** On Darwin's reading of Aristotle, see Allan Gotthelf, "Darwin on Aristotle," *Journal of the History of Biology* 32, no. 1 (1999): 3–30.

17 **"You may recollect me":** James Grece to Charles Darwin, November 12, 1866, Letter 5276, DCP. Aristotle's biological work underwent something of a European revival in the nineteenth century. In his *Histoire des sciences naturelles* of 1841, the French comparative anatomist Georges Cuvier described himself as having been "ravished with astonishment" when he read *The History of Animals.* There were new translations available in German (1811 and 1816), a French translation in 1783, and new English translations of *The History of Animals* in ten volumes in 1862 by Richard Cresswell and of *Parts of Animals* in 1882 by Charles Ogle. George Henry Lewes wrote a book about Aristotle in 1864. A year later, Grece wrote to Darwin again for the last time to claim another favor. He wanted to translate a Dutch grammar book into English for publication. Would Darwin be so kind as to put in a word for him with John Murray? Darwin did. The book was delivered and published in 1874.

2. ARISTOTLE'S EYES

20 **A group of young men wearing finely woven tunics:** See Liba Taub, *Aetna and the Moon: Explaining Nature in Ancient Greece and Rome* (Corvallis: Oregon State University Press, 2008), 87.

21 **The fishermen know one of the philosophers:** See T. E. Rihll, *Greek Science,* New Surveys in the Classics, no. 29 (Oxford: Oxford University Press, 1999), 5.

21 **The older man in the group:** Diogenes describes Aristotle as having long legs and small eyes, fashionable clothes, and rings on his fingers; he also tells us that he shaved. Diogenes, *Lives of the Philosophers,* V:1.

22 **From the sky, the island of Lesbos:** Hugh J. Mason, "Romance in a Limestone Landscape," *Classical Philology* 90, no. 3 (1995): 263–66.

22 **Sixty different species of flower:** P. C. Candargy, *La Végétation de l'île de Lesbos* (Lille: Bigot frères, 1899), 1–39.

22 **Lesbos . . . is also an island of migrants:** The human migrants who arrive on Lesbos today are detained and deported. In the summer of 2009, the new Greek government temporarily closed a notorious detention center on the island built to imprison the hundreds of Afghan or Somalian migrants—men, women, and children—who cross the dangerous nine-mile sea channel between Turkey and Lesbos every month on their route west to Europe fleeing persecution or war zones. What he saw was "worse than Dante's hell," a Greek minister told journalists. Niki Kitsantonis, "Migrants Reaching Greece Despite Efforts to Block Them," *New York Times,* November 18, 2009.

24 **Aristotle had lived a charmed life in Athens:** R. E. Wycherley, "Peripatos: The Athenian Philosophical Scene—II," *Greece and Rome,* 2nd series, vol. 9, no. 1 (1962): 2–21.

24 **"In Athens things that are proper for a citizen":** Anton-Hermann Chroust, *Aristotle: New Light on His Life,* 2 vols. (London: Routledge and Kegan Paul, 1973), 1:158.

24 **Philip of Macedonia laid siege to the important coastal city:** See J. R. Ellis, *Philip II and Macedonian Imperialism* (London: Thames and Hudson, 1976), 95–100.

25 **"an evil-doer and curious person":** Plato's *Apology* from Plato, *Five Great Dialogues of Plato: Euthyphro, Apology, Crito, Meno, Phaedo,* translated by Benjamin Jowett (Claremont, Calif.: Coyote Canyon Press, 2009), 20–21.

25 **his extensive library:** On early libraries, see Jeno Platthy, *Sources on the Earliest Greek Libraries with the Testimonia* (Amsterdam: Hakkert, 1968).

26 **making his way to the Macedonian court in Pella:** Although there is no proof that Aristotle stopped at Pella on his way east, it seems likely to several Aristotle scholars that, given his ambassadorial role in the court at Atarneus and Philip's investment in Aristotle's relationship with Hermias, he would have needed to meet with Philip directly to discuss his future. See in particular Chroust, *Aristotle,* 159.

27 **"the great flood-gates of the wonder-world":** Herman Melville (1851), *Moby-Dick,* ch. 1, "Loomings." On Greek travel by sea and by road, see Lionel Casson, *Travel in the Ancient World* (London: Book Club Associates, 2005), ch. 4.

28 **Aristotle flourished in Hermias' court:** Werner Jaeger, *Aristotle: Fundamentals of the History of His Development,* rev. ed. (Oxford: Clarendon Press, 1948; first published 1934), 114.

29 ***Philosophy starts in wonder and wonderment:*** Aristotle, *The History of Animals,* 982b12ff.

30 **the island of Lesbos:** See also Mason, "Romance in a Limestone Landscape," 263–66. Even the ancient descriptions of the towns of Lesbos are dazzling. The earliest description of Mytilene, the largest city on Lesbos, where Aristotle lived most of the time during his stay there, conveys something of the white, watery light that must have made everything shimmer. Longus, who lived on the island in the second century AD, opens his pastoral tale *Daphnis and Chloe* with a description of Mytilene, the romantic city of his childhood: "There is in Lesbos a large and beautiful city called Mytilene. It is intersected by canals, where the sea flows inland, and decorated with bridges of polished stone. You will think, when you see it, that it is not so much a city as an island." Elsewhere he describes a sacred cave that has nymphs carved into the rock face and water gushing out and flowing away in a stream.

30 **Alcaeus wrote of the "bloom of soft autumn":** Alcaeus 397, 367, 347(a), 345, cited in Peter Green, *Classical Bearings: Interpreting Ancient History and Culture* (Berkeley: University of California Press, 1998), 60.

30 **Sappho described an orchard:** Sappho frr 136, 105(a), 117A, 156, translated by David A. Campbell, cited in Peter Green, *Lesbos and the Cities of Asia Minor* (Austin: Dougherty Foundation, 1984), 21–22.

30 **For two years the beautiful sea lagoon at Pyrra:** For the full story of the significance of the fish of Lesbos and how they have shaped understanding of Aristotle's philosophy, see D'Arcy Wentworth Thompson, *On Aristotle as a Biologist* (Oxford: Clarendon Press, 1913); H.D.P. Lee, "Place-Names and the Date of Aristotle's Biological Works," *Classical Quarterly* 42 (1948): 61–67; and Frank Solmsen, "The Fishes of Lesbos and Their Alleged Significance for the Development of Aristotle," *Hermes* 106 (1978): 467–84.

30 **His zoological questions multiplied here:** Thompson, *On Aristotle as a Biologist,* 50.

31 **the fishermen had built walkways and piers:** Anna Marguerite McCann, "The Harbour and Fishery Remains at Cosa, Italy," *Journal of Field Archaeology* 6, no. 4 (1979): 391–411.

31 **In the introduction to *Parts of Animals*:** Jaeger, *Aristotle*, 337.

31 **"If any person thinks the examination":** Aristotle, *Parts of Animals*, I.5, 645a27–30, translated by William Ogle.

32 **Aristotle would lie for several hours on a small platform:** See Jason A. Tipton, "Aristotle's Observations of the Foraging Interactions of the Red Mullet and Sea Bream," *Archives of Natural History* 35, no. 1 (2008): 164–71, and Tipton, "Aristotle's Study of the Animal World: The Case of the Kobios and the Phucis," *Perspectives in Biology and Medicine* 49, no. 3 (2006): 369–83. The great authority on Greek fish is D'Arcy Wentworth Thompson, *A Glossary of Greek Fishes* (London: Oxford University Press, 1947).

32 **the two fish eat different things, too:** "Other fishes feed habitually on mud or sea-weed or sea-moss or on stalk-weed or growing plants; like the *phucis*, and the goby; and by the way, the only meat that the *phucis* will touch is that of prawns": Aristotle, *The History of Animals*, 591b10, translated by William Ogle.

32 ***So why are two fish so similar:*** D. Balme, "The Place of Biology in Aristotle's Philosophy," in Allan Gotthelf and James G. Lennox, eds., *Philosophical Issues in Aristotle's Biology* (Cambridge: Cambridge University Press, 1987), 301.

33 **Nature itself was perfect:** See John C. Greene (1992), "From Aristotle to Darwin: Reflections on Ernst Mayr's Interpretation in *The Growth of Biological Thought*," *Journal of the History of Biology* 25, no. 2 (1992): 257–84.

33 **Once the pile of collated facts about animals:** See G.E.R. Lloyd, "The Evolution of Evolution: Greco-Roman Antiquity and the Origin of Species," in his *Principles and Practices in Ancient Greek and Chinese Science* (Aldershot: Ashgate Variorum, 2006), 10–11.

33 **"Oviparous fish as a rule spawn only once a year":** Aristotle, *The History of Animals*, 567b12.

33 **Theophrastus was studying the plants of the lagoon:** Decades later in Athens, Theophrastus would write up those plant notes into a book that would be one of the most important works on plants from the ancient world: *Enquiry into Plants* and *On the Causes of Plants*.

34 **He called them "dualisers" or "borderliners":** The Greek word is *epamphoterizein*, which David Balme in Aristotle, *History of Animals*, Books 7–10, edited and translated by D. M. Balme (Loeb Classical Library, Cambridge, Mass., and London: Harvard University Press, 1991) translates as "tend to both sides" but which Geoffrey Lloyd translates, following A. L. Peck, as "dualise" in G.E.R. Lloyd, "Fuzzy Natures," in

his *Aristotelian Explorations* (Cambridge: Cambridge University Press, 1996), 72.

34 **They appeared to straddle the two kingdoms:** See Lloyd, "Evolution of Evolution," 12.

35 **the language he used in his notes wavered a little:** See G.E.R. Lloyd's wonderful essay of 1996, "Fuzzy Natures," 68–83.

35 **"The boundary and the middle, between the non-living and the animals":** Aristotle, *The History of Animals,* 588b5f.

35 **"as regards certain things in the sea":** Ibid., 588b12f.

35 **"how to classify them is unclear":** Aristotle, *Parts of Animals,* 681a28.

35 **A single sponge diver took his place:** This account is based on a detailed description of ancient Greek sponge-diving practice in Oppian, *Halieutica* 5, 612–74, written between AD 170 and 180. I am assuming that sponge-diving practices had changed little between Aristotle and Oppian. Aristotle mentions the use of a sort of air-supply diving bell in his *Problemata:* "In order that these fishers of sponges may be supplied with a facility of respiration, a kettle is let down to them, not filled with water, but with air, which constantly assists the submerged man; it is forcibly kept upright in its descent, in order that it may be sent down at an equal level all around, to prevent the air from escaping and the water from entering," translated by E. S. Forster, cited in Frank J. Frost, "Scyllias: Diving in Antiquity," *Greece and Rome* 15, no. 2 (1968): 180–85, 183.

36 **No one asked the sponge divers questions as Aristotle did:** See Eleni Voultsiadou and Dimitris Vafidis, "Marine Invertebrate Diversity in Aristotle's Biology," *Contributions to Zoology* 76 (2007): 103–20; on diving, see Frost, "Scyllias: Diving in Antiquity," 180–85.

36 **The local people had myths to explain everything:** See Adrienne Mayor, *The First Fossil Hunters: Palaeontology in Greek and Roman Times* (Princeton: Princeton University Press, 2010; first published 2000). Mayor proposes that some of these more bizarre speculations may have come about because of speculations on fossils.

36 **"Coming to be and passing away":** Aristotle, *On Generation and Corruption,* Book 11, ch. 10.

36 **"For becoming starts from non-being":** Aristotle, *The History of Animals,* 741b22–4; cited in Marjorie Grene, *A Portrait of Aristotle* (Chicago: University of Chicago Press, 1963), 64.

37 **the bold propositions of his predecessors:** Lloyd, "Evolution of Evolution," 4–5.

37 **"Empedocles was wrong":** Cited in ibid., 6.

38 **Aristotle was extraordinary and radical:** See Jonathan Barnes, *Greek Philosophers* (Oxford: Oxford University Press, 2001).

39 **In Aristotle's world, all species were fixed:** Ernst Mayr, *The Growth of Biological Thought: Diversity, Evolution, and Inheritance* (Cambridge, Mass., and London: Belknap Press, 1982), 89.

39 **Aristotle was the first to see gradation in nature:** See ibid., 305–7, and Lloyd, *Principles and Practices in Ancient Greek and Chinese Science.*

39 **knowledge could be based only on what was fixed:** See Grene, *Portrait of Aristotle,* 65, 136–37.

40 **Aristotle's philosophical landscape differed profoundly from ours:** See Lloyd, "Evolution of Evolution," 1–15.

40 **After Aristotle died:** I am indebted here to G.E.R. Lloyd's insightful conclusions about evolutionary speculation in ancient Greece and Rome in ibid., 12–15.

3. THE WORSHIPFUL CURIOSITY OF JAHIZ

42 **the port city of Basra:** See Guy Le Strange, *Baghdad During the Abbasid Caliphate: From Contemporary Arabic and Persian Sources* (Oxford: Clarendon Press, 1900). "Mirbad" means literally "kneeling place for camels." In writing this chapter, I depended on the scholarship and knowledge of Professor James Montgomery, professor of classical Arabic at the University of Cambridge, who generously allowed me access to his forthcoming monograph on Jahiz's *Hayawan.*

42 **by the ninth century it was the busiest part of the city:** See A. J. Naji and Y. N. Ali, "The Suqs of Basrah: Commercial Organization and Activity in a Medieval Islamic City," *Journal of the Economic and Social History of the Orient* 24, no. 3 (1981): 298–309. On Basra and the Mirbad, see Charles Pellat, *Le Milieu basrien and la formation de Gahiz* (Paris: Maisonneuve, 1953); Régis Blachère, *Histoire de la littérature arabe des origines à la fin du XVe siècle* (Paris: Maisonneuve, 1964), 3:527; and Hourari Touati, *Islam and Travel in the Middle Ages,* translated by Lydia G. Cochrane (Chicago: University of Chicago Press, 2010).

43 **Jahiz, who may have been of part African descent:** See Charles Pellat, introduction to Charles Pellat, ed., *The Life and Works of Jahiz,* translated by D. M. Hawke (Berkeley: University of California Press, 1969), 2.

43 **listening and talking to the Bedouins in the Mirbad:** "Iraq is the eye of the world," Jahiz wrote, quoting an eighth-century scholar who built a house right in the Mirbad, "Basra is the eye of Iraq and the Mirbad is the eye of Basra": ibid., 192.

43 **Jahiz . . . came close to a theory of evolution and natural selection:** The first scholar to make this claim was the distinguished encyclopedist of science history George Sarton in 1927. In the first of the three volumes of his *An Introduction to the History of Science* (Baltimore: Williams and Wilkins, 1927) he claimed that Jahiz's work contained "the germs of many later theories (evolution, adaptation, animal psychology)" (597). In 1966, the Jahiz scholar and translator Charles Pellat in an entry on "Hayawan" in the *Encyclopaedia of Islam* claimed, "[In the *Kitab al-Hayawan,* Jahiz] sketches a theory of evolution which is not without interest" (3:312); in 1983, Dr. Mehmet Bayrakdar in "Al-Jahiz and the Rise of Biological Evolution," *Islamic Quarterly,* Third Quarter, made a series of significantly larger claims, specifically that Jahiz "described the Struggle for Existence, Transformation of Species into each other and Environmental Factors" (310), that "he profoundly affected the development of zoology and biology" (312), and that "Jahiz and other evolutionist Muslim thinkers influenced Darwin and his predecessors" (313). Bayrakdar's claims were challenged in 2002 by Frank E. Egerton in "A History of the Ecological Sciences, Part 6: Arabic Language Science—Origins and Zoological Writings," *Bulletin of the Ecological Society of America,* 142–46. In 2000, Ahmed Aarab, Philippe Provençal, and Mohamed Idaomar made a more subtle series of claims about the presence of ideas about biodiversity, food chains, animal adaptation, migration, and hibernation in Jahiz's work in "Eco-Ethological Data According to Jahiz Through His Work 'Kitab al-Hayawan,' " *Arabica* 47 (2000): 278–86.

44 **some even allege that Darwin stole Jahiz's work:** One blog writer claims that her history teacher told her that she had once traveled to the British Museum Library in search of Jahiz's *Living Beings* and found the book was missing. Consulting the records, she discovered that the last person to sign out the book had been Darwin and that he had never returned it. The writer concludes that *Origin of Species* is "quite possibly one of the most profound plagiarisms of history." See http://uiforum.uaeforum.org/showthread.php?6582-Early-Islamic-scholars-on-evolution.

46 **A man with pale skin, a flushed face, and bloodshot eyes:** This story was widely circulated in the ninth century and recorded in the tenth century by Al-Nadim, a chronicler of his times and particularly of the Arabian publishing world. Bayard Dodge, ed., *The Fihrist of Al-Nadim: A Tenth-Century Survey of Muslim Culture,* 2 vols. (New York and London: Columbia University Press, 1970), 2:583–84.

46 **the remainder of Aristotle's works were lost:** See Sidney H. Griffith, *The Church in the Shadow of the Mosque* (Princeton: Princeton University Press, 2008).

46 **the desire to rediscover and translate lost knowledge:** Classical scholars Dimitri Gutas and George Saliba disagree about the motivating forces behind the translation movement. Gutas emphasizes the caliphs as directors of the movement, whereas Saliba sees translation as being triggered by social change brought about by competition for patronage among the court officials and administrators.

46 **"And behold," he reported in awe:** Dodge, *Fihrist of Al-Nadim,* 2:585–86; the editor, Bayard Dodge, speculates about this story and the location of the temple: "The building was very likely . . . near Ephesus or Miletus. By the tenth century, the great temple of Apollo Didymaeus at Branchidae near Miletus and the famous library at Pergamum were almost certainly in ruins. It is likely therefore that this library was a second-century building at Ephesus with the famous temple of Diana near by."

47 **what became known as the House of Wisdom:** The concrete realities of the House of Wisdom are frustratingly elusive; for a popular account, see Jonathan Lyons, *The House of Wisdom: How the Arabs Transformed Western Civilisation* (London: Bloomsbury, 2009), and also books by the classical scholars Gutas and Saliba.

47 **competing to commission translations:** For an analysis of the numerous ways in which scholars have described the translation movement and for an astute defense of the plurality of these encounters, see James E. Montgomery, "Islamic Crosspollinations," in Anna Akasoy, James E. Montgomery, and Peter E. Pormann, eds., *Islamic Crosspollinations: Interactions in the Medieval Middle East* (Cambridge: E.J.W. Gibb Memorial Trust, 2007), and Roshdi Rashed, "Greek into Arabic," in James E. Montgomery, ed., *Arabic Theology, Arabic Philosophy: From the Many to the One* (Leuven: Peeters, 2006).

47 **The invention of paper further revolutionized:** Jonathan Bloom, *Paper Before Print: The History and Impact of Paper in the Islamic World* (New Haven and London: Yale University Press, 2001), 49.

47 **Cheap paper meant that books could be produced more easily:** Ibid., 47–56.

49 **Jahiz was an *adib,* a writer of the *adab:*** The literary historian Tarif Khalidi describes the *adab* as "a system for the study of nature and society . . . that eschews narrow specialization in favour of a discursive, multifaceted approach, willing to investigate all natural and social phenomena in a tolerant and sceptical spirit": Tarif Khalidi, *Arabic Historical*

Thought in the Classical Period (Cambridge and New York: Cambridge University Press, 1994), 104.

49 **booksellers in the Mirbad:** On the booksellers' suqs of the Mirbad, see Naji and Ali, "Suqs of Basrah," 303.

49 **Jahiz would rent whole shops for the night:** Dodge, *Fihrist of Al-Nadim*, 2:255.

50 **the zoological gardens of the Abbasid caliphate:** See Qasim Al-Samarrai, "The Abbasid Gardens in Baghdad and Samarra," *Foundation for Science, Technology and Civilisation* 2002, 1–10, www.muslimheritage.com/uploads/ACF9F4.pdf.

50 **now partially translated into Arabic:** On the translation of Aristotle's works into Arabic and Syriac and on the translation movement more generally, see F. E. Peters, *Aristotle and the Arabs: The Aristotelian Tradition in Islam* (New York: New York University Press; London: University of London Press, 1968), and D. Gutas, *Greek Thought, Arabic Culture: The Graeco-Arabic Translation Movement in Baghdad and Early Abbasid Society (2nd–4th/8th–10th Centuries)* (London: Routledge, 1998).

50 **Sometime around 846–47 he hatched the idea:** For the arguments about the date of its composition, see Sa'id H. Mansur, "The World View of al-Jahiz," in *Kitab al-Hayawan* (Alexandria: Dar al-Maare, 1977), 92–96.

51 **There was no other book in the ninth-century Abbasid Empire:** See Charles Pellat, "Hayawan," in *Encyclopaedia of Islam*, 3:305.

51 **the custodians of this knowledge in the Abbasid Empire:** See ibid., 304–15; S. H. Nasr, "Zoology," in his *Islamic Science: An Illustrated Study* (Westerham Press, Kent: World of Islam Festival Publishing, 1976); and Egerton, "History of the Ecological Sciences, Part 6," 142–46.

52 **"to contemplate an entangled bank":** Last paragraph of *Origin of Species* (1859); see also Aarab, Provençal, and Idaomar, "Eco-Ethological Data According to Jahiz," 278–86.

52 **"All you need do is light a fire":** Jahiz, *Hayawan*, 2:110–11; translated by Charles Pellat, ed. *Life and Works of Jahiz*, 142.

53 **"Each species . . . constitutes a food for another species":** Ibid. Translated by Pellat.

54 **"we can see the germs of Darwin's . . . natural selection":** Bayrakdar, "Al-Jahiz and the Rise of Biological Evolution," 307–15.

54 **"every man endowed with reason may know":** Jahiz, *Hayawan*, 2:110, translated by Charles Pellat, ed., *Life and Works of Jahiz*, 142.

55 **"We rarely hear of a statement by a philosopher":** Jahiz, *Hayawan*, 3:268.

55 **In the Wild Beast Park of Baghdad:** On the zoos in the caliphal palace

gardens of Baghdad and Samarra, see Al-Samarrai, "Abbasid Gardens in Baghdad and Samarra," 1–10, www.muslimheritage.com/uploads/ACF9F4.pdf.

55 **"They carry [the pigeons] on their backs":** Jahiz, *Hayawan,* 3:213–14; translated by Charles Pellat, ed., *Life and Works of Jahiz,* 172–73. This knowledge of the ill effects of inbreeding is remarkable; it is of course something that a pigeon breeder would observe over time among his pigeons. It is fundamental for modern genetics. Pigeon fancying and pigeon racing were popular pastimes in medieval Iraq. The Abassids also used a pigeon post.

56 **"We have not devoted a separate chapter to fish":** Jahiz, *Hayawan,* 6:16–17; translation provided by James Montgomery.

57 **caliphal power was beginning to move into a stage of slow decline:** Gutas, *Greek Thought, Arabic Culture,* 124.

57 **Every day, postal envoys brought news of insurrections:** Joel L. Kraemer, "Translator's Foreword," in *The History of al-Tabari,* vol. 34: *Incipient Decline: The Caliphates of Al-Wathiq, Al-Mutawakkil and Al-Muntasir,* AD *841–863* (New York: State University of New York Press, 1989), xv.

57 **Jahiz's principal patron at Baghdad:** A glimpse of their friendship can be seen in Jahiz's teasing letters to his patron in which he chides him for making him tidy his study and bind the pages of his books: Pellat, ed., *Life and Works of Jahiz,* 209–11, 214–15.

57 **But in 847 the caliph al-Wathiq died:** See *History of al-Tabari,* 34: 65–68.

57 **As part of a campaign of retribution:** Ibid., 68.

58 **Jahiz, fearing for his life:** See ibid., 117.

58 **The new caliph Mutawakkil:** Kraemer, "Translator's Foreword," in ibid., xxi.

58 **The rewards for writers and poets in the new court:** See al-Tabari's description of al-Mutawakkil's patronage in "Some Things About al'Mutawakkil and His Way of Life," in ibid., 185.

58 **Fath was a Turkish aristocrat:** Gutas, *Greek Thought, Arabic Culture,* 128.

58 **"The Commander of the Faithful has taken a tremendous liking to you":** Pellat, ed., *Life and Works of Jahiz,* 7–8.

59 **"You will be receiving your monthly allowance":** For translated extracts from Jahiz's work on the Christians, see ibid., 86–89, and on the letter from al-Fath, see 7–8.

59 **To ensure the favor of his new patron:** Kraemer, "Translator's Foreword," in *History of al-Tabari,* 34:xiii.

59 **he died in 869 at the age of ninety-four:** Lawrence I. Conrad, *The Western Medical Tradition* (Cambridge: Cambridge University Press, 2006), 137. At thirty-five or more, the Abbasid Empire had the highest life expectancy of any country in the world at that point.

59 **he was crushed to death:** Pellat, ed., *Life and Works of Jahiz*, 9. Although Jahiz died a century or more before the chronicler Al-Nadim began to write his list of the books, writers, and translators in Baghdad, he was still able to talk to people who had known Jahiz personally. One of them claimed: "I said to Jahiz, 'Do you have an estate at al-Basrah?' He smiled and said 'Verily there is myself, a concubine, the handmaid who serves her, a manservant and a donkey.' " Jahiz then listed his patrons, his books, and how much he was paid for each one, and he concludes his account of himself with the words: "Then I went to Al-Basrah and had an estate which did not require renovation or fertilising." Jahiz's friend remembered his smile when he described his estate. Al-Nadim also recorded that smile; it gives us a glimpse of the pleasure Jahiz felt in returning to Basra. Dodge, *Fihrist of Al-Nadim*, 2:440.

59 **It took the caliphate army two decades:** Amira Bennison, *The Great Caliphs: The Golden Age of the Abbasid Empire* (New York and London: I. B. Tauris, 2009), 89.

60 **But the Catholic Church . . . continued to be deeply suspicious of pagan ideas:** See, in particular, James Hannam, "Heresy and Reason," in his *God's Philosophers: How the Medieval World Laid the Foundations of Modern Science* (London: Icon, 2009), 77–89.

60 **"Neither the works of Aristotle on natural philosophy":** Ibid., 77–81.

60 **In 1423 a Florentine bookseller:** Michael White, *Leonardo: The First Scientist* (London: Abacus, 2000), 42.

4. LEONARDO AND THE POTTER

61 **Sometime in 1493, a family of Italian peasants:** Leonardo tells this story about the peasants arriving from the mountains of Verona in his notebook, the Leicester Codex: Leonardo da Vinci, "Physical Geography," in *The Notebooks of Leonardo da Vinci*, translated by E. MacCurdy (London: Jonathan Cape, 1938), 1:355–56, 359. Michel Jeannert analyzes Leonardo's preoccupation with water and mutability in "Earth Changes: Leonardo Da Vinci," in his *Perpetual Motion: Transforming Shapes in the Renaissance from da Vinci to Montaigne*, translated by Nidra Poller (Baltimore and London: Johns Hopkins University Press, 2001), 50–81. The connection between Leonardo's artistic and intellectual de-

velopment is also discussed, though to a lesser extent, in Pierre Duhem, "Léonard de Vinci, Cardan et Bernard Palissy," *Bulletin Italien* 6, no. 4 (1906): 289–320.

62 **rooms for his assistants:** On Leonardo's assistants in Milan, see Charles Nicholl, *Leonardo da Vinci: The Flights of the Mind* (London: Penguin, 2005), 233–35.

62 **no building could have been better:** Ibid., 248–53.

63 **By 1493 his library contained thirty-seven books:** Ibid., 215; see also Janis Bell, "Color Perspective, ca. 1492," *Achademia Leonardo Vinci* (1992): 64–77; a copy of Aristotle's book appears in a book list from ca. 1490 in the Codex Arundel notebook listed as "meteora d'Aristotle vulgare," and again in a further list in the Codex Arundel from ca. 1490–91 and in the long list in a manuscript in Madrid from 1503–4. On Leonardo's library, see Carlo Maccagni, "Leonardo's List of Books," *Burlington Magazine* 110, no. 784 (1968): 406–10, and Ladislao Reti, "The Two Unpublished Manuscripts of Leonardo da Vinci in the Biblioteca Nacional of Madrid—II," *Burlington Magazine* 110, no. 779 (1968): 81–89.

63 **the collection of natural objects he had gathered:** On the cabinets of curiosities kept by Renaissance princes from the fifteenth century, see Robert Kirkbride, *Architecture and Memory: The Renaissance Studioli of Federico da Montefeltro* (New York: Columbia University Press, 2009).

63 **"astounded . . . fear and desire":** Leonardo, Codex Arundel, 155r, translated by Jean Paul Richter, in Richter, ed., *The Literary Works of Leonardo da Vinci,* 2 vols. (London: Phaidon Press, 1970), 1:1339.

63 **He painted striations of rocky landscapes:** For a dazzling study of the Renaissance fascination with genesis and metamorphosis, see Jeanneret, *Perpetual Motion,* particularly the chapter "Earth Changes: Leonardo da Vinci."

63 **engaged in numerous investigations about the way water:** Nicholl, *Leonardo da Vinci,* 278.

64 **"In the Chiavenna valley . . . are very high barren mountains":** Ibid., 279.

65 **the fresco of *The Last Supper* . . . crumbled, cracked, and faded:** Ibid., 302.

66 **If rocks were no more than forms caught temporarily in time:** On Renaissance ideas of mutability, see Jeanneret, *Perpetual Motion,* particularly the chapter "Earth Changes: Leonardo da Vinci."

67 **"First you must show the smoke of the artillery":** Cited in Nicholl, *Leonardo da Vinci,* 373.

67 **He called battle a "Pazzia Bestialissima":** See Maria Lessing, "Leo-

nardo da Vinci's Pazzia Bestialissima," *Burlington Magazine* 64, no. 374 (1934): 219–31.

67 **a great tangle of human and animal body parts:** For a detailed account of Leonardo's possible borrowings from Greco-Roman sarcophagi, see Kenneth Clark, "Leonardo and the Antique," in C. D. O'Malley, ed., *Leonardo's Legacy* (Berkeley and Los Angeles: University of California Press, 1969), 1–34.

68 **his shoulder a giant xenophora shell:** Richard Preece of the Cambridge Zoology Museum identified this shell. If it is the *Xenophora solaris,* which is not found on Mediterranean shores, Leonardo is most likely to have seen it in one of the many shell collections in the *studioli* of private houses, or he may have picked one up himself from a natural history dealer. On collections, see Patrick Mauries, *Cabinets of Curiosities* (London: Thames and Hudson, 2002). On Renaissance armor, and particularly the use of rams' horns in body armor, see Antonio Domínguez Ortiz, Concha Herrero Carretero, and José A. Godoy, *Resplendence of the Spanish Monarchy: Renaissance Tapestries and Armor from the Patrimonio Nacional* (New York: Metropolitan Museum of Art, 1991); for a further example of a ram's horm used in Renaissance armor around 1520, see 118–19. Leonardo may have seen ancient helmets like these in Rome around 1500 (see arguments by Kenneth Clark in "Leonardo and the Antique") or in Lorenzo's collection of ancient armor—see Mario Scalini, "The Weapons of Lorenzo de Medici," in Robert Held, ed., *Art, Arms and Armour: An International Anthology* (Chiasso, Switzerland: Acquafresca Éditrice, 1979).

68 **"of the winds at Pombino":** Madrid Codices, II, 125r, Biblioteca Nacional, Madrid; cited in Nicholl, *Leonardo da Vinci,* 388.

69 **"On 6 June 1505, on Friday":** Madrid Codices II, 2r; cited in Nicholl, *Leonardo da Vinci,* 390.

70 **he began the Leicester Codex:** The Leicester Codex, one of Leonardo's most important notebooks, was discovered in a chest of manuscripts in Rome in the 1690s and was bought by Thomas Coke, Lord Leicester, remaining in the family until the 1980s, when it was sold to Armand Hammer, who renamed it Codex Hammer. In 1994 it was auctioned at Christie's and bought by Bill Gates for $30 million; he restored the notebook's original name. On Leonardo's fascination with water, see Martin Kemp, *Leonardo* (Oxford: Oxford University Press, 2006), 75–83, and Jeanneret, *Perpetual Motion.*

70 **"With such a rate of motion . . . it would not have travelled":** Leonardo da Vinci, Codex Atlanticus, Ambriosiana Library, Milan, 18v.,

translated by Jean Paul Richter, in Richter, *Literary Works of Leonardo da Vinci,* 1:987.

71 **"The peaks of the Apennines once stood up in a sea":** Leonardo, "Physical Geography," 1:359.

71 **He reread Aristotle and Theophrastus:** All of these writers are referred to in his notebooks.

71 **"It is therefore clear that as time is infinite":** Aristotle, *Meteorologica,* translated by H.D.P. Lee (London: Heinemann, 1952), 119–21.

72 **"He had a very heretical state of mind":** Giorgio Vasari included this passage in the first edition of his *Lives of the Most Excellent Painters Sculptors and Architects* (1550) but excised it in the second edition, thinking it too critical; cited in Nicholl, *Leonardo da Vinci,* 483.

72 **He was also dismissive of the claims of alchemists:** See William R. Newman, *Promethean Ambitions: Alchemy and the Quest to Perfect Nature* (Chicago: University of Chicago Press, 2004), 120–27.

72 **"My concern now . . . is to find subjects and inventions":** Leonardo, Leicester Codex, 2r.

73 **what many Renaissance philosophers and scholars believed to be true:** See, for instance, the Renaissance scholar Marcilio Ficino's *Three Books on Life,* Book III, ch. 1.

73 **"Nothing originates in a spot where there is no sentient":** Richter, *Literary Works of Leonardo da Vinci,* 1:791–92.

73 **He had brilliantly expressed this vision of the earth:** Kemp, *Leonardo,* 148–50.

74 **But he did not contemplate species *evolution:*** One of the reasons for Leonardo's lack of curiosity about species mutation would seem to be his contempt for all alchemical theories. Transmutation was at this point an alchemical notion wrapped up in codes and secrecy and necromancy. Alchemists believed that they could transmute lead to gold and mortal forms to immortal. For Leonardo it was hocus-pocus. No human could transmute flesh into new forms. It was probably this antipathy to the idea of transmutation of any kind that made it impossible for Leonardo, the great philosopher of flux and fluid in the natural world, to countenance the idea that species might have been transmuted from one form to another through vast periods of time by unspecified natural processes.

74 **"The body of the earth, like the bodies of animals":** See Stephen Jay Gould's analysis of Leonardo's fossil writings in "The Upwardly Mobile Fossils," in his *Leonardo's Mountain of Clams and the Diet of Worms* (London: Vintage, 1999).

75 **the most elaborate royal building scheme:** David Thomson, *Renais-*

sance Paris: Architecture and Growth, 1475–1600 (Berkeley and Los Angeles: University of California Press, 1984), 165–75.

75 **The potter, who proudly bore the title of worker of the earth:** For a fascinating analysis of Palissy as an artisan and of how he used his body to embody his art, see Pamela H. Smith, *The Body of the Artisan: Art and Experience in the Scientific Revolution* (Chicago: University of Chicago Press, 2004), 100–106.

76 **Catherine de Médicis, Italian by birth:** On Catherine de Médicis, see Leonie Frieda, *Catherine de Medici: A Biography* (London: Weidenfeld and Nicolson, 2004).

76 **watched her grotto take shape:** An account book survives in the Bibliothèque Nationale in Paris detailing Catherine's expenses on the grotto. It is dated February 22, 1570. The extracts concerning Palissy and the grotto are reprinted in Leonard N. Amico, *Bernard Palissy: In Search of Earthly Paradise* (New York: Flammarion Press, 1996), 231–32.

76 **cast from molds of rocks and shells:** Ibid., 25–26.

77 **they issued from a mind that had become a cauldron:** See William Newman's analysis of the alchemical basis of Palissy's art in Newman, *Promethean Ambitions,* 145–64. See also Jean Céard, "Bernard Palissy et l'alchimie," in Frank Lestringant, ed., *Actes de colloque Bernard Palissy 1510–1590: L'écrivain, le réformé, le céramiste* (Paris: Amis d'Agrippa d'Aubigné, 1992), 157–59.

77 **Palissy's plates . . . had become collectors' pieces:** It is clear from archival records that significant numbers of French aristocrats were already collecting Palissyware during his lifetime; Montmorency in particular had large collections—see Amico, *Bernard Palissy,* appendix 1, documents I and III, 229.

78 **Palissy was fiercely secretive about his art:** Ibid., 41–42.

78 **We know some of the secrets of Palissy's process:** See Hanna Rose Shell, "Casting Life, Recasting Experience: Bernard Palissy's Occupation Between Maker and Nature," *Configurations* 12 (2004): 1–40, and Newman, *Promethean Ambitions,* 158–59.

79 **"There are a great many kinds of ponds":** Bernard Palissy, *The Admirable Discourses of Bernard Palissy,* translated by Aurèle la Rocque (Urbana: University of Illinois Press, 1957), 34–35.

80 **the German-Swiss alchemist Paracelsus in particular:** For more detail about the influence of Paracelsus's work on Palissy, see the remarkable study of Palissy's work in Neil Kamil, *Fortress of the Soul: Violence, Metaphysics, and Material Life in the Huguenots' New World, 1517–1751* (Baltimore: Johns Hopkins University Press, 2005).

81 **Speculation about spontaneous generation:** See Henry Harris, *Things*

Come to Life: Spontaneous Generation Revisited (Oxford: Oxford University Press, 2002), 1–8; and for the link between artificial life and alchemy, see Newman, *Promethean Ambitions,* 164–237.

81 **Pierre Belon and Guillaume Rondelet:** Both Rondelet and Belon had written about rot and spontaneous generation. In *La Nature et diversité des poissons* (1555), Pierre Belon claimed that frogs were generated by both eggs and rot. Belon's vision of spontaneous generation was also continuous with his ideas about the Renaissance. The French naturalist Guillaume Rondelet dedicated a whole chapter of his colossal book on fish, *L'Histoire entière des poissons* (1558), to organisms that live in stagnant marshes, believing them to be "by nature halfway between plants and animals." Palissy also read a French translation of an essay by the great Italian polymath Girolamo Cardano, the son of Leonardo's close friend, entitled "Creatures Born of Putrefaction," in which Cardano claimed: "Frogs spring forth formed out of impure water and sometimes the rain; they are among a certain number of imperfect animals that are born, without seed, from corruption and putrefaction."

81 **"I maintain that shellfish . . . are born on the very spot":** Palissy, *Admirable Discourses,* 244–45.

81 **"I am neither Greek, nor Hebrew, nor Poet":** B. Palissy (1563), "Recepte véritable," *Oeuvres de Bernard Palissy,* edited by A. France (1880), 13; see also a new edition by Frank Lestringant (Paris: Macula, 1996), which has a substantial and very useful introduction.

5. TREMBLEY'S POLYP

86 **The Count of Bentinck's summer residence:** On Dutch gardens in this period, see Erik Jong, *Nature and Art: Dutch Garden and Landscape Architecture, 1650–1740* (Philadelphia: Philadelphia University Press, 2000), and J. W. Vanessa Bezemer-Seller, "The Bentinck Garden at Sorgvliet," in J. D. Hunt, ed., *The Dutch Garden in the Seventeenth Century* (Washington, D.C.: Dumbarton Oaks Research Library and Collection, 1990). I am indebted to Marc Ratcliff of the University of Geneva, author of the masterly study of Enlightenment microscopy, *The Quest for the Invisible: Microscopy in the Enlightenment* (Farnham, Surrey, and Burlington, Vt.: Ashgate, 2009), for giving careful attention to an early draft of this chapter.

87 **their thirty-year-old Genevan tutor, Abraham Trembley:** Abraham Trembley, *Instructions d'un père à ses enfants, sur la nature et sur la religion,* 2 vols (Geneva: Chapuis, 1775), 1:xii. On the Bentinck family, see

Paul-Emile Schazmann, *The Bentincks: The History of a European Family* (London: Weidenfeld and Nicolson, 1976), and Aubrey Le Bond, *Charlotte Sophie, Countess Bentinck: Her Life and Times* (London: Hutchinson, 1912). William Bentinck's correspondence with his sons' tutors and others is in the British Museum: Bentinck Papers. Countess Bentinck later became an intimate friend of Voltaire's and is considered the inspiration for his *Candide.* Trembley gave his own detailed account of his discovery of the polyp in his 1744 work, *Mémoires pour servir à l'histoire d'un genre de polypes d'eau douce.*

88 **"I was surprised," he recalled later:** John R. Baker, *Abraham Trembley of Geneva: Scientist and Philosopher, 1710–1784* (London: Edward Arnold, 1952), 28–29; see also Baker's chapter on Trembley's educational theories, 188–204.

89 **All three men were obsessed with insects:** On seventeenth-century entomology, see Janina Wellmann, "Picture Metamorphosis: The Transformation of Insects from the End of the Seventeenth to the Beginning of the Nineteenth Century," *NTM* 16, no. 2 (2008): 183–211.

90 **he compared his aphid to the virgin Danae:** Bonnet to Réaumur, July 13, 1740, Papers of Réaumur and Bonnet, Archives de l'Académie des Sciences, Paris. The king of Argos imprisoned his daughter Danae in a tower because the Delphic oracle foretold that his death would be at the hands of his daughter's son. But the tower was no protection from the gods. Zeus seduced her disguised as a shower of gold, and she bore Perseus.

90 **a form of asexual reproduction that requires no fertilization:** For a detailed analysis of the discovery of parthenogenesis, see Marc Ratcliff's fine chapter "Insects, Hermaphrodites and Ambiguity" in *Quest for the Invisible,* 57–73.

91 **"These are assuredly observations of great importance":** Réaumur to Bonnet, August 5, 1740; cited in Virginia P. Dawson, *Nature's Enigma: The Problem of the Polyp in the Letters of Bonnet, Trembley and Réaumur* (Philadelphia: Memoirs of the American Philosophical Society, 1988), 80, 114.

92 **"A fact such as the one which aphids presented":** Trembley, *Mémoires,* 18.

92 **"Almost the whole of the month of September 1740 passed":** Baker, *Abraham Trembley of Geneva,* 29.

92 **"It was on the 25th November 1740 that I cut the first polyp":** Ibid., 32.

93 **"Throughout the day I continually observed the points":** Ibid.

94 **"I saw these parts walk, take steps":** Maurice Trembley, *Correspondance inédite,* 28, translated by Virginia P. Dawson in her *Nature's Enigma,* 101–2.

94 **"perhaps one of the most ardent that there is in Nature":** Bonnet to Trembley, December 18, 1740, George Trembley Archives, Toronto, Ontario. See Dawson, *Nature's Enigma,* 89.

94 **"Who knows if one mating might not serve for several generations?":** Trembley to Bonnet, January 27, 1741, Ms Bonnet 24, Bibliothèque Publique et Universitaire de Genève; cited in Dawson, *Nature's Enigma,* 89.

94 **to tell Bonnet about his "little aquatic being":** Ibid., 138.

95 **his "little aquatic Being ought to be regarded":** Bonnet to Trembley, March 24, 1741, George Trembley Archives, Toronto, Ontario; cited in Dawson, *Nature's Enigma,* 138.

95 **the philosophical questions the experiment raised:** On the polyp as a source for materialist and vitalist ideas, see Aram Vartanian, "Trembley's Polyp, La Mettrie and Eighteenth-Century French Materialism," *Journal of the History of Ideas* 11 (1950): 259–80; Jacques Roger, *Les Sciences de la vie dans la pensée française du XVIIIe siècle: La génération des animaux de Descartes à l'Encyclopédie* (Paris: Armand Colin, 1963), 749; Ratcliff, *Quest for the Invisible,* 103–25; Dawson, *Nature's Enigma,* 155–56; Giulio Barsanti, "Les Phénomènes 'étranges' et 'paradoxaux' aux origines de la première révolution biologique (1740–1810)," in Guido Cimino and François Duchesneau, eds., *Vitalisms from Haller to the Cell Theory* (Florence: Olschki, 1997), 67–82; Barbara Maria Stafford, "Images of Ambiguity, Eighteenth-Century Microscopy, and the Neither/Nor," in D. P. Miller and P. H. Reill, eds., *Visions of Empire: Voyages, Botany, and Representations of Nature* (Cambridge: Cambridge University Press, 1997), 230–57; Catherine Wilson, *The Invisible World: Early Modern Philosophy and the Invention of the Microscope* (Princeton: Princeton University Press, 1995), 203; and Brian J. Ford, *Single Lens: The Story of the Simple Microscope* (New York: Harper and Row, 1985), 109–11.

96 **Bonnet . . . began a new series of aphid experiments:** Bonnet to Trembley, March 24, 1741, George Trembley Archives, Toronto, Ontario.

96 **"If this excellent friend had been able to foresee all the evil":** Cited in Dawson, *Nature's Enigma,* 91.

96 **"never did an insect cause so much uproar":** Réaumur to Trembley, August 30, 1741, in Maurice Trembley and Emile Guyénot, eds., *Correspondance inédite entre Réaumur et Abraham Trembley* (Geneva: Georg, 1943), 106.

96 **"The story of the Phoenix who is born from the ashes":** *Histoire de l'Académie Royale des Sciences* (Amsterdam: Pierre Mortier, 1741), 1:46.

96 **"A miserable insect has just shown itself to the world":** [Gilles Auguste Bazin], *Lettres d'Eugène à Clarice* (Strasbourg: Imprimerie du Roi et de Monseigneur le Cardinal de Rohan, 1745); cited in Dawson, *Nature's Enigma,* 186.

97 **Was Descartes then right after all:** Descartes' ideas had been popularized by Bernard de Fontenelle's bestselling *Conversations on the Plurality of Worlds;* they dominated early-eighteenth-century thought.

97 **opportunities to refute this . . . godless way of seeing the world:** Although Descartes' ideas had permeated natural philosophy so far by the eighteenth century that it was second nature for Bonnet, Lyonet, and Trembley to refer to all small organisms as "little machines" and to their workings as "mechanisms," they believed that these mechanisms were operated by God. See Virginia Dawson's chapter "The Ragged Cartesian Fabric of Eighteenth-Century Biology" in *Nature's Enigma,* 25–51.

97 **"All that I ardently wish":** Bonnet to Cramer, June 29, 1741, Ms Suppl. 384, Bibliothèque Publique et Universitaire de Genève; cited in Dawson, *Nature's Enigma,* 141.

97 **"to deal a heavy blow to the System":** Cramer to Bonnet, June 1741, Ms Bonnet 43, Bibliothèque Publique et Universitaire de Genève; cited in Dawson, *Nature's Enigma,* 141.

97 **"Let me breathe a little":** Cramer to Bonnet, December 1741; cited in Dawson, *Nature's Enigma,* 169.

97 **he began to experiment on aquatic worms:** Dawson, *Nature's Enigma,* 143–44.

98 ***"For what end?"* he asked:** Bonnet to Réaumur, November 4, 1741, Papers of Réaumur and Bonnet, Bibliothèque Publique et Universitaire de Genève; cited in Dawson, *Nature's Enigma,* 141.

98 **He proposed a bland explanation:** Réaumur to Bonnet, November 30, 1741, Ms Bonnet 26, Archives de l'Académie des Sciences, Paris.

98 **"I am entirely taken up with dispatching polyps":** Trembley to Martin Folkes, July 16, 1743, Ms Folkes, vol. 4, letter 66; cited in Ratcliff, *Quest for the Invisible,* 12.

98 **It was a brilliant strategy:** Ratcliff, *Quest for the Invisible,* 12.

98 **"Apart from electricity, naturalists did not deal":** Ibid., 13.

99 **She gave him a long report by letter:** Madame Geoffrin to Martin Folkes, January 12, 1743; cited in Harcourt Brown, "Madame Geoffrin and Martin Folkes: Six New Letters," *Modern Language Quarterly* 1 (1940): 219.

99 **"We are no less sensible of your great candour":** Folkes to Trembley, November 30, 1743, Ms Trembley, 91–92; cited in Ratcliff, *Quest for the Invisible,* 21.

100 **Baker published a two-hundred-page account:** Henry Baker, *An Attempt Towards a Natural History of the Polype* (London: R. Dodsley, 1743), 7–10, 209–10.

100 **"The marvelous properties of the new Polyp":** Anonymous, [report of *Philosophical Transactions* 42:467], *Bibliothèque Britannique* 22, no. 1 (1743): 159.

100 **It had become a sensation:** For the early speculation on the polyps and other insects that reproduced after cutting, see Charles Bonnet, "Of Insects Which Are Multiplied, as It Were, by Cutting or Slips," *Philosophical Transactions* 42, no. 470 (1743): 468–88. The discovery was also published in French in Charles Bonnet, *Traité d'insectologie ou observations sur les pucerons* (Amsterdam: Luzac, 1745). See also William Bentinck, "Abstract of Part of a Letter from the Honourable William Bentinck, Esq., F.R.S., to Martin Folkes, Esq., Pr.R.S., Communicating the Following Paper from Mons. Trembley, of the Hague," *Philosophical Transactions* 42, no. 467 (1743): ii (the paper that followed was Abraham Trembley, "Observations and Experiments upon the Freshwater Polypus, by Monsieur Trembley, at the Hague," iii–xi); Duke of Richmond, "Part of a Letter from His Grace the Duke of Richmond to M. Folkes," *Philosophical Transactions* 42, no. 470 (1743): 510–13; Baker, *Attempt Towards a Natural History of the Polype;* and Thomas Lord, "Concerning Some Worms Whose Parts Live After They Have Been Cut Asunder," *Philosophical Transactions* 42, no. 470 (1743): 522–23.

102 **The polyp appeared to be the point of passage:** See Dawson, *Nature's Enigma,* 167–68.

102 **God had originally created a multitude of germs:** This idea, held by others in the eighteenth century, was called preformism. See Bentley Glass, Owsei Temkin, and William L. Straus, eds., *The Forerunners of Darwin: 1745–1859* (Baltimore: Johns Hopkins University Press, 1959), 164–69.

102 **If there was progress, he argued:** Ibid., 168.

102 **"Nature is assuredly admirable in the conservation of individuals":** C. Bonnet, *Considérations sur les corps organisés,* in *Oeuvres d'histoire naturelle et de philosophie,* 8 vols. (Neuchâtel, 1779–83; first published 1762), 3:90; cited in Bentley Glass, "Heredity and Variation in the Eighteenth Century Concept of Species," in Glass, Temkin, and Straus, *Forerunners of Darwin,* 164.

102 **Man was immortal:** See Charles Otis Whitman, "The Palingenesia and the Germ Doctrine" and "Bonnet's Theory of Evolution—A System of Negation," in his *Biological Lectures* (Woods Hole, Mass.: Marine Biological Laboratory, 1894), 205–72.

102 **"Mr. de Buffon claims to explain nearly everything":** Abraham Trembley to William Bentinck, January 9/20, 1750, Ms Egerton 1726, British Library, London; cited in Dawson, *Nature's Enigma*, 187.

102 **Now that so many men and women had seen the polyp regenerate:** For a fascinating discussion of what Marc Ratcliff calls the Trembley effect, see his fine essay *L'Effet Trembley, ou la naissance de la zoologie marine* (Geneva: La Baconnière, 2010).

103 **Ponds, seabeds, and rock pools . . . were terra incognita:** Baker, *Attempt Towards a Natural History of the Polype*, 207.

103 **In the four vignettes commissioned for Trembley's *Memoir*:** The polyp diagrams in the *Memoir* were engraved by Pierre Lyonet, the four vignettes by the Dutch draftsman Cornelius Pronk, a protégé of William Bentinck's.

104 **Despite the emptiness portrayed in the picture:** Baker, *Attempt Towards a Natural History of the Polype*, 37.

105 **they may have been leaders of a group of freemasons:** Margaret C. Jacob makes these claims in *Living the Enlightenment: Freemasonry and Politics in Eighteenth-Century Europe* (New York: Oxford University Press, 1991). Her claims have been disputed by several scholars, including Marc Ratcliff.

105 **They had strong links with radical publishing networks:** Margaret C. Jacob, *The Radical Enlightenment: Pantheists, Freemasons and Republicans* (London: Allen and Unwin, 1981), 245–47. See also Jacob, *Living the Enlightenment*, 129–30.

105 **The freethinkers of The Hague:** Paul Hazard, *Le Crise de la conscience européenne, 1680–1714* (Paris: Boivin, 1935), and Margaret C. Jacob, "Hazard Revisited," in Phyllis Mack, ed., *Politics and Culture in Early Modern Europe: Essays in Honour of H. G. Koenigsberger* (Cambridge: Cambridge University Press, 1987), 250–72.

106 **Trembley's discovery spawned a new age of natural philosophical speculation:** Charles Bonnet continued to struggle with his eyesight for the rest of his life; he seems rarely to have left the country house where he lived outside Geneva with his wife. Though childless, they raised his wife's adopted nephew, who became the celebrated physicist and Alpine traveler Horace-Bénédict de Saussure. Unable to undertake close microscopal work, Bonnet wrote books of an increasingly philosophical

and metaphysical nature that supported his developing theory of preexistent germs. Trembley left the Bentinck household in 1747; he traveled through Europe on the Grand Tour as the tutor and companion of the fifteen-year-old 3rd Duke of Richmond from 1750 to 1755. Later he married and dedicated the rest of his life to raising and educating his children and writing books on educational methods.

6. THE CONSUL OF CAIRO

107 **the French consul, Benoît de Maillet:** See Harriet Dorothy Rothschild, "Benoît de Maillet's Cairo Letters," *Studies on Voltaire and the Eighteenth Century* 169 (1977): 134, and Paul Masson, *Histoire du commerce français dans le Levant au XVIIIe siècle* (Paris: Hachette, 1911); Louis Phélypeaux, Comte de Pontchartrain, was one of a dynasty of French politicians who ran the Conseil de la Marine, all of whom Maillet reported to: Louis was secretary from 1690 until 1699; his son Jérôme succeeded him from 1699 to 1714; Jérôme's son Jean was secretary of a newly formed Conseil de la Marine from 1723 to 1737. Louis Phélypeaux was a family friend from Lorraine and appointed Maillet to the position in Egypt in 1692. Despite his dislike for Jérôme, Maillet remained loyal to the family of his patrons for more than forty years. I have largely composed this picture of Maillet's life in Egypt, Leghorn, Paris, and Marseilles from the letters he left behind, now in the Archives Nationales, Correspondance Consulaire, which have been examined in Rothschild, "Benoît de Maillet's Cairo Letters," 115–85; Rothschild, "Benoît de Maillet's Letters to the Marquis de Caumont," *Studies on Voltaire and the Eighteenth Century* 60 (1977): 311–38; Rothschild, "Benoît de Maillet's Leghorn Letters," *Studies on Voltaire and the Eighteenth Century* 30 (1964): 351–76; and Rothschild, "Benoît de Maillet's Marseilles Letters," *Studies on Voltaire and the Eighteenth Century* 37 (1965): 109–45.

108 **Cairo, once a great capital:** Of the twenty-five trading posts that flourished in Egypt in 1709, only eight or nine survived by 1724. Letter from Maillet to the Comte de Maurepas, March 19, 1724, in Rothschild, "Benoît de Maillet's Marseilles Letters," 120.

110 **the first sustained attempt to prove that species had mutated:** Lester G. Crocker, "Diderot and Eighteenth-Century French Transformism," in Glass, Temkin, and Straus, *Forerunners of Darwin*, 123–24.

110 **"One cannot but admire the beauty of these domes":** Benoît de Maillet, *Description de l'Egypte*, 1735 ed., 1:200–201.

110 **Maillet began his investigations into Egyptian history:** See Claudine

Cohen, "Benoît de Maillet et la diffusion de l'histoire naturelle à l'aube des lumières," *Revue d'Histoire des Sciences* 44, no. 3–4 (1991): 334.

111 **In *Histories,* Maillet read Herodotus' excited observations:** Almost certainly the 1677 edition published in Paris by G. de Luyne.

111 **Benoît de Maillet set out to find Memphis:** Maillet, *Telliamed,* 1750, 100 (I have used the English edition of 1750, which is a very close translation of the original French edition of 1748); see also the annotated translation and detailed study of the various manuscript versions undertaken by Albert V. Carozzi: Carozzi, ed., *Telliamed* (Champaign: University of Illinois Press, 1968). A league is around three miles or an hour's walk; hence at this point Memphis was seventy-five miles from the sea.

112 **"this fame Flux and Reflux":** Maillet, *Telliamed,* 92.

113 **he knew he would have to publish anonymously:** On the history of anonymity in publishing, see John Mullan, *Anonymity: A Secret History of English Literature* (London: Faber and Faber, 2008).

115 **"What if they were clever enough to navigate":** Fontenelle, *Conversations on the Plurality of Worlds,* translated by Elizabeth Gunning (London: Hurst, 1803; first published in 1686), 65.

115 **"Were the sky only a blue arch":** Ibid., 112.

116 **the theory that stars had been born from a vortex:** Claude Gadrois, *Discours sur les influences des astres selon les principes de M. Descartes* (Paris: J.-B. Coignard, 1671).

116 **an approving description of "the simple potter":** For further material on Maillet's use of Palissy, see Carozzi, *Telliamed,* 335–36.

116 **new species were being produced constantly:** Maillet, *Telliamed,* 276–77.

116 **After leaving Egypt in 1708 and settling in Leghorn:** See Rothschild, "Benoît de Maillet's Leghorn Letters," 360–63.

117 **Here they found a hollow log:** Maillet, *Telliamed,* 50–51.

118 **Khayyám's ideas about the diminution of the sea:** Ibid., 159.

118 **In 1717, Maillet returned from Leghorn:** Rothschild, "Benoît de Maillet's Marseilles Letters," 125.

118 **Maillet set up house in the rue de Rome:** Ibid., 133.

118 **it was difficult for him to focus on his book:** Ibid., 113.

119 **A clandestine book trade had burgeoned in Paris:** See Jane McLeod, "Provincial Book Trade Inspectors in Eighteenth-Century France," in *French History* 12, no. 2 (1998): 127–48, and Robert Darnton, "A Police Officer Sorts His Files," in his *The Great Cat Massacre and Other Episodes in French Cultural History* (London: Vintage, 1985), 145–89.

119 **But what was he to do with his most dangerous book?:** See Miguel

Benítez, "Benoît de Maillet et la littérature clandestine: Étude de sa correspondance avec l'abbé Le Mascrier," *Studies on Voltaire and the Eighteenth Century* 183 (1980): 143.

120 **Conversation 3, "On the Origin of Species":** Rothschild, "Benoît de Maillet's Letters to the Marquis de Caumont," 315.

120 **"Telliamed has all the trouble in the world":** Ibid., 315–16.

121 **"The Transformation of a Silk-worm or a Caterpillar":** Maillet, *Telliamed*, 225.

121 **"Your Histories read . . . that in the Year 592 of your Era":** Ibid., 230–31.

122 **records of sea people sightings:** Ibid., 232–44.

123 **"I was informed that there was a Courtesan":** Ibid., 249–50.

123 **what he described as "troublous times":** Rothschild, "Benoît de Maillet's Marseilles Letters," 117.

124 **He offered his eighty-volume oriental library:** Ibid., 136.

124 **The French capital overwhelmed his imagination:** Rothschild (1968), "Benoît de Maillet's Letters to the Marquis de Caumont," 332.

125 **his five notebooks on Ethiopia and Coptic Christianity:** Ibid., 334.

125 **a bestselling heretical book written by two Dutchmen:** On *Ceremonies*, see Lynn Hunt, Margaret C. Jacob, and Wijnand Mijnhardt, *The Book That Changed Europe: Picart and Bernard's "Religious Ceremonies of the World"* (Cambridge, Mass., and London: Belknap Press, 2010).

125 **As Maillet lay dying in his house in Marseilles:** Rothschild, "Benoît de Maillet's Letters to the Marquis de Caumont," 335.

126 **"Even after centuries of petrification":** Letter cited in Benítez, "Benoît de Maillet et la littérature clandestine," 153–54.

126 ***Telliamed* at last found its way into print:** On Mascrier's changes, see Albert V. Carozzi's edition of the text and study of the changes: Carozzi, *Telliamed*, 26–30.

127 **"What a folly in this author to substitute Telliamed":** A. J. Dézallier d'Argenville, *L'Histoire naturelle éclaircie dans une de ses parties principales. . . .* (Paris: Debure l'aîné, 1757), 74.

127 **"This consul Maillet was one of those charlatans":** Voltaire (J.F.M. Arouet) in *Cabales* (1772), in *Oeuvres complètes,* 70 vols. (Paris: Firmin-Didot, 1875), 2:749.

128 **"He was a Jesuit for a long time":** Darnton, "A Police Officer Sorts His Files," 158; I have used Darnton's translation of this passage but have translated the French titles into English.

129 **"I am bound to read it":** Darwin to Isaac Anderson-Henry, May 22, 1867, Letter 5545, DCP.

7. THE HOTEL OF THE PHILOSOPHERS

130 **They had come to interrogate the man:** Arthur M. Wilson, *Diderot* (New York: Oxford University Press, 1972), 55–56.

131 **"In the presence of the said Diderot":** Paul Bonnefon, "Diderot prisonnier à Vincennes," *Revue d'Histoire Littéraire de la France* 6 (1899): 204–5.

132 **"passed his early years in debauchery":** Ibid., 203.

133 **He impounded the manuscript he found:** See Wilson, *Diderot,* 63–64.

134 ***Letter on the Blind* had been written by an atheist:** There were deaths and births in these years. Denis and Nanette buried their six-week-old firstborn child, a daughter—Angélique—in 1744; Nanette gave birth to a son in 1746, but the boy was sickly; Diderot's mother died in 1748. Diderot began a passionate affair with a younger and married woman—a writer, Madeleine d'Arsant de Puisieux—while Nanette was pregnant with the boy. On the emergence of Diderot's atheism, see Aram Vartanian, "From Deist to Atheist: Diderot's Philosophical Orientation, 1746–1749," *Diderot Studies* 1 (1949): 31–51.

135 **Paris savants began discussing Trembley's polyp:** See May Spanger, "Science, philosophie et littérature: Le polype de Diderot," *Recherches sur Diderot et sur l'Encyclopédie* 23 (1997): 89–107.

135 **you had only to look at the complexity of design:** Diderot makes this argument in *Philosophical Thoughts* (1746).

136 **For Maupertuis, the polyp proved:** The historian of evolution Ernst Mayr claims that Maupertuis was not an evolutionist and that he was more of a cosmologist than a biologist. Instead, Mayr argues, he was one of the pioneers of genetics. Maupertuis proposed a theory of pangenesis, arguing that particles from both mother and father were responsible for the characteristics of the child. Bowler credits him with conducting studies on heredity, with elucidating the natural origin of human races, and with the originating idea that forms of life may have changed with time. His philosophy was basically materialist. Mayr, *The Growth of Biological Thought,* 328–29.

136 **no spiritual or supernatural presence in the universe:** Vartanian, "From Deist to Atheist," 31–51, and Crocker, "Diderot and Eighteenth-Century French Transformism," 117.

137 **an acquaintance of Diderot's and a man he much admired:** Jacques Roger, *Buffon: A Life in Natural History,* edited by L. Pearce Williams, translated by Sarah Lucille Bonnefoi (Ithaca and London: Cornell University Press, 1997), 199.

137 **Although Buffon believed that species were fixed:** Arthur O. Lovejoy, "Buffon and the Problem of Species," in Glass, Temkin, and Straus, *Forerunners of Darwin,* 68. Ernst Mayr claims that though Buffon was not an evolutionist, "it is nonetheless true that he was the father of evolutionism": Mayr, *Growth of Biological Thought,* 330.

137 **Diderot also read Benoît de Maillet's *Telliamed:*** Vartanian, "From Deist to Atheist," 59. Lester Crocker in "Diderot and Eighteenth-Century French Transformism" argues that Maillet's book had no apparent influence on Diderot's *Letter on the Blind.*

138 **his prison room in Vincennes:** See Wilson, *Diderot,* 105.

138 **Diderot had no books with him other than:** P. N. Furbank, *Diderot: A Critical Biography* (London: Secker and Warburg, 1992), 153; Wilson, *Diderot,* 109.

138 **"The detention of M. Diderot":** Bonnefon, "Diderot prisonnier à Vincennes," 206.

138 **he had become a celebrity:** Wilson, *Diderot,* 117.

138 **In the long entry "Animal" written for the first volume:** Diderot, "Animal," *Encyclopédie ou Dictionnaire raisonné des sciences, des arts et des métiers,* edited by Denis Diderot and Jean le Rond d'Alembert, 17 vols. (Paris: 1751–72), 1:469. See also Mary Efrosni Gregory, *Diderot and the Metamorphosis of Species* (London: Routledge, 2008), 109–10.

139 **The Jansenists declared it a work of heresy:** See Roger, *Buffon,* 187–89.

139 ***"Sur la scène du monde, je m'avance masqué":*** Otis Fellows, "Buffon's Place in the Enlightenment," *Studies on Voltaire and the Eighteenth Century* 25 (1963): 613.

139 **in 1751 the first volume appeared:** Wilson, *Diderot,* 7.

140 **There is no record of their first meeting:** Wilson argues that d'Holbach and Diderot did not meet until around the autumn of 1751, but Kors argues for an earlier date. Discussing the events of 1751, Rousseau says d'Holbach was already "linked for a long time with Diderot": Alan Charles Kors, *D'Holbach's Coterie: An Enlightenment in Paris* (Princeton: Princeton University Press, 1976), 14.

140 **an elegant and spacious six-story house:** Baron d'Holbach bought the house on the rue Royale in 1759, but, as he was holding his salons there before that date, he may have rented it for the first few years. The address today is 8, rue des Moulins. See Wilson, *Diderot,* 175; André Billy, *Diderot: Sa vie, son oeuvre* (Paris: A. Cresson, 1972), 314–15, quotes the undated bill of sale.

141 **"Let us hasten," he wrote, "to make philosophy popular":** Wilson, *Diderot,* 198.

142 **Rousseau described him as a fervent recruiter of freethinkers:** Rous-

seau writes: "A natural repugnancy prevented me a long time from answering his advances. One day, when he asked me the reason of my unwillingness, I told him he was too rich. He was, however, resolved to carry his point, and at length succeeded. My greatest misfortune proceeded from my being unable to resist the force of marked attention. I have ever had reason to repent of having yielded to it." *Confessions,* Book 1.

142 **He had a great deal to offer his guests:** See Max Pearson Cushing, *Baron d'Holbach: A Study of Eighteenth Century Radicalism in France* (New York: Columbia University Press, 1914), 21. A catalog of the baron's library was published by Deburé in Paris in 1789.

142 **"[D'Holbach] pursued the incredulity of Diderot":** Dominique-Joseph Garat, *Mémoires historiques sur la vie de M. Suard, sur les écrits, et sur le XVIIIe siècle,* 2 vols. (Paris, 1820), 1:208–9.

142 **D'Holbach was devastated:** For a description of d'Holbach's grief, see Rousseau's *Confessions,* Book 8.

143 **the furious evangelism of his atheism:** See Michael Bush, introduction to the 1999 Clinamen Press reprint of M. Mirabaud's 1797 translation of d'Holbach's *System of Nature,* ix. D'Holbach's first child, Charles Marius, was born in August 1757; his two daughters were born in 1758 and 1760, and a further son whose birth date has not survived for the historical record was born some time afterward. See T. C. Newland, "D'Holbach, Religion, and the 'Encyclopédie,'" *Modern Language Review* 69, no. 3 (1974): 523–33.

143 **an unrepentant and unwavering atheist:** The Abbé Morellet, a twenty-five-year-old theology student in 1753 and a friend of Diderot's, described him as having "had in mind the desire to gain proselytes, not precisely to atheism, but to philosophy and reason. . . . He defended them [his atheistic ideas] without acrimony, and without looking unfavourably upon those who did not share them." Morellet also described various attempts by young curés to convert the now famously atheist Diderot and the theological wranglings that took place at the rue de l'Estrapade when they turned up to preach. Morellet, *Mémoires,* 1:29–30, 34–35.

143 **he published a further attempt to answer those questions:** See Wilson, *Diderot,* 194.

143 **"May it not be," he wrote, "that, just as an individual organism":** Cited in ibid., 194–95.

144 ***May it not be . . . If the faith had not taught us otherwise?:*** See Arthur O. Lovejoy, "Some Eighteenth-Century Evolutionists," *Popular Science Monthly* 65 (1904): 238–51.

144 **opaque, incomprehensible, and at best obscure:** See Wilson, *Diderot,* 196–98.

144 **"He who resolves to apply himself to the study of philosophy":** Cited in ibid., 198.

145 **"in the middle of the persecution incited against philosophy":** Cited in Roger, *Buffon,* 338. On Buffon and transformism, see Mayr, *Growth of Biological Thought,* 330–37; Lovejoy, "Buffon and the Problem of Species"; and Mary Efrosni Gregory, *Evolutionism in Eighteenth-Century French Thought* (New York: Peter Lang, 2008), 69–92.

146 **"have you ever thought seriously about what living means?":** Diderot to Sophie Volland, October 15, 1759, Denis Diderot, *Diderot's Letters to Sophie Volland: A Selection,* translated by Peter France (London: Oxford University Press, 1972), 37–38. The English translation lists it as October 17, but the original French edition (edited by Ernest Babelon) gives it as October 15. Cited in Kors, *D'Holbach's Coterie,* 99.

146 **life had become more dangerous within the d'Holbach circle:** Cushing, *Baron d'Holbach,* 26.

147 **"The pedlar, the pedlar's wife, and the apprentice":** Letter to Sophie Volland, October 8, 1768, Diderot, *Diderot's Letters to Sophie Volland,* 180–81.

147 **"It is raining bombs in the house of the Lord":** Ibid., 189.

148 **"So you can see me . . . surrounded by engravings":** Cited in Wilson, *Diderot,* 559.

148 **"everything is bound up with everything else":** Ibid., 568; on the influences on *Rêve,* see Crocker, "Diderot and Eighteenth-Century French Transformism," 137–43.

148 **Yes, he answered, man, "but not as he is":** Denis Diderot, *Oeuvres complètes de Diderot,* edited by Jules Assezat and Maurice Tourneux, 20 vols. (Paris: Garnier, 1875), 4:94–96.

149 **Laurence Sterne's anarchic *Tristram Shandy:*** He called *Tristram Shandy* "the craziest, wisest and gayest of all books. . . . This book that is so mad, so wise and so gay, is the Rabelais of the English." Cited in Wilson, *Diderot,* 457.

149 **"It is the height of extravagance":** Letter to Sophie Volland, August 31, 1769, Diderot, *Diderot's Letters to Sophie Volland,* x.

149 **"the whole is constantly changing":** Diderot, *Rêve,* 180–81.

150 **"Let us assume a long succession of armless generations":** Ibid., 180.

150 **"Stop thinking about individuals and answer me this":** Ibid., 181.

151 **living webs of connectedness:** Diderot's vision of connectedness is essentially Rabelaisian.

151 ***"tout est en un flux perpetuel":*** Buffon had used the idea of flux for the

first time in volume 9 (1761) to describe a stream of events in nature: "Nature, I declare, is in a movement of continual flux [*de flux continue*]; but it is enough for man to seize it in the instant of his century, and to throw a few glances behind and ahead, in order to try to catch a glimpse of what it might have been, and what it could become in time": "Des animaux communs aux des continents," in *Histoire naturelle, générale et particulière* (Paris: Imprimerie Royale, 1749–67).

152 **Diderot claimed later that he had indeed destroyed it:** Jean Philibert Damiron, *Mémoires sur Naigeon et accessoirement sur Sylvain Maréchal et Delalande* (Paris: Durand, 1857), 409.

152 **d'Holbach . . . set off his own incendiary device:** Kors, *D'Holbach's Coterie,* 235–43.

153 **"[I have] acquired a taste for solitude":** Letter to Sophie Volland, November 28, 1770, Diderot, *Diderot's Letters to Sophie Volland,* 206–7.

153 **"a chaos, a great moral sickness":** Voltaire, *Oeuvres,* 66:394.

153 **"I think that nothing has debased our century more":** Cushing, *Baron d'Holbach,* 58.

153 **"I hardly ever go out":** Letter to Sophie Volland, July 22, 1773, Diderot, *Diderot's Letters to Sophie Volland,* 209.

154 **"My old age does not leave me the time":** Roger, *Buffon,* 432, and Mayr, *The Growth of Biological Thought,* 336.

154 **Buffon's *Epochs of Nature* was immediately attacked:** Jean Stengers, "Buffon et la Sorbonne," in Roland Mortier and Hervé Hasquin, eds., *Études sur le XVIIIe siècle* (Brussels: Éditions de l'Université de Bruxelles, 1974), 113–24.

154 **"The people need a religion":** Roger, *Buffon,* 423.

155 **"When I become ill and feel my end approaching":** Quoted in Georges Louis Leclerc Buffon, *Correspondance inédite de Buffon,* edited by H. Nadault de Buffon, 2 vols. (Paris: Hachette, 1860), 2:615.

8. ERASMUS UNDERGROUND

157 **a lead mine called Tray Cliff:** Tray Cliff Cavern is now known as Treak Cliff Cavern. Erasmus Darwin described his cave visits in a letter to Josiah Wedgwood on July 2, 1767, cited in Desmond King-Hele, ed., *The Collected Letters of Erasmus Darwin* (Cambridge: Cambridge University Press, 2007), 44. See John Whitehurst, *An Inquiry into the Original State and Formation of the Earth* (London: Bent, 1778). Whitehurst's biographer Maxwell Craven confirms that although *An Inquiry* was not published until 1778, a full sketch of its observations and conclusions was in circulation as early as 1763. It seems to have been finished by 1767 and

was being circulated among members of the Lunar Society in the 1760s. Erasmus Darwin may well have read this early draft. He would certainly have heard Whitehurst recounting his theories. I am grateful to the historian of science Patricia Fara for a careful and insightful reading of this chapter.

157 **the rare fluorite rock called Blue John:** See Trevor B. Ford, *Treak Cliff Cavern and the Story of Blue John Stone* (Castleton: Harrison Taylor, 1992).

158 **a temple of mysteries, the altar of the goddess Nature:** For evidence of Erasmus Darwin's neoclassical imagination in relation to caves, see the geological notes to his *Temple of Nature.* See also Irwin Primer, "Erasmus Darwin's *Temple of Nature:* Progress, Evolution, and the Eleusinian Mysteries," *Journal of the History of Ideas* 25, no. 1 (1964): 58–76.

159 **aristocratic collectors such as . . . Sir Ashton Lever:** For Ashton Lever's collection, see Richard Daniel Altick, *The Shows of London* (Cambridge, Mass., and London: Belknap Press, 1978), 30.

159 **Others declared them to be the remains of the deluge:** See John Woodward, for instance, in *An Essay Toward a Natural History of the Earth and Terrestrial Bodies, Especially Minerals, etc.* (1695), *Brief Instructions for Making Observations in All Parts of the World* (1696), and *An Attempt Towards a Natural History of the Fossils of England,* 2 vols. (1728–29).

159 **But John Whitehurst had other ideas:** Jenny Uglow, *The Lunar Men: The Friends Who Made the Future, 1730–1810* (London: Faber and Faber, 2003), 150–51; Maxwell Craven, *John Whitehurst of Derby: Clockmaker and Scientist, 1713–88* (Ashbourne: Mayfield, 1996).

160 **"a rarity, the like whereof has not been observ'd before in this Island":** William Stukeley, "An Account of the Impression of the Almost Entire Skeleton of a Large Animal in a Very Hard Stone, Lately presented the Royal Society, from Nottinghamshire," *Philosophical Transactions* 30, no. 360 (1719): 963–68.

161 **a box of giant fossilized bones and tree trunks and rocks:** Josiah Wedgwood described the fossils in a letter to Thomas Bentley, April 2, 1767; cited in Eliza Meteyard, *The Life of Josiah Wedgwood,* 2 vols. (London: Hurst and Blackett, 1865), 1:501.

161 **"These various strata . . . seem from various circumstances":** Ibid., 500–502.

161 **"I have lately travel'd two days into the bowels of the earth":** Erasmus Darwin to Josiah Wedgwood, July 2, 1767; cited in King-Hele, *Collected Letters of Erasmus Darwin,* 44.

161 **"I want to see you and Dr Small much if you will fix a Day":** Erasmus Darwin to Matthew Boulton, July 29, 1767; cited in King-Hele, *Collected Letters of Erasmus Darwin,* 45.

163 **"I do not mean to attack the Christian Religion":** Erasmus Darwin to Richard Gifford, September 4, and October 15, 1768; cited in King-Hele, *Collected Letters of Erasmus Darwin,* 91–97.

163 **an anonymous satirical poem entitled "Omnis e Conchis":** Cited in Desmond King-Hele, *Erasmus Darwin: A Life of Unequalled Achievement* (London: Giles de la Mare, 1999), 9.

164 **"Don't kill them all, leave me one":** Ibid., 92.

164 **Seventeen-year-old Mary Parker:** Mary Parker arrived on July 26, 1770; her first child, Susan, was born in May 1772 and her second child, Mary, in May 1774; see King-Hele, *Erasmus Darwin: A Life of Unequalled Achievement,* 106–7.

165 **Erasmus's two daughters continued to live with him:** When his two daughters came of age he set up a school for them in Ashbourne and wrote a book on education for them.

165 **The Scottish geologist James Hutton came to stay with Erasmus:** King-Hele, *Collected Letters of Erasmus Darwin,* 153–54.

165 **"What shall I send you in return for these?"** Ibid., 137–38.

166 **he was falling in love with a beautiful married woman:** Erasmus met Elizabeth Pole for the first time (probably) as early as 1771 (King-Hele, *Erasmus Darwin: A Life of Unequalled Achievement,* 127) and began treating her children in 1775, possibly earlier. She lost one of her twin babies in 1774—it is probable that he attended the family during the child's illness.

166 **Erasmus sent her unsigned love poems:** King-Hele, *Collected Letters of Erasmus Darwin,* 139–40.

166 **an experimental garden—"tangled and sequestered":** Anna Seward, *Memoirs of the Life of Dr. Darwin* (London: J. Johnson, 1804), 125–32; the remains of the garden and bathhouse are still to be seen in the grounds of a school for children with dyslexia called Maple Hayes School on Abnalls Lane.

167 **"The Linnaean system . . . is unexplored poetic ground":** Seward, *Memoirs of the Life of Dr. Darwin,* 130–31.

167 **promoting Linnaeus's ideas about plant sentience and sexuality:** Ernst Mayr addresses the part that Linnaeus played in the development of evolutionary thought in *Growth of Biological Thought,* 340–41. He argues that though Linnaeus is often considered the archfoe of evolutionism, his consistent opposition to evolutionary ideas brought the problem into scientific recognition and his system drew attention to the discontinuities in the natural world.

167 **coining new words to describe the sexual parts of plants:** For the coining of words, see Darwin's letter to Joseph Banks, September 29,

1781, King-Hele, *Collected Letters of Erasmus Darwin,* 189–91, and his letter to Josiah Wedgwood, October 4, 1781, ibid., 192–93. For a detailed study of *The Loves of the Plants,* see Janet Browne, "Botany for Gentlemen: Erasmus Darwin and *The Loves of the Plants,*" *Isis* 80, no. 4 (1989): 601.

168 **"manuscript has undergone as many alterations":** Wedgwood to Bentley, October 24, 1778; cited in Craven, *John Whitehurst of Derby,* 94–95; Wedgwood to Bentley, November 4, 1778, cited in Uglow, *Lunar Men,* 300–301.

169 **He was still thinking through his earlier ideas:** See Erasmus Darwin, letter to unknown recipient [summer 1782?], King-Hele, *Collected Letters of Erasmus Darwin,* 204–5.

169 ***The Loves of the Plants* was quickly taking shape:** Erasmus Darwin to Joseph Johnson, May 23, 1784, King-Hele, *Collected Letters of Erasmus Darwin,* 235. Erasmus wrote and published *The Loves of the Plants* first, although *The Economy of Vegetation* is the first volume in the chronology of the whole.

169 **"I would not have my name affix'd to the work":** Erasmus Darwin to Joseph Johnson, May 23, 1784, King-Hele, *Collected Letters of Erasmus Darwin,* 235.

170 **"vast and massive, his head almost buried in his shoulders":** *The Biography of Mrs Schimmel Penninck,* 177; cited in Uglow, *Lunar Men,* 424.

170 **began to put together an ambitious fossil collection:** Erasmus's letters from 1788 onward are full of references to fossils.

170 **The critical response to the poem was rapturous:** For a detailed account of the reception of *The Loves of the Plants,* see Desmond King-Hele, *The Life and Genius of Erasmus Darwin* (London: Faber and Faber, 1977), 197–98.

171 **"You are such an infidel in religion":** James Keir to Erasmus Darwin, March 15, 1790, in James Keir, *Sketch of the Life of James Keir* (London: R. E. Taylor, 1868), 111; the term "oxyde hydro-carbonneux" is a reference to the work of the French chemist Lavoisier.

171 **"I have some medico-philosophical works in MS":** Erasmus Darwin to James Watt, January 19, 1790, King-Hele, *Collected Letters of Erasmus Darwin,* 358. In February 1792 he wrote to Robert Darwin repeating his intention to publish *Zoonomia:* "I intend to write no more verse and to try a medico-philosophical work next, called Zoonomia," ibid., 364.

171 **It was a dangerous time, certainly:** For a powerful and detailed account of this time, see Uglow, *Lunar Men,* 440–44.

173 **"the highroads for full half a mile of the house":** Robert K. Dent, *Old and New Birmingham: A History of the Town and Its People* (Wakefield: EP Publishing, 1972–73; first published 1879), 229.

173 **"I am now too old and harden'd to fear a little abuse":** King-Hele, *Collected Letters of Erasmus Darwin*, 399.

173 **"This is the most important crisis in the history of British liberty":** King-Hele, *Erasmus Darwin: A Life of Unequalled Achievement*, 292.

174 **Erasmus was no longer getting away with it:** Ibid., 293.

175 **"one of the most important productions of the age":** Ibid., 291.

176 **"I have a profess'd spy shoulders us on the right":** Erasmus Darwin to R. L. Edgeworth, March 15, 1796. The spy was Mr. Upton—recruited to watch activities in the house by John Reeve's Association for Preserving Liberty and Property Against Republicans and Levellers, founded in 1792.

176 **"America is the only place of safety":** King-Hele, *Collected Letters of Erasmus Darwin*, 472.

177 **Prison beckoned:** King-Hele, *Erasmus Darwin: A Life of Unequalled Achievement*, 314–17; on Johnson's imprisonment, see Jane Worthington Smyser, "The Trial and Imprisonment of Joseph Johnson, Bookseller," *Bulletin of the New York Public Library* 77 (1974): 418–35.

177 **the less provocative *The Temple of Nature*:** Martin Priestman, "Darwin's Early Drafts for the Temple of Nature," in C.U.M. Smith and Robert Arnott, eds., *The Genius of Erasmus Darwin* (Aldershot: Ashgate, 2005), 311.

179 **his evolution emerged from his medical knowledge:** This observation was made in detail in Maureen McNeil, *Under the Banner of Science: Erasmus Darwin and His Age* (Manchester: Manchester University Press, 1987).

179 **There was not a single good review:** Norton Garfinkle, "Science and Religion in England, 1790–1800: The Critical Response to the Work of Erasmus Darwin," *Journal of the History of Ideas* 16, no. 3 (1955): 385.

179 **The poet Samuel Taylor Coleridge:** Letter to William Wordsworth, May 30, 1815, in Earl Leslie Griggs, ed., *The Collected Letters of Samuel Taylor Coleridge*, 6 vols. (Oxford: Clarendon Press, 1956–71), 4:574–75; Samuel Taylor Coleridge, "Notes on Stillingfleet," *Athenaeum*, March 27, 1875, 2474:423.

181 **Mary Godwin, Shelley's brilliant and intellectually voracious lover:** Ellen Moers, "Female Gothic," reprinted in George Levine and U. C. Knoepflmacher, eds., *The Endurance of Frankenstein: Essays on Mary Shelley's Novel* (Berkeley: University of California Press, 1982), 83–84.

9. THE JARDIN DES PLANTES

183 **the botanical garden of the Jardin des Plantes:** For fascinating accounts of the Jardin, see R. W. Burkhardt, "The Leopard in the Garden: Life in Close Quarters at the Muséum d'Histoire Naturelle," *Isis* 98, no. 4 (2007): 675–94, Dorinda Outram, "Le Muséum National d'Histoire Naturelle après 1793: Institution scientifique ou champ de bataille pour les familles et les groupes d'influence?" in C. Blanckaert, Claudine Cohen, Pietro Corsi, and Jean-Louis Fischer, eds., *Le Muséum au premier siècle de son histoire* (Paris: Muséum National d'Histoire Naturelle, 1997), 25–30, and Dorinda Outram, *Georges Cuvier: Science, Vocation and Authority in Post-Revolutionary France* (Manchester: Manchester University Press, 1984); for the earlier history of the Jardin des Plantes, see Emma Spary, *Utopia's Garden: French Natural History from Old Regime to Revolution* (Chicago: University of Chicago Press, 2000); for the politics of the Jardin, see Pietro Corsi, *The Age of Lamarck: Evolutionary Theories in France, 1790–1830* (Berkeley: University of California Press, 1988), Toby A. Appel, *The Cuvier-Geoffroy Debate: French Biology in the Decades Before Darwin* (Oxford: Oxford University Press, 1987), and R. W. Burkhardt, "Lamarck, Evolution and the Politics of Science," *Journal of the History of Biology* 3 (1970): 275–96; for a contemporary account, see Joseph Deleuze, *Histoire et description du Muséum Royal d'Histoire Naturelle, ouvrage rédigé d'après les ordres de l'administration du Muséum* (Paris: Royer, 1823; translated into English in 1823). In writing, revising, and correcting this chapter, I owe a debt of gratitude to the generous scholars Dorinda Outram and Richard Burkhardt.

185 **Three professors; three different versions of nature:** A. S. Packard, *Lamarck: The Founder of Evolution* (New York: Longmans, Green, 1901), 42–43. The key biographies on these three men are: on Lamarck—Raphaël Bange and Pietro Corsi, "Chronologie de la vie de Jean-Baptiste Lamarck," Centre National de la Recherche Scientifique (online); Ludmilla Jordanova, *Lamarck* (Oxford: Oxford University Press, 1984); R. W. Burkhardt, *The Spirit of System: Lamarck and Evolutionary Biology* (Cambridge, Mass., and London: Harvard University Press, 1977), and Pietro Corsi, *Lamarck, philosophe de la nature* (Paris: Presses Universitaires de France, 2006); on Cuvier—Dorinda Outram, *Georges Cuvier: Science, Vocation and Authority in Post-Revolutionary France,* and Toby Appel, *The Cuvier-Geoffroy Debate;* on Geoffroy—Théophile Cahn, *La Vie et l'oeuvre d'Étienne Geoffroy Saint-Hilaire* (Paris: Presses Universitaires de France, 1962), and Hervé Le Guyader, *Geoffroy Saint-Hilaire: A Visionary Natural-*

ist, translated by Marjorie Grene (Chicago: University of Chicago Press, 2004).

186 **When Cuvier first arrived in Paris from Germany:** See Outram, *Georges Cuvier: Science, Vocation and Authority in Post-Revolutionary France,* 166–68.

186 **his own museum of comparative anatomy:** Ibid., 176.

186 **Soon Cuvier's Museum of Comparative Anatomy:** On Cuvier's early years in the Museum of Natural History, see Dorinda Outram, "Uncertain Legislator: Georges Cuvier's Laws of Nature in Their Intellectual Context," *Journal of the History of Biology* 19, no. 3 (1986): 323–68.

187 **Lamarck . . . must have seemed an old-fashioned generalist:** This portrait of Lamarck's rather obsessive personality comes from Henri-Marie Ducrotay de Blainville, *Histoire des sciences de l'organisation et de leurs progrès comme base de la philosophie, rédigée etc. par F.L.M. Maupied,* 3 vols. (Paris, 1845), 3:358.

187 **Georges Cuvier was a plain facts man:** See Appel, *Cuvier-Geoffroy Debate, 53–59,* and Outram, *Georges Cuvier: Vocation, Science and Authority in Post-Revolutionary France,* 128.

188 **Lamarck and Cuvier belonged to different traditions:** Corsi, *Age of Lamarck,* 64–65.

188 **Hundreds of young lawyers, medical students:** Appel, *Cuvier-Geoffroy Debate,* 34–37.

188 **They were mostly French, although there were also students:** For detailed information about the nationalities and biographies of the young men who attended Lamarck's lectures, see Professor Pietro Corsi's valuable database of auditors at www.lamarck.cnrs.fr/auditeurs/presentation.php?lang=en.

188 **Lamarck . . . had changed his mind:** Various reasons have been proposed for Lamarck's conversion from fixity to flux. See in particular Richard Burkhardt, "The Inspiration of Lamarck's Belief in Evolution," *Journal of the History of Biology* 5 (1972): 413–38, and Burkhardt, *Spirit of System,* ch. 5.

188 **The naturalist had to be a philosopher:** Lamarck first defined his role as naturalist-philosopher in this lecture in 1800. See Burkhardt, "Lamarck, Evolution and the Politics of Science," 285.

189 **he began to describe nature as being in a state of perpetual flux:** Corsi, *Age of Lamarck,* 93.

189 **"Little by little," he wrote, "nature has reached the state":** Cited in ibid., 100.

189 **It was a picture that would lend itself easily to parody:** Ibid., 93–94. Corsi describes a discovery made by Richard Burkhardt in the

archives—a passage that Cuvier had excised from a draft of *Discourse on the Revolutions of the Earth* in which he mercilessly mocked Lamarck's description of metamorphosing birds.

190 **"would reduce all natural history to . . . variable forms":** Cuvier, "Mémoire sur les espèces d'éléphants tant vivantes que fossiles," 12; cited in Appel, *Cuvier-Geoffroy Debate,* 51.

190 **his poor health would not allow him to finish:** Burkhardt, "Lamarck, Evolution and the Politics of Science," 287.

190 **Lamarck expanded and defended his transformist theories:** See Corsi, *Age of Lamarck,* 122.

190 **"It is not organs—that is, the nature and shape of an animal's body":** Jean-Baptiste Lamarck, *Recherches sur l'organisation des corps vivants et particulièrement sur son origine, sur la cause de ses développements et des progrès de sa composition, et sur celle qui, tendant continuellement à la detruire dans chaque individu, amène nécessairement sa mort; précédé du discours d'ouverture du cours de zoologie, donné dans le Muséum National d'Histoire Naturelle* (Paris: Maillard, 1802), 50.

191 **"In this imperceptibly slow process, the sea is constantly breaking up":** Jean-Baptiste Lamarck, *Hydrogéologie* (Paris: Chez l'Auteur, Agasse et Maillard, 1802), 54; cited in Corsi, *Age of Lamarck,* 106.

191 **"utterly transcended man's capacity to calculate":** Lamarck, *Hydrogéologie,* 88. Corsi explains the contradictions between Lamarck's developing history of the earth and the history of life in *Age of Lamarck,* 115–17.

191 **"To the examples I cited," he added as an afterthought:** Lamarck, *Recherches,* 208. He used the giraffe much more extensively in his *Philosophie Zoologique* of 1809.

191 **the numbers of students attending his lectures almost doubled:** See Corsi's list of student numbers at www.lamarck.cnrs.fr/.

192 **In 1802 Lamarck was a rising star among these students:** The largest audiences in the Jardin were for Desfontaines' uncontroversial botany lectures. He regularly had audiences of between 500 and 600 auditors; these had a much higher proportion of women. In comparison, Lamarck's audiences were small. Cuvier often attracted between 200 and 300 students to his lectures. See Deleuze, *Histoire et description du Muséum Royal d'Histoire Naturelle.*

192 **Following Lamarck, they claimed that the first life-forms:** For a useful summary and assessment of Lamarck's key ideas, see Mayr, *Growth of Biological Thought,* 359.

192 **The idea of species change:** See Pietro Corsi, "Before Darwin: Transformist Concepts in European Natural History," *Journal of the History of Biology* 38 (2005): 167–83.

193 **Geoffroy Saint-Hilaire had been professor of vertebrates:** Appel, *Cuvier-Geoffroy Debate,* 20–21.

193 **"It seems that nature is confined within certain limits":** Geoffroy, 1796; cited in Le Guyader, *Geoffroy Saint Hilaire,* 21.

195 **"I have never seen such water birds":** Nina Burleigh, *Mirage: Napoleon's Scientists and the Unveiling of Egypt* (New York: Harper, 2007), 189.

196 **"demand no less than the throne of anatomy":** Geoffroy, *Lettres écrite d'Egypte,* 95–96; cited in Appel, *Cuvier-Geoffroy Debate,* 75.

196 **"I am so overwhelmed with business":** Burleigh, *Mirage,* 197.

196 **But in this atmosphere of political volatility:** Robert Solé, *Les Savants de Bonaparte* (Paris: Editions du Seuil, 1999), 160.

196 **"The bombing, the fires, the ambushes":** Étienne Geoffroy Saint-Hilaire, *Études progressives d'un naturaliste pendant les années 1834 et 1835* (Paris), 149–51; cited in Appel, *Cuvier-Geoffroy Debate,* 78.

197 **a "very vast theory" that would revolutionize science:** Geoffroy, *Lettres écrites d'Egypte,* 205; cited in Appel, *Cuvier-Geoffroy Debate,* 76.

198 **Geoffroy locked his manuscripts and his great theory away:** Appel, *Cuvier-Geoffroy Debate,* 81.

198 **"One cannot master the transports of one's imagination":** Burleigh, *Mirage,* 190.

200 **Three thousand years was nothing in the age of the earth:** Appel, *Cuvier-Geoffroy Debate,* 82.

200 **"true science should be sought on a broader and higher plane":** Isidore Geoffroy Saint-Hilaire, *Vie, travaux et doctrine scientifique d'Étienne Geoffroy Saint-Hilaire* (Paris and Strasbourg, 1847), 116–17; cited in Appel, *Cuvier-Geoffroy Debate,* 83.

200 **Geoffroy . . . begged Cuvier to intervene on his behalf:** Appel, *Cuvier-Geoffroy Debate,* 84.

200 **Isidore, who would become an eminent and important zoologist:** Isidore Geoffroy Saint-Hilaire also promoted transformist ideas; Darwin included him in his "Historical Sketch."

200 **Cuvier married a widow:** On Cuvier's four stepchildren and the four children he had with Madame Duvaucel, see Mrs. R. Lee, *Memoirs of Baron Cuvier* (London: Longman, Rees, Orme, Brown, Green and Longman, 1833), 18–19.

200 **The baby lived only a few weeks:** On the death of Cuvier's firstborn son, christened Georges, see ibid., 19.

200 **his theatrical and flamboyant lecturing style:** See Isidore Bourdon, *Illustres Médecins et naturalistes des temps modernes* (Paris: Comptoir des Imprimeurs-Unis, 1844), 116–17.

201 **only the beauty and goodness of the Creator were proper objects:**

François-René Chateaubriand, *The Genius of Christianity,* 5 vols. (Paris: Migueret, 1802), Part 3, Book 2, ch. 2; cited in Outram, "Uncertain Legislator," 335.

201 **geology need not be antireligious:** On Cuvier's religious beliefs, see Outram, *Georges Cuvier,* 141–60; on his geology lectures, see Martin J. S. Rudwick, *Georges Cuvier, Fossil Bones and Geological Catastrophe: New Translations and Interpretations of the Primary Texts* (Chicago: University of Chicago Press, 1997), 74–88.

204 **When Lamarck died in 1829:** See Packard, *Lamarck,* 57–61.

204 **There was very little money to spare in the museum coffers:** For the history of the Jardin and specifically the history of the Museum of Comparative Anatomy, see Deleuze, *Histoire et description du Muséum Royal d'Histoire Naturelle.*

204 **what Baron Cuvier now had to say in his obituary:** An English translation, probably by Robert Jameson, was published in 1836: Georges Cuvier, "Elegy of Lamarck," *Edinburgh New Philosophical Journal* 20 (January 1836), 1–22. This elegy was read to the French Institut National in Paris by the Baron Silvestre (Cuvier had recently died) on November 26, 1832. It was intended to follow an elegy to M. Volta on June 27, 1831, but was postponed. It was published (after an unaccountable delay) in France in *Mémoires de l'Académie Royale des Sciences de l'Institut de France,* vol. 13 (Paris, 1835), i–xxxi.

205 **every organism, he declared, belonged to one of four branches:** Cuvier's system demolished the idea of nature that had dominated natural philosophy for centuries—the Great Chain of Being, the belief that nature was arranged like a ladder with the simplest organisms on the bottom and the most complex at the top in a continuous sequence. For the classic text on this subject, see Arthur O. Lovejoy, *The Great Chain of Being: A Study of the History of an Idea* (Cambridge, Mass.: Harvard University Press, 1936).

205 **his stepdaughter and assistant, Sophie Duvaucel:** Sophie Duvaucel was now his chief, but unacknowledged, collaborator on the *Règne animal* volumes and his senior assistant and illustrator. She managed the group of assistants assembled to help on the project, who worked in a room adjoining Cuvier's study in the Gallery of Comparative Anatomy. See M. Orr, "Keeping It in the Family: The Extraordinary Case of Cuvier's Daughters," in Cynthia Burek and Bettie Higgs, eds., *The Role of Women in the History of Geology* (London: Geological Society of London, 2007), Special Publications, 281:277–86.

205 **Cuvier rose to his feet and demanded a retraction:** Appel, *Cuvier-Geoffroy Debate,* 146.

206 **the German school of philosophical natural history, the *Naturphilosophie*:** On this important group of natural philosophers, see Robert J. Richards, *The Romantic Conception of Life: Science and Philosophy in the Age of Goethe* (Chicago: University of Chicago Press, 2002), and Iain Hamilton Grant, *Philosophies of Nature After Schelling* (New York and London: Continuum, 2006).

206 **"The volcano has come to an eruption":** Johann Peter Eckermann, *Conversations of Goethe with Eckermann and Soret*, translated by John Oxenford (London: George Bell and Sons, 1874), 121–22. Goethe appeared on Darwin's list as a supporter of a form of transformism.

207 **His savage obituary of Lamarck:** Dorinda Outram, "The Language of Natural Power: The Funeral Éloges of Georges Cuvier," *History of Science* 16 (1978): 153–78.

208 **One such young man was the brilliant young soldier:** Corsi, *Age of Lamarck*, 179.

208 **All through his years of exile he published books and papers:** Ibid., 223.

209 **Transformism . . . had always been political:** See Adrian Desmond, *The Politics of Evolution: Morphology, Medicine, and Reform in Radical London* (Chicago and London: University of Chicago Press, 1991).

209 **the French madman:** On Rafinesque, see C. T. Ambrose, "Darwin's Historical Sketch—An American Predecessor," *Archives of Natural History* 37, no. 2 (2010): 191–202, and Jim Endersby, " 'The Vagaries of a Rafinesque': Imagining and Classifying American Nature," *Studies in History and Philosophy of Science Part C: Studies in History and Philosophy of Biological and Biomedical Sciences* 40, no. 3 (2009): 168–78.

10. THE SPONGE PHILOSOPHER

211 **the harbor port of Leith:** For the social history of Leith, see James Scott Marshall, *The Life and Times of Leith* (Edinburgh: John Donald, 1986); Sue Mowat, *The Port of Leith: Its History and Its People* (Edinburgh: John Donald in association with the Forth Ports, 1994); and Joyce M. Wallace, *Traditions of Trinity and Leith* (Edinburgh: John Donald, 1997).

213 **Robert Jameson, Regius professor of natural history:** For further information on Professor Jameson, see James A. Secord, "Edinburgh Lamarckians: Robert Jameson and Robert E. Grant," *Journal of the History of Biology* 24 (1991): 1–18.

214 **he had fallen upon *Zoonomia*:** For more on the work of Erasmus Darwin, see King-Hele, *Erasmus Darwin: A Life of Unequalled Achievement*, and McNeil, *Under the Banner of Science.*

214 **Grant was twenty-two when he arrived in Paris:** See Simona Pakenham, *In the Absence of the Emperor: London-Paris, 1814–15* (London: Cresset Press, 1968).

215 **had made Paris the heart of the new medicine:** See Roy Porter, *The Greatest Benefit to Mankind: A Medical History of Humanity from Antiquity to the Present* (London: Fontana, 1997), 306–14.

215 **Grant did not attend:** We presume Grant did not attend because he did not sign the register. See Pietro Corsi's fascinating database documenting the auditors of Lamarck's lectures: www.lamarck.cnrs.fr/auditeurs/presentation.php?lang=en.

216 **"All facts known about the sponge":** Robert Grant, "Observations and Experiments on the Structure and Functions of the Sponge," *Edinburgh Philosophical Journal* 13, no. 25 (1825): 99.

216 **"It is pleasing to observe that our forefathers":** Ibid., 97.

217 **He kept his work to himself, however:** The most detailed research into Grant's life and work has been undertaken by Adrian Desmond in *The Politics of Evolution,* in Desmond, "Robert E. Grant's Later Views on Organic Development," *Archives of Natural History* 11 (1984): 395–413, and in Desmond, "Robert E. Grant: The Social Predicament of a Pre-Darwinian Transmutationist," *Journal of the History of Biology* 17, no. 2 (1984): 189–223. A short book about Grant's life has also been put together by Sarah E. Parker: *Robert Edmond Grant (1793–1894) and His Museum of Zoology and Comparative Anatomy* (London: Grant Museum of Zoology, 2006). Grant, trained in Parisian medical research techniques, took notes about everything he dissected, all his ideas, particular trains of thought, and critical conversations. But all of this has disappeared. He died unmarried with no close relatives, and although his library survived, these valuable journals and bundles of notes and letters did not. So tracing his intellectual and physical journeys is a matter of detective work. A long biographical essay written by his friend Thomas Wakley in 1850 for *The Lancet,* probably based on interviews with Grant, survives, but almost nothing else, apart from the dozens of essays he published in the 1820s and 1830s: Thomas Wakley, "Biographical Sketch of Robert Edmond Grant, M.D.," *Lancet* 2 (1850): 686–95. For a full and detailed list of all of Grant's publications on the sponge, see Parker, *Robert Edmond Grant (1793–1894) and His Museum of Zoology and Comparative Anatomy.*

218 **"On moving the watch-glass":** Robert Grant, "Observations and Experiments on the Structure and Functions of the Sponge," *Edinburgh Philosophical Journal* 13 (1825): 102.

219 **the hole could be described as a "fecal orifice":** Robert Grant, "Obser-

vations and Experiments on the Structure and Functions of the Sponge," *Edinburgh New Philosophical Journal* 2 (1826): 126. While Grant was publishing his sponge sequence, the *Edinburgh Philosophical Journal* was renamed the *Edinburgh New Philosophical Journal.*

220 **"This animal," Grant wrote, "seems eminently calculated":** Ibid., 136.

220 **"I have plunged portions of the branched and sessile sponges":** Robert Grant, "Observations and Experiments on the Structure and Functions of the Sponge," *Edinburgh Philosophical Journal* 14, no. 27 (1826): 123.

220 **It shared both plant and animal characteristics:** The sponge is in fact an animal and not a transitional organism. Sponges are an ancient group of animals that diverged from other metazoans more than 600 million years ago. This divergence required the evolution of mechanisms for cell division, growth, specialization, adhesion, and death. In modern genetics they play a central role in the search for the origins of metazoan multicellular processes. I am grateful to the geneticist Kate Downes for conversations about the sponge.

220 **"My praise is altogether an unclean thing":** John Hutton Balfour, *Biography of the Late John Coldstream* (London: J. Nisbet, 1865), 6.

221 **a certain mysterious sense of disgust about his body:** The first historian to suggest that Robert Grant might have been homosexual was Adrian Desmond in Desmond, *Archetypes and Ancestors: Palaeontology in Victorian London, 1850–1875* (London: Blond and Biggs, 1982), as a conjecture based upon accounts of Grant's reputation from the zoology department at University College London. Grant never married and continued to have intense friendships with men throughout his life; he traveled with many of them for several months abroad. If sexual feelings arose between Coldstream and Grant, this may account for the level of self-loathing that Coldstream expressed in his diaries during the time he worked alongside Grant, particularly given the intensity of Coldstream's religious beliefs. It may also, as Desmond points out, be one possible factor among many in the decline of Grant's reputation in London and Charles Darwin's eventual distancing from him. I am grateful to Adrian Desmond for discussions on this matter.

222 **the boundary between the animal and vegetable kingdoms:** I am again grateful to Adrian Desmond for generous assistance and advice on this matter some years ago. At the University of Heidelberg, Grant had met the young professor of anatomy and physiology Frederick Tiedemann. Lamarck, for all his transformism, still believed in an absolute demarcation between animal and vegetable kingdoms. Tiedemann,

however, believed that in the most simple and ancient life-forms the boundaries between these kingdoms were not fixed.

222 **Plinian Natural History Society meetings:** See J. H. Ashworth, "Charles Darwin as a Student in Edinburgh, 1825–1827," *Proceedings of the Royal Society of Edinburgh* 55 (1935): 97–113. For this period of Darwin's life, see the fascinating account by Adrian Desmond and James Moore, *Darwin* (Harmondsworth: Penguin, 1992); P. Helveg Jespersen, "Charles Darwin and Dr. Grant," *Lychnos* (1948–49): 159–67; George Sheppersen, "The Intellectual Background of Charles Darwin's Student Years at Edinburgh," in M. Banton, ed., *Darwinism and the Study of Society* (London: Tavistock Publications; Chicago: Quadrangle Books, 1961), 17–35; and J. H. Ashworth, "Charles Darwin as a Student in Edinburgh," *Proceedings of the Royal Society of Edinburgh* 55 (1935): 97–113.

223 **"the lower animals possess every faculty & propensity of the human mind":** Plinian Minutes Ms, 1ff., 34–36, Edinburgh University Library, Dc.2.53.

224 **"How far this law is general with zoophytes":** Grant, "Observations on the Spontaneous Motions of the Ova of Zoophytes," *Edinburgh New Philosophical Journal* 1 (1826): 156.

224 **"Having procured some specimens of the Flustra Carbocea":** Edinburgh Notebook, listed as DAR 118 in the Darwin archive in Cambridge University Library, 56.

225 **"One frequently finds sticking to oyster and other old shells":** Ibid.

225 **Darwin gave a paper on the ova of the *Flustra:*** Cited in Ashworth, "Charles Darwin as a Student in Edinburgh," 105.

225 **"I then made him repeat what he had told me before":** Jespersen, "Charles Darwin and Dr. Grant," 164–65. The note that Jespersen refers to has now been lost, so one should read this reminiscence with care. However, later when Grant worked in London he was notorious for his priority disputes and guarded about his research findings: see Desmond, *The Politics of Evolution.*

227 **"he was troubled with doubts":** Balfour, *Biography of the Late John Coldstream,* 38.

227 **"In our day the majority of naturalists, I fear are infidels":** Ibid., 69.

227 **"A fair exterior covers a perfect sink of iniquity":** Ibid.

227 **"no pursuit is more becoming for a physician than Nat. Hist.":** Letter from John Coldstream to Charles Darwin, February 28, 1829, Letter 58, DCP.

227 **"Be so good as to write me again soon":** Ibid.

227 **"What a fellow that D. is for asking questions":** Cited in Desmond and Moore, *Darwin,* 82.

228 **"It strikes me, that all our knowledge":** Charles Darwin to William Darwin Fox, July 9, 1831, Letter 101, DCP.

228 **"Natural History . . . is very suitable to a Clergyman":** Charles Darwin, *Autobiography with original omissions restored; edited with appendix and notes by his grand-daughter, Nora Barlow* (London: Collins, 1958), 71–72.

228 **Coldstream drew careful instructions and diagrams:** Coldstream soon returned to some kind of peace in his rock pool explorations. He married in 1835 and settled down to family life and his growing medical practice, writing occasional encyclopedia entries on jellyfish, limnoria, and barnacles.

229 **he defiantly promoted his friend:** For a detailed account of the Wakley-Grant alliance and a brilliant analysis of politics and science in the early nineteenth century, see Desmond, *Politics of Evolution,* 122–23. Wakley published Grant's entire lecture course in *The Lancet* in 1833–34, further radicalizing Grant's ideas by association.

229 **"While myriads of individuals appear and disappear":** Robert Grant, Lecture 55, *Lancet* 2 (1833–34): 1001.

231 **"He appears to have allowed himself to be frightened":** From Carl Gustav Carus, *On the State of Medicine in Britain in 1844;* cited in Desmond, *Politics of Evolution,* 258.

231 **The backlash came, perhaps inevitably, from close quarters:** Adrian Desmond, "Richard Owen's Reaction to Transmutation in the 1830's," *British Journal for the History of Science* 18, no. 1 (1985): 25–50.

231 **set out to demolish Lamarckian transmutation:** Pietro Corsi, "The Importance of French Transformist Ideas for the Second Volume of Lyell's *Principles of Geology," British Journal for the History of Science* 11, no. 3 (1978): 221–44.

232 **to "[degrade] Man from his high Estate":** Desmond, *Politics of Evolution,* 328.

232 **It worked. Grant was ousted:** For a brilliant account of Grant's struggles in London, see Desmond, "Robert E. Grant: The Social Predicament," 189–223.

233 **"I have found the world to be chiefly composed of knaves and harlots":** John Beddoe, *Memories of Eighty Years* (Bristol: Arrowsmith, 1910), 32–33.

11. THE ENCYCLOPEDIST

238 **"violence held rule almost everywhere":** William Chambers, *Memoir of Robert Chambers and Autobiographical Reminiscences of William Cham-*

bers (Edinburgh and London: W. & R. Chambers, 1872), 50. I am indebted to James A. Secord not only for his generously reading, correcting, and providing feedback on a draft of this chapter but for his masterly book *Victorian Sensation: The Extraordinary Publication, Reception, and Secret Authorship of "Vestiges of the Natural History of Creation"* (Chicago: University of Chicago Press, 2000), without which this chapter and indeed parts of other chapters could not have been written.

238 **Chambers had taken refuge in the local bookshop:** C. H. Layman, ed., *Man of Letters: The Early Life and Love Letters of Robert Chambers* (Edinburgh: Edinburgh University Press, 1990), 57.

239 **"all my spare time was spent beside the chest":** Ibid., 58–59.

241 **William was learning the book trade:** Chambers, *Memoir of Robert Chambers,* 77.

241 **"He was great in electricity":** Layman, *Man of Letters,* 84–85.

241 **a literary journal called *Kaleidoscope:*** Cited in Chambers, *Memoir of Robert Chambers,* 242–43.

242 **"This fervour is as fatal to literature as the irruption of the Goths":** Chambers to Scott, March 30, 1830; cited in James A. Secord, "Behind the Veil: Robert Chambers and *Vestiges,*" in James R. Moore, ed., *History, Humanity and Evolution: Essays for John C. Greene* (Cambridge and New York: Cambridge University Press, 1989), 169.

243 **eighty thousand in bookshops around the country:** Secord, *Victorian Sensation,* 234.

243 **"All previous hardships and experiences":** Ibid., 241.

243 **"It has been a matter of congratulation":** Ibid., 238.

244 **"On all hands," William wrote, "we were beset with requests":** Ibid., 246.

244 **"The shepherds, who are scattered there":** Layman, *Man of Letters,* 177.

245 **The *Journal* became more outspoken by the year:** For information on the increasingly outspoken nature of the *Journal,* see Robert J. Scholnick, " 'The Fiery Cross of Knowledge': *Chambers's Edinburgh Journal,* 1832–1844," *Victorian Periodicals Review* 32, no. 4 (1999): 324–58; unfortunately, Scholnick fails to register the significance of the fiery cross reference, however, arguing instead that it is evidence of Chambers's increasingly "messianic" tone, which it is not. It has a specific political meaning in the context of Scottish history.

245 **"I believe this liberal view is advancing":** Chambers to Ireland, no date; cited in Secord, "Behind the Veil," 171.

246 **"When we reflect . . . that some of the forms of heathenism":** Rob-

ert Chambers to George Combe, November 24, 1835, National Library of Scotland Ms 7234; cited in Secord, *Victorian Sensation,* 87.

247 **He denounced the "revolutionary ruffians":** George Combe, *Constitution of Man Considered in Relation to External Objects* (Edinburgh: J. Anderson Jr., 1828), 301.

247 **"The acute and anatomical knowledge of the Doctor":** Robert Chambers, "Natural History: Animals with a Backbone," *Chambers's Edinburgh Journal,* November 24, 1832, 338.

247 **"man himself, Socrates, Shakespeare and Newton":** Robert Chambers, "Popular Information on Science: Transmutation of Species," *Chambers's Edinburgh Journal,* September 26, 1835, 273–74; cited in Secord, *Victorian Sensation,* 93.

247 **he started writing a treatise on phrenology:** In 1837, Robert Chambers wrote to friends that all his spare moments in the previous two years had been spent on a manuscript treatise on "the philosophy of phrenology." That book became *Vestiges.* Secord, "Behind the Veil," 174.

249 **"work at his secret with all the security of a criminal unrecognized":** Eliza Priestly, *The Story of a Lifetime* (London: Kegan, Paul, Trench, Trubner, 1908), 43.

249 **"I do not think Churchill is likely to boggle":** Robert Chambers to Alexander Ireland, June 30, 1844; cited in Secord, *Victorian Sensation,* 114.

249 **he proposed publishing a thousand copies:** Secord, *Victorian Sensation,* 115.

250 **"pointed to his house in which he had eleven children":** R. C. Lehmann, *Memories of Half a Century* (London: Smith, Elder, 1908), 7; Lehmann is quoting here from the memoirs of Frederick Lehmann, who married one of the Chambers daughters in 1852.

250 **"they can but suspect and surmise":** Cited in Secord, *Victorian Sensation,* 376.

251 **"The great plot comes out here":** Ibid., 104.

252 **"Every effort is made that reason and common sense would at all admit of":** Robert Chambers to Alexander Ireland, June 30, 1844, National Library of Scotland; cited in Secord, "Behind the Veil," 171.

252 **"I am happy to say that I have been able at the end":** Robert Chambers to Alexander Ireland [1844], National Library of Scotland; cited in Secord, "Behind the Veil," 170–71.

254 **"I think you could smash him and I wish you would":** George W. Featherstonhaugh to the Reverend Adam Sedgwick, November 16, 1844; cited in Secord, *Victorian Sensation,* 222.

254 **"You have no conception what mischief the book has done":** Adam

Sedgwick to Macvey Napier, May 4, 1845; cited in Secord, *Victorian Sensation,* 240.

254 **"iron heel upon the head of the filthy abortion":** Adam Sedgwick to Macvey Napier, April 10, 1845, in Macvey Napier, ed., *Selections from the Correspondence of the Late Macvey Napier* (London: Macmillan, 1879), 492.

255 **"comes before them with a bright, polished, and many-coloured surface":** Cited in Secord, *Victorian Sensation,* 246.

255 **praised Sedgwick's "masterly essay":** *Christian Remembrancer,* June 1845, 612.

255 **"We seem to be standing on the verge of a vast volcano":** [James McCosh], "Periodicals for the People," *Lowe's Edinburgh Magazine,* January 1847, 200.

255 **"the lower levels of society had sunk into a miasmatic marsh":** Hugh Miller, "The People Their Own Best Portrait Painters," *Witness,* December 5, 1849, 2.

256 **"Infidelity is of the spirit of the present":** Scottish Association for Opposing Prevalent Errors, *Report of the Proceedings of the First Public Meeting of the Scottish Association for Opposing Prevalent Errors, Held in the Salon of Gibb's Royal Hotel, Princes Street, Edinburgh on Tuesday 9th March, 1847,* cited in Secord, *Victorian Sensation,* 289.

257 **"The thing that was most curious of all":** Cited in Secord, *Victorian Sensation,* 444.

257 **"Mr Chambers studiously excludes all religious subjects":** Ibid., 294–95.

258 **"somewhat less amused at it":** Joseph Hooker to Charles Darwin, December 30, 1844, Letter 804, DCP.

259 **the group he called "us transmutationists":** Charles Darwin to Joseph Hooker, April 18, 1847, Letter 1012, DCP.

259 **"almost as unorthodox about species as *Vestiges* itself":** Charles Darwin to T. H. Huxley, September 2, 1854, Letter 1587, DCP.

12. ALFRED WALLACE'S FEVERED DREAMS

262 **He had published a paper about the species question:** Alfred Russel Wallace, "On the Law Which Has Regulated the Introduction of New Species," dated Sarawak, Borneo, *Annals and Magazine of Natural History,* 2nd series, 16 (1855): 184–96. I am indebted to the fine biographer Peter Raby for providing generous feedback on this chapter, for talking over the finer points of what is known about Wallace's life, and for his masterly biography of Wallace.

263 **the feverish bolt-from-the-blue moment of realization:** Wallace was always a little foggy about where the fever-induced idea occurred. He claimed it was in Ternate, a small town across the water on the island of Ternate that served as his base camp, but his field notes indicate that he was staying in Dodinga on the island of Gilolo for all of February. As his biographer Ross A. Slotten points out, this is likely to have been because most of his readers would have heard of Ternate but Gilolo was too small to have any geographical significance in the West. Ross A. Slotten, *The Heretic in Darwin's Court: The Life of Alfred Russel Wallace* (New York: Columbia University Press, 2004), 509n6.

263 **the "checks" that stopped populations from growing:** A. R. Wallace, *My Life: A Record of Events and Opinions,* 2 vols. (London: Chapman and Hall, 1905), 1:361.

263 **"it occurred to me to ask the question, *Why do some die and some live?*:** Ibid., 362.

264 **But out here in this hut in the Malay Archipelago, there was only Ali:** Wallace wrote about Ali in his autobiography—he employed him in 1855 and trained him to hunt and prepare specimens; on Ali, see also Jane R. Camerini, "Wallace in the Field," in H. Kuklick and R. Kohler, eds., "Science in the Field," *Osiris* 11 (1996): 55–56. Wallace's other main assistant, Charles Allen, who had sailed with him from England, had left him in 1856 in order to enter a monastery. He rejoined Wallace in 1860.

264 **Ali was smart, a fine collector, a hard worker:** In 1862, when Wallace sailed from Sumatra by mail steamer for Singapore, he left Ali a rich man, giving him money and his two guns, ammunition, stores, and tools. In presenting Ali's photograph in his autobiography, he described him as the faithful companion of nearly all of his journeys in the East and the best native servant he ever had.

265 **"I said that I hoped," he remembered later:** Wallace, *My Life,* 1:363.

266 **he had been diving for it for years:** As Jim Endersby has pointed out in his review essay of 2003, Wallace is rarely presented as anything other than the also-ran in the history of evolution and his life rarely seen as anything but a shadow in the Darwin story—see Jim Endersby, "Escaping Darwin's Shadow," *Journal of the History of Biology* 36, no. 2 (2003): 385–403; of the many biographies of Wallace available, Peter Raby's is still the finest—see Raby, *Alfred Russel Wallace: A Life* (Princeton: Princeton University Press, 2001). See also Slotten, *Heretic in Darwin's Court.*

266 **Wallace had been interested in . . . geographical borders:** For these insights into Wallace's early interest in borders and borderlines, I am indebted to James Moore's important essay of 1997, "Wallace's Malthu-

sian Moment: The Common Context Revisited," in Bernard Lightman, ed., *Victorian Science in Context* (Chicago: University of Chicago Press, 1997), 290–311.

266 **the "little Saxon":** Wallace, *My Life,* 1:29.

267 **Daniel Defoe's *History of the Great Plague:*** Wallace also read, at around the same time, Thomas Hood's moral tale of a gang of robbers breaking into the deserted houses of plague evacuees, "The Tale of the Great Plague."

267 **Defoe described the three weeks in 1665:** Defoe was only four years old when the plague struck; the journal he published in 1722 as the journal of a single eyewitness was part fictionalized, part assembled from multiple eyewitness accounts.

267 **"Nothing but the immediate finger of God":** Daniel Defoe, *Journal of the Plague Year,* 282.

268 **London gave the young Wallace a political education:** Wallace, *My Life,* 1:80.

268 **"a pervading spirit of scepticism":** Ibid., 227.

269 **"the word 'atheist' had always been," he wrote, "used with bated breath":** Ibid., 226.

269 **Wallace would later call the Enclosure Acts land theft:** Wallace makes a long and eloquent attack on the principle of enclosure in his autobiography: ibid., 78–84; he was later active in the land nationalization movement.

269 **mobs of young farmers and agricultural workers:** These were called the Rebecca Riots (1839–43) after the biblical Rebecca. Many of the agitators, who were mostly agricultural workers and farmers, attacked tollbooths at night in gangs disguised as women. See David Williams, *The Rebecca Riots: A Study in Agrarian Discontent* (Cardiff: University of Wales Press, 1955).

269 **bringing a mechanics' institute and science library to Neath:** Wallace wrote his first essay in these years in Neath, "The South-Wales Farmer," which was later published in *My Life,* 1:207–22; it is now anthologized in Jane R. Camerini, ed., *The Alfred Russel Wallace Reader: A Selection of Writings from the Field* (Baltimore: Johns Hopkins University Press, 2001), 18–60.

270 **"It was," Wallace wrote later, "the first work I had yet read":** Wallace, *My Life,* 1:232. He read Malthus in 1844 and his understanding of natural selection came in 1858, so the time lapse was actually less than he remembered.

271 **a "wild, neglected park with the ruins of a mansion":** Ibid., 237.

272 **The scandalous and much discussed red-leathered book:** R. Elwyn

Hughes, "Alfred Russel Wallace: Some Notes on the Welsh Connection," *British Journal for the History of Science* 22, no. 4 (1989): 401–18.

272 **"I do not consider it as a hasty generalization":** Wallace to Bates, December 28, 1845, reproduced in Wallace, *My Life,* 1:254.

273 **Wallace plunged into preparations for a journey:** Raby, *Alfred Russel Wallace,* 28.

274 **"I begin to feel rather dissatisfied with a mere local collection":** Wallace, *My Life,* 1:256–57.

274 **"The more I see of this country, *the more I want to*":** Cited in David Quammen, "The Man Who Knew Islands," in his *The Song of the Dodo: Island Biogeography in an Age of Extinctions* (London: Hutchinson, 1996), 65.

274 **"Wherever I go dogs bark, children scream":** From Wallace's Ms Journal, 39, 54, Linnaean Society archives; cited in Camerini, "Wallace in the Field," 53.

275 **"And now everything was gone," he wrote, "and I had not one specimen":** Cited in Quammen, "The Man Who Knew Islands," 71.

275 **"the limits beyond which certain species never passed":** Cited in ibid., 73–74.

276 **the sheer labor of collecting and surviving:** He had developed routines and rituals, patterns for his days, that made him a highly successful collector: he was always up at half past five, arranging and drying insects and preparing equipment for the day; from nine until three he would collect out in the jungle or fields or forests, returning to kill and pin insects with his assistants; then he would have dinner at four and work again until six or later depending on the weight of the day's catch, and then he would have a few hours of reading or talking before bed. Details from a letter Wallace sent from Singapore on May 28, 1854, cited by Wallace *My Life,* 1:337–38.

277 **he decided to nail his colors to the mast and publish a paper:** Wallace's essay "On the Law Which Has Regulated the Introduction of New Species" appeared in 1855 in the *Annals and Magazine of Natural History.*

277 **"forty black, naked, mop-headed savages":** Alfred Russel Wallace, *The Malay Archipelago: The Land of the Orang-Utan, and the Bird of Paradise. A Narrative of Travel, with Studies of Man and Nature,* 2 vols. (London: Macmillan & Co., 1869), 2:176, 177, 179.

278 **he roamed across these scattered islands:** Wallace's descriptions of his time on the Aru Islands are particularly fascinating—see ibid., 2:196–298.

278 **"I was startled at first to see you already ripe":** Henry Bates to A. R.

Wallace, November 19, 1856; cited in Slotten, *Heretic in Darwin's Court,* 135.

278 **"By your letter & even still more by your paper in Annals":** Charles Darwin to A. R. Wallace, May 1, 1857, Letter 2086, DCP.

279 **"As this subject is now attracting much attention among naturalists":** Cited in Slotten, *Heretic in Darwin's Court,* 139.

280 **"I had discovered the exact boundary line between the Malay and Papuan races":** Wallace, *Malay Archipelago,* 2:20; cited in Moore, "Wallace's Malthusian Moment: The Common Context Revisited," 296.

281 **"Your words have come true with a vengeance":** Charles Darwin to Charles Lyell, June 18, 1858, Letter 2285, DCP.

282 **"I would far rather burn my whole book":** Charles Darwin to Charles Lyell, June 25, 1858, Letter 2294, DCP.

283 **Darwin asked others to make the necessary decisions:** Some historians have suggested that Darwin behaved in a less than gentlemanly way; others that he actively manipulated the situation to his own advantage; still others that he stole some of Wallace's ideas. John Langdon Brookes has pointed out that Wallace's paper could have arrived as early as May 18 and thus Darwin could have had it for at least two weeks and perhaps even a month before he noted its receipt in his private journal. That would have provided him with enough time to "appropriate" some of Wallace's ideas. Arnold Brackman suggests that there was a conspiracy between Darwin, Hooker, and Lyell and that class was an issue. Another view is that the two theories are not the same, that Wallace's version of natural selection acted at the level of varieties and species while Darwin's acted at an individual level—see Slotten, *Heretic in Darwin's Court,* 159.

283 **"The interest excited was intense":** Hooker wrote this account for Francis Darwin, ed., *The Life and Letters of Charles Darwin,* 3 vols. (London: John Murray, 1887), 2:126.

283 **"Allow me in the first place sincerely to thank yourself":** A. R. Wallace to Joseph Hooker, October 6, 1858, cited in Slotten, *Heretic in Darwin's Court,* 160.

285 **"I was stopped in the street one day":** Thomas Barbour, *Naturalist at Large* (London: Scientific Book Club, 1950), 36; cited in Camerini, "Wallace in the Field," 55.

Bibliography

GENERAL

Bowler, Peter. *Evolution: The History of an Idea.* Berkeley, Los Angeles, and London: University of California Press, 1989.

Corsi, Pietro. "Before Darwin: Transformist Concepts in European Natural History," *Journal of the History of Biology* 38: 67–83.

Glass, Bentley, Owsei Temkin, and William L. Straus, eds. *The Forerunners of Darwin: 1745–1859.* Baltimore: Johns Hopkins Univeristy Press, 1959.

Grant, Edward. *A History of Natural Philosophy from the Ancient World to the Nineteenth Century.* Cambridge: Cambridge University Press, 2007.

Larson, E. J. *Evolution: The Remarkable History of a Scientific Legacy.* New York: Modern Library, 2004.

Lovejoy, Arthur O. "Some Eighteenth-Century Evolutionists," *Popular Science Monthly* 65 (1904): 238–51, 323–40.

———. *The Great Chain of Being: A Study of the History of an Idea.* Cambridge, Mass.: Harvard University Press, 1936.

Mayr, Ernst. *The Growth of Biological Thought: Diversity, Evolution and Inheritance.* Cambridge, Mass.: Harvard University Press, 1982.

Osborn, H. F. *From the Greeks to Darwin.* New York: Macmillan, 1894.

Zirkle, Conway. "Natural Selection Before *The Origin of Species*," *Proceedings of the American Philosophical Society* 84, no. 1 (1941): 71–123.

1. DARWIN'S LIST

Browne, Janet. *Charles Darwin.* 2 vols. London: Jonathan Cape, 1995–2002.

Corsi, Pietro. *Science and Religion: Baden Powell and the Anglican Debate, 1800–1860.* Cambridge and New York: Cambridge University Press, 1988.

Desmond, Adrian, and James Moore. *Darwin.* London: Penguin, 1992.

Gotthelf, Allan. "Darwin in Aristotle," *Journal of the History of Biology* 32, no. 1 (1999): 3–30.

Johnson, Curtis N. "The Preface to Darwin's *Origin of Species:* The Curious History of the 'Historical Sketch,'" *Journal of the History of Biology* 40 (2007): 529–56.

2. ARISTOTLE'S EYES

Balme, D. "The Place of Biology in Aristotle's Philosophy." In A. Gotthelf and James G. Lennox, eds, *Philosophical Issues in Aristotle's Biology.* Cambridge: Cambridge University Press, 1987, 9–20.

Barnes, Jonathan. *Aristotle.* Oxford: Oxford University Press, 1982.

———, ed. *Cambridge Companion to Aristotle.* Cambridge: Cambridge University Press, 1995.

———. *Greek Philosophers.* Oxford: Oxford University Press, 2001.

———. *Coffee with Aristotle.* London: Duncan Baird, 2008.

Candargy, P. C. *La Végétation de l'île de Lesbos.* Lille: Bigot frères, 1899.

Casson, Lionel. *Travel in the Ancient World.* London: Book Club Associates, 2005.

Chroust, Anton-Hermann. "Aristotle Leaves the Academy," *Greece and Rome* 14, no. 1 (1967): 39–43.

———. *Aristotle: New Light on His Life.* 2 vols. London: Routledge and Kegan Paul, 1973.

Ellis, J. R. *Philip II and Macedonian Imperialism.* London: Thames and Hudson, 1976.

Frost, Frank J. "Scyllias: Diving in Antiquity," *Greece and Rome* 15, no. 2 (1968): 180–85.

Green, Peter. *Lesbos and the Cities of Asia Minor.* Austin: Dougherty Foundation, 1984.

Greene, John C. "From Aristotle to Darwin: Reflections on Ernst Mayr's Interpretation in *The Growth of Biological Thought*," *Journal of the History of Biology* 25, no. 2 (1992): 257–84.

Grene, Marjorie. *A Portrait of Aristotle.* Chicago: University of Chicago Press, 1963.

———. "Aristotle and Modern Biology," *Journal of the History of Ideas* 33, no. 3 (1972): 395–424.

Jaeger, Werner. *Aristotle: Fundamentals of the History of his Development.* Oxford: Clarendon Press, 1934.

Lee, H.D.P. "Place-Names and the Date of Aristotle's Biological Works," *Classical Quarterly* 42 (1948): 61–67.

Lloyd, G.E.R. *Aristotelian Explorations.* Cambridge: Cambridge University Press, 1996.

———. *Aristotle: The Growth and Structure of His Thought.* Cambridge: Cambridge University Press, 1968.

———. *Early Greek Science: Thales to Aristotle.* London: Chatto and Windus, 1970.

———. "The Evolution of Evolution: Greco-Roman Antiquity and the Origin of Species." In G.E.R. Lloyd, *Principles and Practices in Ancient Greek and Chinese Science.* Aldershot: Ashgate Variorum, 2006, 1–15.

———. *Magic, Reason and Experience: Studies in the Origin and Development of Greek Science.* Cambridge: Cambridge University Press, 1979.

Mason, Hugh J. "Romance in a Limestone Landscape," *Classical Philology* 90, no. 3 (1995): 263–66.

Mayor, Adrienne. *The First Fossil Hunters: Palaeontology in Greek and Roman Times.* Rev. ed. Princeton: Princeton University Press, 2010. First published 2000.

Mayr, Ernst. *The Growth of Biological Thought: Diversity, Evolution and Inheritance.* Cambridge, Mass., and London: Belknap Press, 1982.

Rihll, T. E. *Greek Science.* New Surveys in the Classics, no. 29. Oxford: Oxford University Press, 1999.

Solmsen, Frank. "The Fishes of Lesbos and Their Alleged Significance for the Development of Aristotle," *Hermes* 106 (1978): 467–84.

Taub, Liba. *Aetna and the Moon: Explaining Nature in Ancient Greece and Rome.* Corvallis: Oregon State University Press, 2008.

Thompson, D'Arcy Wentworth. *A Glossary of Greek Fishes.* London: Oxford University Press, 1947.

———. *On Aristotle as a Biologist.* Herbert Spencer Lecture. Oxford: Clarendon Press, 1913.

Tipton, Jason A. "Aristotle's Observations of the Foraging Interactions of the Red Mullet and Sea Bream," *Archives of Natural History* 35, no. 1 (2008): 164–71.

———. "Aristotle's Study of the Animal World: The Case of the Kobios and the Phucis," *Perspectives in Biology and Medicine* 49, no. 3 (2006): 369–83.

Voultsiadou, Eleni, and Dimitris Vafidis. "Marine Invertebrate Diversity in Aristotle's Biology," *Contributions to Zoology* 76 (2007): 103–20.

Wycherley, R. E. "Peripatos: The Athenian Philosophical Scene—II." *Greece and Rome,* 2nd series, vol. 9, no. 1 (1962): 2–21.

3. THE WORSHIPFUL CURIOSITY OF JAHIZ

Aarab, Ahmed, Philippe Provençal, and Mohamed Idaomar. "Eco-Ethological Data According to Jahiz Through His Work, Kitab al-Hayawan," *Arabica* 47 (2000): 278–86.

Ahsan, Muhammad Manazir. *Social Life Under the Abbasids, 170–289 AH, 786–902 AD.* London and New York: Librarie du Liban, 1979.

Al-Samarrai, Qasim. "The Abbasid Gardens in Baghdad and Samarra," *Foundation for Science, Technology and Civilisation* (2002): 1–10.

Bayrakdar, Mehmet. "Al-Jahiz and the Rise of Biological Evolutionism," *Islamic Quarterly,* Third Quarter (1983): 307–15.

Bennison, Amira. *The Great Caliphs: The Golden Age of the Abbasid Empire.* New York and London: I. B. Tauris, 2009.

Blachère, Régis. *Histoire de la littérature arabe des origines à la fin du XVe siècle.* 3 vols. Paris: Maisonneuve, 1964.

Bloom, Jonathan. *Paper Before Print: The History and Impact of Paper in the Islamic World.* New Haven and London: Yale University Press, 2006.

Dodge, Bayard, ed. *The Fihrist of Al-Nadim: A Tenth-Century Survey of Muslim Culture.* 2 vols. New York: Columbia University Press, 1970.

Egerton, Frank E. "A History of the Ecological Sciences, Part 6: Arabic Language Science—Origins and Zoological Writings," *Bulletin of the Ecological Society of America* (2002): 142–46.

Grant, Edward. *A History of Natural Philosophy from the Ancient World to the Nineteenth Century.* Cambridge: Cambridge University Press, 2007.

———. *Science and Religion, 400 BC to AD 1550: From Aristotle to Copernicus.* Johns Hopkins University Press, 2006.

Griffith, Sidney H. *The Church in the Shadow of the Mosque.* Princeton: Princeton University Press, 2008.

Gutas, D. *Greek Thought, Arabic Culture: The Graeco-Arabic Translation Movement in Baghdad and Early Abbasid Society (2nd–4th/8th–10th Centuries).* London: Routledge, 1998.

Hannam, James. *God's Philosophers: How the Medieval World Laid the Foundations of Modern Science.* London: Icon, 2009.

Heinemann, Arnim, Manfred Kropp, Tarif Khalidi, and John Lash Meloy. *Al-Jahiz: A Muslim Humanist for Our Time.* Würzburg and Beirut: Ergon Verlag, 2009.

Kennedy, Hugh. *The Court of the Caliphs: When Baghdad Ruled the Muslim World*. London: Phoenix, 2004.

Khalidi, Tarif. *Arabic Historical Thought in the Classical Period*. Cambridge and New York: Cambridge University Press, 1994.

Kraemer, Joel L. "Translator's Foreword." In *The History of al-Tabari*, vol. 34: *Incipient Decline: The Caliphates of Al-Wathiq, Al-Mutawakkil and Al-Muntasir, AD 841–863*. New York: State University of New York Press, 1989, xi–xxiv.

Kruk, R. "A Frothy Bubble: Spontaneous Generation in the Medieval Islamic Tradition," *Journal of Semitic Studies* 35 (1990): 265–82.

Le Strange, Guy. *Baghdad During the Abbasid Caliphate: From Contemporary Arabic and Persian Sources*. Oxford: Clarendon Press, 1900.

Lindsay, James E. *Daily Life in the Medieval Islamic World*. Westport, Conn.: Greenwood Press, 2005.

Lyons, Jonathon. *The House of Wisdom: How the Arabs Transformed Western Civilisation*. London: Bloomsbury, 2009.

Montgomery, James E. "Al Jahiz." In Shawkat M. Toorawa and Michael Cooperson, eds., *Dictionary of Literary Biography: Arabic Literary Culture, 500–925*, Farmington Hills, Mich.: Gale Press, 2005, 231–42.

———. "Jahiz's *Kitab al-Bayan wa-I-Tabyin*." In Julia Bray, ed., *Writing and Representation: Muslim Horizons*. London: Routledge, 2006, 91–152.

———. "Al-Jahiz and Hellenizing Philosophy." In C. d'Ancona, ed., *The Libraries of the Neoplatonists*. Leiden: Brill, 2007.

———. "Islamic Crosspollinations." In Anna Akasoy, James E. Montgomery, and Peter E. Pormann, eds., *Islamic Crosspollinations: Interactions in the Medieval Middle East*. Cambridge: E.J.W. Gibb Memorial Trust, 2007.

———, ed. *Arabic Theology, Arabic Philosophy: From the Many to the One*. Leuven: Peeters, 2006.

Naji, A. J., and Y. N. Ali. "The Suqs of Basrah: Commercial Organization and Activity in a Medieval Islamic City," *Journal of the Economic and Social History of the Orient* 24:3 (1981): 298–309.

Nasr, S. H. *Islamic Science: An Illustrated Study*. Westerham Press, Kent: World of Islam Festival Publishing, 1976.

———. *Science and Civilisation in Islam*. Cambridge, Mass.: Harvard University Press, new edn: Islamic Texts Society, 1968.

Osborn, H. F. *From the Greeks to Darwin*. New York: Macmillan, 1894.

Pellat, Charles. "Al-Jahiz." In Julia Ashtiany, T. M. Johnstone, J. D. Latham, and R. B. Serjeant, eds., *The Cambridge History of Arabic Literature: Abbasid Belles-Lettres*. Cambridge: Cambridge University Press, 1990.

———. "Hayawan," *Encyclopaedia of Islam* 3 (1966): 304–15.

———, ed. *The Life and Works of Jahiz*. Translated by D. M. Hawke. Berkeley: University of California Press, 1969.

———. *Le Milieu basrien et la formation de Gahiz*. Paris: Maisonneuve, 1953.

Peters, F. E. *Aristotle and the Arabs: The Aristotelian Tradition in Islam*. New York: New York University Press; London: University of London Press, 1968.

Rashed, Roshdi. "Greek into Arabic." In James E. Montgomery, ed., *Arabic Theology, Arabic Philosophy: From the Many to the One*. Leuven: Peeters, 2006.

Rosenthal, Franz. *Greek Philosophy in the Arab World: A Collection of Essays*. Aldershot: Variorum, 1990.

———. "The Stranger in Medieval Islam," *Arabica* 44 (1997): 35–75.

Sabra, A. I. "The Appropriation and Subsequent Naturalisation of Greek Science in Medieval Islam: A Preliminary Statement." *History of Science* 25 (1987): 223–43.

Saliba, George. *Islamic Science and the Making of the European Renaissance*. Cambridge, Mass., and London: MIT Press, 2007.

Sarton, George. *Introduction to the History of Science*. Baltimore: Williams and Wilkins, 1927–31.

Savage-Smith, E. "Attitudes Towards Dissection in Medieval Islam," *Journal of the History of Medicine and Allied Sciences* 50 (1995): 67–110.

Silverstein, Adam J. *Postal Systems in the Pre-Modern Islamic World*. Cambridge: Cambridge University Press, 2007.

Touati, Hourari. *Islam and Travel in the Middle Ages*. Translated by Lydia G. Cochrane. Chicago: University of Chicago Press, 2010.

Turner, Howard R. *Science in Medieval Islam: An Illustrated Introduction*. Austin: University of Texas Press, 1997.

Walbridge, John. *The Leaven of the Ancients: Suhrawardhi and the Heritage of the Greeks*. Albany: State University of New York Press, 2000.

Young, M.J.L., J. D. Latham, and R. B. Serjeant. *Religion, Learning and Science in the Abbasid Period*. Cambridge: Cambridge University Press, 1990.

Zirkle, Conway. "Natural Selection Before *The Origin of Species*," *Proceedings of the American Philosophical Society* 84, no. 1 (1941): 71–123.

4. LEONARDO AND THE POTTER

Allbutt, Thomas Clifford. "Palissy, Bacon and the Revival of Natural Science," *Proceedings of the British Academy* (1913–14): 234–47.

Amico, Leonard N. *Bernard Palissy: In Search of Earthly Paradise*. New York: Flammarion Press, 1996.

Bell, Janis. "Color Perspective, c. 1492," *Achademia Leonardo Vinci* 5 (1992): 64–77.

Céard, Jean. "Bernard Palissy et l'alchimie." In Frank Lestringant, ed., *Actes de colloque Bernard Palissy, 1510–1590: L'écrivain, le réformé, le céramiste*. Paris: Amis d'Agrippa d'Aubigné, 1992, 157–59.

Clark, Kenneth. *Leonardo da Vinci.* Edited by M. Kemp. London: Penguin, 1993.

———. "Leonardo and the Antique." In C. D. O'Malley, ed., *Leonardo's Legacy.* Berkeley and Los Angeles: University of California Press, 1969, 1–34.

Duhem, Pierre. *Études sur Léonard de Vinci.* Paris: A. Hermann, 1906.

———. "Léonard de Vinci, Cardan et Bernard Palissy," *Bulletin Italien* 6, no. 4 (1906): 289–320.

Frieda, Leonie. *Catherine de Medici: A Biography.* London: Weidenfeld and Nicolson, 2003.

Gould, Stephen Jay. "The Upwardly Mobile Fossils." In Stephen Jay Gould, *Leonardo's Mountain of Clams and the Diet of Worms.* London: Vintage, 1999.

Harris, Henry. *Things Come to Life: Spontaneous Generation Revisited.* Oxford: Oxford University Press, 2002.

Huppert, George. *Style of Paris: Renaissance Origins of the French Enlightenment.* Bloomington: Indiana University Press, 1999.

Jeanneret, Michel. *Perpetual Motion: Transforming Shapes in the Renaissance from da Vinci to Montaigne.* Translated by Nidra Poller. Baltimore and London: Johns Hopkins University Press, 2001.

Johnson, Jerah. "Bernard Palissy, Prophet of Modern Ceramics," *Sixteenth Century Journal* 14, no. 4 (1983): 399–410.

Kamil, Neil. *Fortress of the Soul: Violence, Metaphysics, and Material Life in the Huguenots' New World, 1517–1751.* Baltimore: Johns Hopkins University Press, 2005.

Kemp, Martin. *Leonardo da Vinci: The Marvellous Works of Nature and Man.* Rev. ed. Oxford: Oxford University Press, 2006.

Kirkbride, Robert. *Architecture and Memory: The Renaissance Studioli of Federico da Montefeltro.* New York: Columbia University Press, 2008.

Kirsop, Allace. "The Legend of Bernard Palissy," *Ambix* 9 (1961): 136–94.

Leonardo da Vinci. Leicester Codex, in "Physical Geography," in *The Notebooks of Leonardo da Vinci,* translated by E. MacCurdy. 2 vols. London: Jonathan Cape, 1938.

Lessing, Maria. "Leonardo da Vinci's Pazzia Bestialissima," *Burlington Magazine* 64, no. 374 (May 1934): 219–31.

Maccagni, Carlo. "Leonardo's List of Books," *Burlington Magazine* 110, no. 784 (July 1968): 406–10.

Mauries, Patrick. *Cabinets of Curiosities.* London: Thames and Hudson, 2002.

Newman, William R. *Promethean Ambitions: Alchemy and the Quest to Perfect Nature.* Chicago: University of Chicago Press, 2004.

Nicholl, Charles. *Leonardo da Vinci: The Flights of the Mind.* London: Penguin, 2005.

Ogilvie, Brian W. *Science of Describing: Natural Science in Renaissance Europe.* Chicago: University of Chicago Press, 2006.

Ortiz, Antonio Domínguez, Concha Herrero Carretero, and José A. Godoy. *Resplendence of the Spanish Monarchy: Renaissance Tapestries and Armor from the Patrimonio Nacional.* New York: Metropolititan Museum of Art, 1996.

Reti, Ladislao. "The Two Unpublished Manuscripts of Leonardo da Vinci in the Biblioteca Nacional of Madrid—II," *Burlington Magazine* 110, no. 779 (February 1968): 81–89.

Richter, Jean Paul, ed. *The Literary Works of Leonardo da Vinci.* 2 vols. London: Phaidon Press, 1970.

Scalini, Mario. "The Weapons of Lorenzo de Medici." In Robert Held, ed., *Art, Arms and Armour: An International Anthology.* Chiasso, Switzerland: Acquafresca Editrice, 1979.

Shell, Hanna Rose. "Casting Life, Recasting Experience: Bernard Palissy's Occupation Between Maker and Nature," *Configurations* 12 (2004): 1–40.

Smith, Pamela H. *The Body of the Artisan: Art and Experience in the Scientific Revolution.* Chicago: University of Chicago Press, 2004.

Thomson, David. *Renaissance Paris: Architecture and Growth, 1475–1600.* Berkeley and Los Angeles: University of California Press, 1984.

Thompson, H. R. "The Geographical and Geological Observations of Bernard Palissy the Potter." *Annals of Science* 10:2 (1954): 149–65.

White, Michael. *Leonardo: The First Scientist.* London: Abacus, 2000.

5. TREMBLEY'S POLYP

Baker, John R. *Abraham Trembley of Geneva: Scientist and Philosopher, 1710–1784.* London: Edward Arnold, 1952.

Baker, Henry. *An Attempt Towards a Natural History of the Polype.* London: R. Dodsley, 1743.

Barsanti, Giulio. "Les Phénomènes 'étranges' et 'paradoxaux' aux origines de la première révolution biologique (1740–1810)." In Guido Cimino and François Duchesneau, eds., *Vitalisms from Haller to the Cell Theory.* Florence: Olschki, 1997.

Bezemer-Seller, Vanessa J. W. "The Bentinck Garden at Sorgvliet." In J. D. Hunt, ed., *The Dutch Garden in the Seventeenth Century.* Washington, D.C.: Dumbarton Oaks Research Library and Collection, 1990.

Brown, Harcourt. "Madame Geoffrin and Martin Folkes: Six New Letters," *Modern Language Quarterly* 1 (1940): 219.

Dawson, Virginia P. *Nature's Enigma: The Problem of the Polyp in the Letters of Bonnet, Trembley and Réaumur.* Philadelphia: Memoirs of the American Philosophical Society, 1988.

Ford, Brian J. *Single Lens: The Story of the Simple Microscope.* New York: Harper and Row, 1985.

Glass, Bentley. "Heredity and Variation in the Eighteenth Century Concept of Species." In Bentley Glass, Owsei Temkin, and William L. Straus, eds., *The Forerunners of Darwin: 1745–1859*. Baltimore: Johns Hopkins University Press, 1959, 144–73.

Hazard, Paul. *Le Crise de la conscience européenne, 1680–1714*. Paris: Boivin, 1935.

Jacob, Margaret C. "Hazard Revisited." In Phyllis Mack, ed., *Politics and Culture in Early Modern Europe: Essays in Honour of H. G. Koenigsberger.* Cambridge: Cambridge University Press, 1987, 250–72.

———. *Living the Enlightenment: Freemasonry and Politics in Eighteenth-Century Europe.* New York: Oxford University Press, 1991.

———. *The Radical Enlightenment: Pantheists, Freemasons and Republicans.* London: George Allen and Unwin, 1981.

Jong, Erik. *Nature and Art: Dutch Garden and Landscape Architecture, 1650–1740.* Philadelphia: Philadelphia University Press, 2000.

Le Bond, Aubrey. *Charlotte Sophie, Countess Bentinck: Her Life and Times.* London: Hutchinson, 1912.

Ratcliff, Marc J. *L'Effet Trembley ou la naissance de la zoologie marine.* Geneva: La Baconnière, 2010.

———. *The Quest for the Invisible: Microscopy in the Enlightenment.* Farnham, Surrey, and Burlington, Vt.: Ashgate, 2009.

———. "Trembley's Strategy of Generosity and the Scope of Celebrity in the Mid-Eighteenth Century." *Isis* 95, no. 4 (2004): 555–75.

Ratcliff, Marc J., and Marian Fournier, "Abraham Trembley's Impact on the Construction of Microscopes." In Dario Generali and Marc J. Ratcliff, eds., *From Makers to Users: Microscopes, Markets and Scientific Practices in the Seventeenth and Eighteenth Centuries.* Florence: Olschki, 2007.

Roger, Jacques. *Les sciences de la vie dans la pensée française du XVIIIe siècle: La génération des animaux de Descartes à l'Encyclopédie.* Paris: Armand Colin, 1963.

Schazmann, Paul-Emile. *The Bentincks: The History of a European Family.* London: Weidenfeld and Nicolson, 1976.

Stafford, Barbara Maria. "Images of Ambiguity, Eighteenth-Century Microscopy, and the Neither/Nor." In D. P. Miller and P. H. Reill, eds., *Visions of Empire: Voyages, Botany, and Representations of Nature.* Cambridge: Cambridge University Press, 1997.

Trembley, Abraham. *Instructions d'un père à ses enfants, sur la nature et sur la religion.* 2 vols. Geneva: Chapuis, 1775.

———. *Mémoires pour servir à l'histoire d'un genre de polypes d'eau douce.* Paris, 1744.

Trembley, Maurice, and Émile Guyénot, eds. *Correspondance inédite entre Réaumur et Abraham Trembley.* Geneva: Georg, 1943.

Vartanian, Aram. "Trembley's Polyp, La Mettrie and Eighteenth-Century French Materialism," *Journal of the History of Ideas* 11 (1950): 259–80.

Wellmann, Janina. "Picture Metamorphosis: The Transformation of Insects from the End of the Seventeenth to the Beginning of the Nineteenth Century," *NTM* 16, no. 2 (2008): 183–211.

Wilson, Catherine. *The Invisible World: Early Modern Philosophy and the Invention of the Microscope.* Princeton: Princeton University Press, 1995.

6. THE CONSUL OF CAIRO

Allen, Don Cameron. "The Predecessors of Champollion," *Proceedings of the American Philosophical Society* 104, no. 5 (1960): 527–47.

Benítez, Miguel. "Benoît de Maillet et la littérature clandestine: Étude de sa correspondance avec l'abbé Le Mascrier," *Studies on Voltaire* 183 (1980): 133–59.

———. "Benoît de Maillet et l'origine de la vie dans la mer: Conjecture amusante ou hypothèse scientifique?" *Revue de Synthèse,* 3rd series, 113–14 (1984): 37–54.

———. *La Face cachée des Lumières: Recherches sur les manuscrits philosophiques clandestins de l'âge classique.* Paris: Voltaire Foundation, 1996.

———. "Fixisme et évolutionnisme au temps des Lumières: Le *Telliamed* de Benoît de Maillet." *Rivista di Storia della Filosofia* 45 (1990): 247–68.

Carozzi, Albert V., ed. *Telliamed.* Champaign: University of Illinois Press, 1968.

Carré, Jean-Marie. *Voyageurs et écrivains français en Egypte.* Rev. and corrected ed. Cairo: Institut français d'archéologie orientale du Caire. 1990. Facsimile of 1956 edition.

Cohen, Claudine. "L' 'Anthropologie' de *Telliamed,*" *Bulletins et Mémoires de la Société d'Anthropologie de Paris* 1, no. 3–4 (1989): 45–56.

———. "Benoît de Maillet et la diffusion de l'histoire naturelle à l'aube des Lumières," *Revue d'histoire des sciences* 44, no. 3–4 (1991): 325–42.

———. *La Genèse de "Telliamed": Théorie de la terre et histoire naturelle à l'aube des Lumières.* Paris: Presses Universitaires de France, 1989.

———. *Science, libertinage et clandestinité à l'aube des Lumières.* Paris: Presses Universitaires de France, 2011.

Darnton, Robert. "A Police Officer Sorts His Files." In Robert Darnton, *The Great Cat Massacre and Other Episodes in French Cultural History.* London: Vintage, 1985.

Hunt, Lynn, Margaret C. Jacob, and Wijnand Mijnhardt. *The Book That Changed Europe: Picart and Bernard's Religious Ceremonies of the World.* Cambridge, Mass., and London: Belknap Press, 2010.

Masson, Paul. *Histoire du commerce français dans le Levant au XVIIIe siècle.* Paris: Hachette, 1911.

McLeod, Jane. "Provincial Book Trade Inspectors in Eighteenth-Century France," *French History* 12, no. 2 (1998): 127–48.

Mézin, Anne. *Les Consuls de France au siècle des Lumières: 1715–1792.* Paris: La Documentation française, 1997.

Rothschild, Harriet Dorothy. "Benoît de Maillet's Cairo Letters," *Studies on Voltaire and the Eighteenth Century* 169 (1977): 115–85.

———. "Benoît de Maillet's Leghorn Letters," *Studies on Voltaire and the Eighteenth Century* 30 (1964): 351–75.

———. "Benoît de Maillet's Letters to the Marquis de Caumont," *Studies on Voltaire and the Eighteenth Century* 60 (1968): 311–38.

———. "Benoît de Maillet's Marseilles Letters," *Studies on Voltaire and the Eighteenth Century* 37 (1965): 109–45.

7. THE HOTEL OF THE PHILOSOPHERS

Billy, André. *Diderot: Sa vie, son oeuvre.* Paris: A. Cresson, 1949.

Bonnefon, Paul. "Diderot prisonnier à Vincennes," *Revue d'Histoire Littéraire de la France* 6 (1899): 200–24.

Buffon, Georges Louis Leclerc. *Correspondance inédite de Buffon.* Edited by H. Nadault de Buffon. 2 vols. Paris: Hachette, 1860.

———. *The Embattled Philosopher: A Biography of Denis Diderot.* London: Neville Spearman, 1955.

Crocker, Lester G. "Diderot and Eighteenth-Century French Transformism." In Bentley Glass, Owsei Temkin, and William L. Straus, eds., *The Forerunners of Darwin, 1745–1859.* Baltimore: Johns Hopkins University Press, 1959, 114–43.

Cushing, Max Pearson. *Baron d'Holbach: A Study of Eighteenth Century Radicalism in France.* New York: Columbia University Press, 1914.

Darnton, Robert. *The Literary Underground of the Old Regime.* Cambridge, Mass.: Harvard University Press, 1982.

———. "A Police Officer Sorts His Files." In Robert Darnton, *The Great Cat Massacre and Other Episodes in French Cultural History.* London: Vintage, 1985.

Diderot, Denis. *Diderot's Letters to Sophie Volland: A Selection.* Translated by Peter France. London: Oxford University Press, 1972.

———. *Oeuvres complètes de Diderot.* Edited by Jules Assezat and Maurice Tourneux. 20 vols. Paris: Garnier, 1875.

———. *Rameau's Nephew and D'Alembert's Dream.* Translated by Leonard Tancock. Harmondsworth and New York: Penguin, 1976.

———, and Jean le Rond d'Alembert, eds. *Encyclopédie ou Dictionnaire raisonné des sciences, des arts et des métiers.* 17 vols. Paris, 1751–72. See online project at encyclopedie.uchicago.edu/.

Fellows, Otis. "Buffon's Place in the Enlightenment," *Studies on Voltaire and the Eighteenth Century* 25 (1963): 603–29.

Furbank, P. N. *Diderot: A Critical Biography.* London: Secker and Warburg, 1992.

Gregory, Mary Efrosni. *Diderot and the Metamorphosis of Species.* London: Routledge, 2007.

———. *Evolutionism in Eighteenth-Century French Thought.* New York: Peter Lang, 2008.

Hanley, W. "The Policing of Thought in Eighteenth-Century France." *Studies on Voltaire and the Eighteenth Century* 183 (1980): 279–84.

Hill, Emita. "Materialism and Monsters in Diderot's Rêve d'Alembert," *Diderot Studies* 10 (1968): 67–93.

Kafker, Frank A., and Jeff Loveland. "Diderot et Laurent Durand, son éditeur principal." *Recherches sur Diderot et sur l'Encyclopédie* 39 (2005): 29–40.

Kors, Alan Charles. "The Atheism of d'Holbach and Naigeon." In Michael Hunter and David Wooton, eds., *Atheism from the Reformation to the Enlightenment.* Oxford: Clarendon Press, 1992.

———. *D'Holbach's Coterie: An Enlightenment in Paris.* Princeton: Princeton University Press, 1976.

Llana, James. "Natural History and the *Encyclopaedie*," *Journal of the History of Biology* 33 (2000): 1–25.

Lovejoy, Arthur O. "Buffon and the Problem of Species." In Bentley Glass et al., *Forerunners of Darwin, 1745–1859.* Baltimore: Johns Hopkins University Press 1959.

———. "Some Eighteenth-Century Evolutionists," *Popular Science Monthly* 65 (1904): 238–51.

Newland, T. C. "D'Holbach, Religion and the 'Encyclopédie.' " *Modern Language Review* 69, no. 3 (1974): 523–33.

Roger, Jacques. *Buffon: A Life in Natural History.* Edited by L. Pearce Williams, translated by Sarah Lucille Bonnefoi. Ithaca and London: Cornell University Press, 1997.

———. "Diderot et Buffon en 1749," *Diderot Studies* 4 (1963): 221–36.

———. *The Life Sciences in Eighteenth-Century French Thought.* Edited by Keith R. Benson. Translated by Robert Ellrich. First published 1963. Stanford: Stanford University Press, 1997.

Spanger, May. "Science, philosophie et littérature: Le polype de Diderot," *Recherches sur Diderot et sur l'Encyclopédie* 23 (1997): 89–107.

Stengers, Jean. "Buffon et la Sorbonne." In Roland Mortier and Hervé Hasquin, eds., *Études sur le XVIIIe siècle*. Brussels: Éditions de l'Université de Bruxelles, 1974, 113–24.

Topazio, Virgil W. *D'Holbach's Moral Philosophy: Its Background and Development*. Geneva: Institut et Musée Voltaire, 1956.

Vartanian, Aram. "From Deist to Atheist: Diderot's Philosophical Orientation, 1746–1749," *Diderot Studies* 1 (1949): 46–63.

———. *Science and Humanism in the French Enlightenment*. Charlottesville, Va: Rookwood, 1999.

———. "Trembley's Polyp, La Mettrie and Eighteenth-Century French Materialism," *Journal of the History of Ideas* 2, no. 3 (1950): 259–86.

Voltaire (J.F.M. Arouet). *Cabales* (1772). In *Oeuvres complètes,* 70 vols. Paris: Firmin-Didot, 1875, 2.

Wickwar, W. H. *Baron d'Holbach: A Prelude to the French Revolution*. London: George Allen and Unwin, 1935.

Wilson, Arthur M. *Diderot*. New York: Oxford University Press, 1972.

8. ERASMUS UNDERGROUND

Barlow, Nora. "Erasmus Darwin FRS 1731–1802," *Notes and Records of the Royal Society of London* 14, no. 1 (1959): 85–98.

Bewell, Alan. "Erasmus Darwin's Cosmopolitan Nature," *ELH* 76, no. 1 (2009): 19–48.

Browne, Janet. "Botany for Gentlemen: Erasmus Darwin and *The Loves of the Plants*," *Isis* 80, no. 4 (1989): 593–621.

Coleridge, Samuel Taylor. *The Collected Letters of Samuel Taylor.* Edited by Earl Leslie Griggs, 6 vols. Oxford: Clarendon Press, 1956–71.

Craven, Maxwell. *John Whitehurst of Derby: Clockmaker and Scientist, 1713–88*. Ashbourne: Mayfield, 1996.

Darwin, Erasmus. *Zoonomia, or the Laws of Organic Life*. London: J. Johnson, 1794–96.

Dean, Bashford. "Two Letters of Dr. Darwin: The Early Date of His Evolutionary Writings." *Science* 23, no. 600 (1906): 986–87.

Dent, Robert K. *Old and New Birmingham: A History of the Town and Its People*. Wakefield: EP Publishing, 1972–73. Reprint of 1878–80 edition.

Elliott, Paul. "Erasmus Darwin, Herbert Spencer and the Origins of the Evolutionary Worldview in British Provincial Scientific Culture." *Isis* 94, no. 1 (2003): 1–29.

Ford, Trevor B. *Treak Cliff Cavern and the Story of Blue John Stone*. Castleton: Harrison Taylor, 1992.

Garfinkle, Norton. "Science and Religion in England, 1790–1800: The Critical

Response to the Work of Erasmus Darwin." *Journal of the History of Ideas* 16, no. 3 (1955): 376–88.

Harrison, James. "Erasmus Darwin's View of Evolution." *Journal of the History of Ideas* 32, no. 2 (1971): 247–64.

Hassler, Donald M. *The Comedian as the Letter D: Erasmus Darwin's Comic Materialism*. The Hague: Martinus Nijhoff, 1973.

Keir, James. *Sketch of the Life of James Keir.* London: R. E. Taylor, 1868.

King-Hele, Desmond. *The Collected Letters of Erasmus Darwin*. Cambridge: Cambridge University Press, 2006.

———. *Erasmus Darwin: A Life of Unequalled Achievement*. London: Giles de la Mare, 1999.

———. *Erasmus Darwin and the Romantic Poets*. Basingstoke: Macmillan, 1986.

———. "Erasmus Darwin's Life at Lichfield: Fresh Evidence," *Notes and Records of the Royal Society of London* 49, no. 2 (1995): 231–43.

———. *Letters of Erasmus Darwin*. Cambridge: Cambridge University Press, 1981.

———. *The Life and Genius of Erasmus Darwin*. London: Faber and Faber, 1977.

McNeil, Maureen. *Under the Banner of Science: Erasmus Darwin and His Age.* Manchester: Manchester University Press, 1987.

Meteyard, Eliza. *A Group of Englishmen (1795 to 1815): Being Records of the Younger Wedgwoods and Their Friends*. London: Longmans, Green, 1871.

———. *The Life of Josiah Wedgwood*. 2 vols. London: Hurst and Blackett, 1865.

Moers, Ellen. "Female Gothic." In George Levine and U. C. Knoepflmacher, eds., *The Endurance of Frankenstein: Essays on Mary Shelley's Novel.* Berkeley: University of California Press, 1982. First published 1976.

Palmer, Stanley. *Police and Protest in England and Ireland, 1780–1850*. Cambridge: Cambridge University Press, 1989.

Porter, Roy. "Erasmus Darwin: Doctor of Evolution?" In James Moore, ed., *History, Humanity and Evolution: Essays for John C. Greene.* New York and Cambridge: Cambridge University Press, 1989, 39–69.

Posner, E. "Erasmus Darwin and the Sisters Parker," *History of Medicine* 6, no. 2 (1975): 39–43.

Priestman, Martin. *Romantic Atheism: Poetry and Freethought, 1780–1830*. Cambridge: Cambridge University Press, 1999.

Priestman, Martin. "Darwin's Early Drafts for the *Temple of Nature*." In C.U.M. Smith and Robert Arnott, eds., *The Genius of Erasmus Darwin*. Aldershot: Ashgate, 2005, 307–19.

Primer, Irwin. "Erasmus Darwin's *Temple of Nature:* Progress, Evolution, and the Eleusinian Mysteries," *Journal of the History of Ideas* 25, no. 1 (1964): 58–76.

Seward, Anna. *Memoirs of the Life of Dr. Darwin*. London: J. Johnson, 1804.

Smith, C.U.M., and Robert Arnott, eds. *The Genius of Erasmus Darwin.* Aldershot: Ashgate, 2005.

Smyser, Jane Worthington. "The Trial and Imprisonment of Joseph Johnson, Bookseller," *Bulletin of the New York Public Library* 77 (1974): 418–35.

Stukeley, William. "An Account of the Impression of the almost Entire Skeleton of a large Animal in a very hard Stone, lately presented the Royal Society, from Nottinghamshire." *Philosophical Transactions* 30, no. 360 (1719): 963–68.

Taylor, David. *Crime, Policing and Punishment in England, 1750–1914.* Basingstoke: Macmillan, 1998.

Uglow, Jenny. *The Lunar Men: The Friends Who Made the Future, 1730–1810.* London: Faber and Faber, 2003.

Whitehurst, John. *An Inquiry into the Original State and Formation of the Earth.* London: Bent, 1778.

9. THE JARDIN DES PLANTES

Ambrose, C. T. "Darwin's Historical Sketch—An American Predecessor," *Archives of Natural History* 37, no. 2 (2010): 191–202.

Appel, Toby A. *The Cuvier-Geoffroy Debate: French Biology in the Decades Before Darwin.* Oxford: Oxford University Press, 1987.

Bange, Raphaël, and Pietro Corsi. "Chronologie de la vie de Jean-Baptiste Lamarck." Centre National de la Recherche Scientifique, 1997. Online at www.lamarck.cnrs.fr/chronologie.

Blainville, Henri-Marie Ducrotay de. *Histoire des sciences de l'organisation et de leurs progrès comme base de la philosophie, rédigée etc. par F.L.M. Maupied.* 3 vols. Paris, 1845.

Bourdier, Frank. "Le Prophète Geoffroy Saint-Hilaire, Georges Sand et les Saint-Simoniens," *Histoire et Nature* 3 (1973): 47–66.

Burkhardt, R. W. "The Inspiration of Lamarck's Belief in Evolution," *Journal of the History of Biology* 5 (1972): 413–38.

———. "Lamarck, Evolution and the Politics of Science," *Journal of the History of Biology* 3 (1970): 275–96.

———. "The Leopard in the Garden: Life in Close Quarters at the Muséum d'Histoire Naturelle," *Isis* 98, no. 4 (2007): 675–94.

———. *The Spirit of System: Lamarck and Evolutionary Biology.* Cambridge, Mass., and London: Harvard University Press, 1977.

Burleigh, Nina. *Mirage: Napoleon's Scientists and the Unveiling of Egypt.* New York: Harper, 2007.

Cahn, Théophile. *Vie et l'oeuvre d'Étienne Geoffroy Saint-Hilaire.* Paris: Presses Universitaires de France, 1962.

Corsi, Pietro. *The Age of Lamarck: Evolutionary Theories in France, 1790–1830.* Berkeley: University of California Press, 1988.

———. "Before Darwin: Transformist Concepts in European Natural History," *Journal of the History of Biology* 38, no. 1 (2005): 167–83.

———. *Lamarck, philosophe de la nature.* Paris: Presses Universitaires de France, 2006.

Cuvier, Georges. "Elegy of Lamarck." *Edinburgh New Philosophical Journal* 20 (January 1836): 21–22.

Deleuze, Joseph. *Histoire et description du Muséum Royal d'Histoire Naturelle, ouvrage rédigé d'après les ordres de l'administration du Muséum.* Paris: Royer, 1823. Translated into English 1823.

Desmond, Adrian. *The Politics of Evolution: Morphology, Medicine and Reform in Radical London.* Chicago and London: University of Chicago Press, 1989.

Endersby, Jim. " 'The Vagaries of a Rafinesque': Imagining and Classifying American Nature," *Studies in History and Philosophy of Science Part C: Studies in History and Philosophy of Biological and Biomedical Sciences* 40, no. 3 (2009): 168–78.

Gould, Stephen Jay. "A Tree Grows in Paris: Lamarck's Division of Worms and the Division of Nature." In Stephen Jay Gould, *The Lying Stones of Marrakech: Penultimate Reflections in Natural History.* New York: Harmony Books, 2000, 115–43.

Grant, Iain Hamilton. *Philosophies of Nature After Schelling.* New York and London: Continuum, 2006.

Gregory, Mary Efrosni. *Evolutionism in Eighteenth-Century French Thought.* New York: Peter Lang, 2009.

Henry, Freeman G. "Rue Cuvier, Rue Geoffroy-Saint-Hilaire, Rue Lamarck: Politics and Science in the Streets of Paris," *Nineteenth-Century French Studies* 35, no. 3–4 (2007): 513–25.

Jordanova, Ludmilla. *Lamarck.* Oxford: Oxford University Press, 1984.

Lee, Mrs. R. *Memoirs of Baron Cuvier.* London: Longman, Rees, Orme, Brown, Green and Longman, 1833.

Le Guyader, Hervé. *Geoffroy Saint Hilaire: A Visionary Naturalist.* Translated by Marjorie Grene. Chicago: University of Chicago Press, 2004.

Loveland, Jeff. "Daubenton's Lions: From Buffon's Shadow to the French Revolution," *New Perspectives on the Eighteenth Century* 1 (2004): 29–47.

Orr, M. "Keeping It in the Family: The Extraordinary Case of Cuvier's Daughters." In Cynthia Burek and Bettie Higgs, eds., *The Role of Women in the History of Geology.* London: Geological Society of London, Special Publications, 2007, 281:277–86.

Outram, Dorinda. "The Language of Natural Power: The Funeral Éloges of Georges Cuvier," *History of Science* 16 (1978): 153–78.

———. "Le Muséum National d'Histoire Naturelle après 1793: Institution scientifique ou champ de bataille pour les familles et les groupes d'influence?" In Claude Blanckaert, Claudine Cohen, Pietro Corsi, and Jean-Louis Fischer, *Le Muséum au premier siècle de son histoire.* Paris: Muséum National d'Histoire Naturelle, 1997, 25–30.

———. *Science, Vocation and Authority in Post-Revolutionary France: Georges Cuvier.* Manchester: Manchester University Press, 1984.

———. "Uncertain Legislator: Georges Cuvier's Laws of Nature in Their Intellectual Context," *Journal of the History of Biology* 19, no. 3 (1986): 323–68.

Packard, A. S. *Lamarck: The Founder of Evolution.* New York: Longmans, Green, 1901.

Richards, Robert J. *The Romantic Conception of Life: Science and Philosophy in the Age of Goethe.* Chicago: University of Chicago Press, 2002.

Rudwick, Martin J. S. *Bursting the Limits of Time: The Reconstruction of Geohistory in the Age of Revolution.* Chicago: University of Chicago Press, 2005.

———. *Georges Cuvier, Fossil Bones and Geological Catastrophe: New Translations and Interpretations of the Primary Texts.* Chicago: University of Chicago Press, 1997.

Solé, Robert. *Les Savants de Bonaparte.* Paris: Editions du Seuil, 1999.

Spary, Emma. *Utopia's Garden: French Natural History from Old Regime to Revolution.* Chicago: University of Chicago Press, 2000.

Strathern, Paul. *Napoleon in Egypt: The Greatest Glory.* London: Jonathan Cape, 2007.

10. THE SPONGE PHILOSOPHER

Ashworth, J. H. "Charles Darwin as a Student in Edinburgh, 1825–27." *Proceedings of the Royal Society of Edinburgh,* 55 (1935): 97–113.

Balfour, John Hutton. *Biography of the Late John Coldstream.* London: J. Nisbet, 1865.

Beddoe, John. *Memories of Eighty Years.* Bristol: Arrowsmith, 1910.

Corbin, Alain. *The Lure of the Sea: The Discovery of the Seaside, 1750–1840.* Harmondsworth: Penguin, 1995.

Corsi, Pietro. "The Importance of French Transformist Ideas for the Second Volume of Lyell's *Principles of Geology,*" *British Journal for the History of Science* 11, no. 3 (1978): 221–44.

Darwin, Charles. *Autobiography with original omissions restored; edited with appendix and notes by his grand-daughter, Nora Barlow.* London: Collins, 1958.

Desmond, Adrian. *Archetypes and Ancestors: Palaeontology in Victorian London, 1850–1875.* London: Blond & Biggs, 1982.

———. *The Politics of Evolution: Morphology, Medicine and Reform in Radical London.* Chicago: University of Chicago Press, 1989.

———. "Richard Owen's Reaction to Transmutation in the 1830s." *British Journal for the History of Science* 18, no. 1 (1985): 25–50.

———. "Robert E. Grant's Later Views on Organic Development." *Archives of Natural History* 11 (1984): 395–413.

———. "Robert E. Grant: The Social Predicament of a Pre-Darwinian Transmutationist." *Journal of the History of Biology* 17, no. 2 (1984): 189–223.

———, and James Moore. *Darwin*. Harmondsworth: Penguin, 1992.

Grant, Robert. "Observations and Experiments on the Structure and Functions of the Sponge." *Edinburgh Philosophical Journal* 13, no. 25 (1825): 99.

Jespersen, P. Helveg. "Charles Darwin and Dr. Grant." *Lychnos* (1948–49): 159–67.

Marshall, James Scott. *The Life and Times of Leith*. Edinburgh: John Donald, 1986.

Mowat, Sue. *The Port of Leith: Its History and Its People*. Edinburgh: John Donald in association with the Forth Ports, 1994.

Pakenham, Simona. *In the Absence of the Emperor: London-Paris, 1814–15*. London: Cresset Press, 1968.

Parker, Sarah E. *Robert Edmond Grant (1793–1894) and His Museum of Zoology and Comparative Anatomy*. London: Grant Museum of Zoology, 2006.

Porter, Roy. *The Greatest Benefit to Mankind: A Medical History of Humanity from Antiquity to the Present*. London: Fontana, 1997.

Royle, Edward. *Victorian Infidels: The Origins of the British Secularist Movement, 1791–1866*. Manchester: Manchester University Press, 1974.

Secord, James A. "Edinburgh Lamarckians: Robert Jameson and Robert E. Grant." *Journal of the History of Biology* 24 (1991): 1–18.

Sheppersen, George. "The Intellectual Background of Charles Darwin's Student Years at Edinburgh." In M. Banton, ed., *Darwinism and the Study of Society*. London: Tavistock Publications; Chicago: Quadrangle Books, 1961, 17–35.

Stevenson, Sara. *Hill and Adamson's "The Fishermen and Women of the Firth of Forth."* Edinburgh: Scottish National Portrait Gallery, 1991.

Wakley, Thomas. "Biographical Sketch of Robert Edmund Grant, M.D." *Lancet* 2 (1850): 686–95.

Wallace, Joyce M. *Traditions of Trinity and Leith*. Edinburgh: John Donald, 1997.

11. THE ENCYCLOPEDIST

Chambers, Robert. "Natural History: Animals with a Backbone." *Chambers's Edinburgh Journal*, November 24, 1832.

———. "Popular Information on Science: Transmutation of Species." *Chambers's Edinburgh Journal*, September 26, 1835.

———. *Vestiges of the Natural History of Creation and Other Evolutionary Writings.* Edited by J. A. Secord. Chicago: University of Chicago Press, 1994.

Chambers, William. *Memoir of Robert Chambers and Autobiographical Reminiscences of William Chambers.* Edinburgh and London: W. & R. Chambers, 1872.

Combe, George. *Constitution of Man Considered in Relation to External Objects.* Edinburgh: J. Anderson Jr., 1828.

Layman, C. H., ed. *Man of Letters: The Early Life and Love Letters of Robert Chambers.* Edinburgh: Edinburgh University Press, 1990.

Lehmann, R. C. *Memories of Half a Century.* London: Smith, Elder, 1908.

Scholnick, Robert J. " 'The Fiery Cross of Knowledge': *Chambers's Edinburgh Journal,* 1832–1844." *Victorian Periodicals Review* 32, no. 4 (1999): 324–58.

Secord, Anne. "Corresponding Interests: Artisans and Gentlemen in Nineteenth-Century Natural History." *British Journal for the History of Science* 27 (1994): 383–408.

———. "Science in the Pub." *History of Science* 32 (1994): 269–315.

Secord, James A. "Behind the Veil: Robert Chambers and Vestiges." In James R. Moore, ed., *History, Humanity and Evolution: Essays for John C. Greene.* Cambridge and New York: Cambridge University Press, 1989, 165–94.

———. *Controversy in Victorian Geology: The Cambrian-Silurian Debate.* Princeton: Princeton University Press, 1986.

———. "The Discovery of a Vocation: Darwin's Early Geology." *British Journal for the History of Science* 24 (1991): 133–57.

———. *Victorian Sensation: The Extraordinary Publication, Reception, and Secret Authorship of "Vestiges of the Natural History of Creation."* Chicago: University of Chicago Press, 2000.

12. ALFRED WALLACE'S FEVERED DREAMS

Brooks, John Langdon. *Just Before the Origin: Alfred Russel Wallace's Theory of Evolution.* New York: Columbia University Press, 1984.

Camerini, Jane R. "Remains of the Day: Early Victorians in the Field." In Bernard Lightman, ed., *Victorian Science in Context.* Chicago: University of Chicago Press, 1997, 354–77.

———. "Wallace in the Field." In H. Kuklick and R. Kohler, eds., "Science in the Field." *Osiris* 11 (1996): 44–65.

———, ed. *The Alfred Russel Wallace Reader: A Selection of Writings from the Field.* Baltimore: Johns Hopkins University Press, 2001.

Endersby, Jim. "Escaping Darwin's Shadow." *Journal of the History of Biology* 36, no. 2 (2003): 385–403.

Hodge, Jonathan, and Gregory Radick. "The Place of Darwin's Theories in the

Intellectual Long Run." In Jonathan Hodge and Gregory Radick, eds., *The Cambridge Companion to Darwin*. 2nd ed. Cambridge: Cambridge University Press, 2009, 246–73.

Hughes, R. Elwyn. "Alfred Russel Wallace (1823–1913): The Making of a Scientific Non-Conformist." *Proceedings of the Royal Institution* 63 (1991): 175–83.

———. "Alfred Russel Wallace: Some Notes on the Welsh Connection." *British Journal for the History of Science* 22, no. 4 (1989): 401–18.

Macdougall, Ian. *All Men Are Brethren: French, Scandinavian, Italian, German, Dutch, Belgian, Spanish, Polish, West Indian, American and Other Prisoners of War in Scotland During the Napoleonic Wars, 1803–1814*. Edinburgh: John Donald, 2008.

McKinney, H. Lewis. *Wallace and Natural Selection*. New Haven and London: Yale University Press, 1972.

Moore, James. "Wallace's Malthusian Moment: The Common Context Revisited." In Bernard Lightman, ed., *Victorian Science in Context*. Chicago: University of Chicago Press, 1997, 290–311.

Quammen, David. *The Song of the Dodo: Island Biogeography in an Age of Extinctions*. London: Hutchinson, 1996.

Raby, Peter. *Alfred Russel Wallace: A Life*. Princeton: Princeton University Press, 2001.

Slotten, Ross A. *The Heretic in Darwin's Court: The Life of Alfred Russel Wallace*. New York: Columbia University Press, 2004.

Wallace, A. R. *My Life: A Record of Events and Opinions*. London: Chapman and Hall, 1908.

———. "On the Law Which Has Regulated the Introduction of New Species." Dated Sarawak, Borneo. *Annals and Magazine of Natural History*, 2nd series, 16 (1855): 184–96.

Williams, David. *The Rebecca Riots: A Study in Agrarian Discontent*. Cardiff: University of Wales Press, 1955.

Index

Page numbers in *italics* refer to illustrations.

A Conversation with Rebecca Stott

John Williams of *The New York Times*: The very first sentence of your book is: "I grew up in a Creationist household." How much did that drive your interest in Darwin?

Rebecca Stott: Darwin was described as the mouthpiece of Satan in the fundamentalist Christian community in which I was raised. His ideas were censored, and of course censorship can act as a kind of provocation to curiosity. The school library had a good encyclopedia with several pages on Darwin. I can't say I understood much of his ideas back then, but I understood enough to be mute with fascination. It was extraordinarily different from the biblical version of how things had come to be—but no less strange.

JW: You've written about Darwin before. What led you to concentrate on this aspect of his story?

RS: In writing *Darwin and the Barnacle,* I had come to respect the kinds of risks Darwin took in asking these dangerous questions about the origins of species. But I also knew there had been others before him who entertained similar ideas, and I wanted to know if they had had to take similar risks.

JW: Alfred Russel Wallace was independently reaching the same conclusions as Darwin around the same time, and Darwin felt compelled to rush his book to publication to establish his primacy. Wallace responded to the situation with incredible equanimity. Why didn't he fight for more turf?

RS: I don't think it occurred to him to do that. There were subtle class issues that determined his place in the question of priority for him, I think. Wallace had long looked up to Darwin and [the geologist] Charles Lyell and [the botanist] Joseph Hooker—they were gentlemen of science, whereas he thought of himself as a collector. Other things mattered to him more than fame—he was determined to play his part in the collection of proof about evolution. He was deeply proud to have been part of that, and also probably relieved to be able to slip away from the politics and the fuss and get back into the field.

JW: Aside from Wallace, who came closest to scientifically (as opposed to metaphorically) figuring out natural selection before Darwin?

RS: Jean-Baptiste Lamarck was one of the first men of science to have access to enough fossil and living animal specimens and bones to really gather the weight of evidence that would be needed to understand the ways in which species evolve. Lamarck worked in the Museum of Natural History in Paris, which in 1800 had the most remarkable collection of natural history specimens in the world—Napoleon Bonaparte had stolen hundreds of famous European natural history collections during the Napoleonic Wars and brought them all to Paris.

JW: The book's roster of notable predecessors starts with Aristotle. As brilliant as he was, that's a very early time for thoughts about evolution. What did he know or intuit that makes him a part of this intellectual lineage?

RS: What is remarkable about Aristotle is that he was the first to practice empirical science, rather than to settle for large-scale hypothetical theories about natural laws or cosmologies. He insisted on gathering facts; only facts he had verified with his own eyes. By trying to gather together all the information on all the animal species in the world, he was asking questions about species diversity and adaptation that would lead later scientists to evolutionary speculations. But he was not an evolutionist. He believed in the fixity of species.

JW: You start in 344 B.C. Then you hop forward to A.D. 850. And then to the late fifteenth century. What accounts for such large gaps between periods of progress in this subject?

RS: I wish I knew. Perhaps certain thinkers or schools of thought have been lost to history. Perhaps in the West it was due to the dominance of Christianity, and particularly Catholicism, over intellectual inquiry. Some of the periods of acceleration in the history of evolutionary thought were caused by material changes—the development of the printing press or of the microscope, growth in literacy rates, the gradual opening up of libraries and natural history collections to the public—but it always strikes me as salutary that one of the greatest periods of acceleration in evolutionary speculation took place in post-Revolutionary Paris between 1790 and 1815, when the priests had been banished and the professors had been given license to pursue any question they liked. That's when evolutionary ideas really came into their own.

JW: The polyp is a small organism that plays a large role in the story. What about it captured people's imaginations?

RS: I am particularly fond of the polyp. It was first "discovered" by a Swiss naturalist called Abraham Trembley in The Hague in the 1730s. Under a powerful microscope, he found that if you cut the "animal" in half, it could regenerate itself. The discovery caused a sensation amongst European naturalists, philosophers and theologians because it seemed to challenge all natural laws—animals cannot regenerate themselves.

Why could a simple organism like the polyp have such powers and not humans?

JW: Darwin would add people to a list of acknowledgments, then cross them off. He criticized a predecessor in one edition of *Origin* and then struck that criticism from the next edition. What drove his anxiety about the list and his fiddling with it the way he did?

RS: I am more and more convinced that assembling that list of predecessors was a kind of political act as well as a public relations exercise for Darwin. He was effectively saying: "Look, I'm not the first. Here are the men who have made this claim before me. We are all responsible." He wanted to have sane, respectable, ordinary people on that list to try to persuade his readers that evolution wasn't a mad, French, radical, anti-establishment idea. So it seems no coincidence that he worked particularly hard at finding hardworking respectable people like himself to put on his list, and that a large proportion of those people are British.

PHOTO: © BRUCE ROBERTSON

Rebecca Stott is a professor of English literature and creative writing at the University of East Anglia and an affiliated scholar at the department of the history and philosophy of science at Cambridge University. She is the author of several books, including *Darwin and the Barnacle* and the novels *Ghostwalk* and *The Coral Thief*. She lives in Cambridge, England.

ABOUT THE TYPE

This book was set in Monotype Dante, a typeface designed by Giovanni Mardersteig (1892–1977). Conceived as a private type for the Officina Bodoni in Verona, Italy, Dante was originally cut only for hand composition by Charles Malin, the famous Parisian punch cutter, between 1946 and 1952. Its first use was in an edition of Boccaccio's *Trattatello in laude di Dante* that appeared in 1954. The Monotype Corporation's version of Dante followed in 1957. Though modeled on the Aldine type used for Pietro Cardinal Bembo's treatise *De Aetna* in 1495, Dante is a thoroughly modern interpretation of that venerable face.